W0041972

The Animal Kingdom

VOLUME 5:
SYNOPSIS OF THE SPECIES OF THE CLASS
MAMMALIA AND THE CLASS REPTILIA

GEORGES CUVIER
EDITED AND TRANSLATED BY
EDWARD GRIFFITH

CAMBRIDGE
UNIVERSITY PRESS

CAMBRIDGE UNIVERSITY PRESS

Cambridge, New York, Melbourne, Madrid, Cape Town,
Singapore, São Paolo, Delhi, Mexico City

Published in the United States of America by Cambridge University Press, New York

www.cambridge.org
Information on this title: www.cambridge.org/9781108049580

© in this compilation Cambr dge University Press 2012

This edition first published 1828
This digitally printed version 2012

ISBN 978-1-108-04958-0 Paperback

This book reproduces the text of the original edition. The content and language reflect
the beliefs, practices and terminology of their time, and have not been updated.

Cambridge University Press wishes to make clear that the book, unless originally published
by Cambridge, is not being republished by, in association or collaboration with, or
with the endorsement or approval of, the original publisher or its successors in title.

CAMBRIDGE LIBRARY COLLECTION

Books of enduring scholarly value

Life Sciences

Until the nineteenth century, the various subjects now known as the life sciences were regarded either as arcane studies which had little impact on ordinary daily life, or as a genteel hobby for the leisured classes. The increasing academic rigour and systematisation brought to the study of botany, zoology and other disciplines, and their adoption in university curricula, are reflected in the books reissued in this series.

The Animal Kingdom

Georges Cuvier (1769–1832), made a peer of France in 1819 in recognition of his work, was perhaps the most important European scientist of his day. His most famous work, Le Règne Animal, was published in French in 1817; Edward Griffith (1790–1858), a solicitor and amateur naturalist, embarked in 1824, with a team of colleagues, on an English version which resulted in this illustrated sixteen-volume edition with additional material, published between 1827 and 1835. Cuvier was the first biologist to compare the anatomy of fossil animals with living species, and he named the now familiar 'mastodon' and 'megatherium'. However, his studies convinced him that the evolutionary theories of Lamarck and St Hilaire were wrong, and his influence on the scientific world was such that the possibility of evolution was widely discounted by many scholars both before and after Darwin. Volume 5 is an overview of mammals and reptiles.

Cambridge University Press has long been a pioneer in the reissuing of out-of-print titles from its own backlist, producing digital reprints of books that are still sought after by scholars and students but could not be reprinted economically using traditional technology. The Cambridge Library Collection extends this activity to a wider range of books which are still of importance to researchers and professionals, either for the source material they contain, or as landmarks in the history of their academic discipline.

Drawing from the world-renowned collections in the Cambridge University Library and other partner libraries, and guided by the advice of experts in each subject area, Cambridge University Press is using state-of-the-art scanning machines in its own Printing House to capture the content of each book selected for inclusion. The files are processed to give a consistently clear, crisp image, and the books finished to the high quality standard for which the Press is recognised around the world. The latest print-on-demand technology ensures that the books will remain available indefinitely, and that orders for single or multiple copies can quickly be supplied.

The Cambridge Library Collection brings back to life books of enduring scholarly value (including out-of-copyright works originally issued by other publishers) across a wide range of disciplines in the humanities and social sciences and in science and technology.

THE

ANIMAL KINGDOM

ARRANGED IN CONFORMITY WITH ITS
ORGANIZATION,

BY THE BARON CUVIER,

MEMBER OF THE INSTITUTE OF FRANCE, &c. &c. &c.

WITH

ADDITIONAL DESCRIPTIONS

OF

ALL THE SPECIES HITHERTO NAMED, AND OF
MANY NOT BEFORE NOTICED,

BY

EDWARD GRIFFITH, F.L.S., A.S., *&c.*
AND OTHERS.

———

VOLUME THE FIFTH.

———

LONDON:

PRINTED FOR GEO. B. WHITTAKER,

AVE-MARIA-LANE.

———

MDCCCXXVII.

SYNOPSIS OF THE SPECIES

OF THE

CLASS MAMMALIA,

AS ARRANGED WITH

REFERENCE TO THEIR ORGANIZATION,

BY

CUVIER,

AND OTHER NATURALISTS.

WITH

SPECIFIC CHARACTERS, SYNONYMA,
&c. &c.

VOLUME THE FIFTH.

LONDON:

PRINTED FOR GEO. B. WHITTAKER,

AVE-MARIA-LANE.

MDCCCXXVII.

LONDON:
Printed by WILLIAM CLOWES,
Charing Cross.

A

SYNOPSIS OF THE SPECIES

OF THE CLASS

MAMMALIA.

THE preceding supplemental essays on the text of our
author, like the text itself, by no means furnish even a
sketch of all the species hitherto described, but as the ex-
amination of the most ingenious machinery, however interesting in the detail of all its parts, is but an idle amusement, unless the final object and utility of the machine itself
be borne in mind, so the study of the various peculiarities of
organized nature is but a profitless pursuit unless the characters, habits, and relative situations of the several animals themselves be considered.

Hence some additional biographical matter seemed absolutely necessary to the utility of the present undertaking, as
otherwise that harmonious adaptation of means to ends of
the works of creation, which becomes the more apparent in
proportion to the pains bestowed on its investigation,
might be neglected or forgotten.

One considerable difficulty in the insertion of additional
descriptions and biographical matter is, however, to know
where to stop—to feel satisfied that the English reader has
been introduced into the hitherto neglected arcana of zoological science sufficiently to enable him, by drawing conclu-
sions in his own mind from the premises stated, to derive
both profit and amusement from the pursuit; and at the

1

same time fairly to presume that neither his patience nor his purse have been unnecessarily or excessively drawn upon.

To accomplish this object as far as possible, the following tabular view* is appended. This is intended to present, not merely a list of all the species hitherto known, but also their various synonyma with references to the first zoologists who have described them under each name given, and to the best figures, with occasional observations, particularly where any uncertainty seems to arise as to the real distinctiveness of any particular species. In presenting this list, however, with the authorities for the several species enumerated, it is by no means pretended to vouch for the accuracy or propriety of the specific separation of each; but they will be translated from the best monographs on the subject, revised by actual observations, as far as the collections of England, and the remarks of foreign writers, will permit.

This table is constructed according to the arrangement of the ' Règne Animal' of Cuvier. To the student in systematic zoology it will matter little whether he divides the orders with that author into a few genera, and each genus into several sub-genera or groups, or whether, with the more modern and refined zoologists, he treats each group or subdivision as a distinct genus. The present table will avail equally either way.

Class I.—MAMMALIA.

Animals covered with hair, having a back bone or vertebral column, red warm blood, and respiring air, by means of lungs, floating in a peculiar cavity. The fœtus, sustained by the mother in the womb, and born with signs of life. The young nourished with milk from the mammæ or teats of its mother.

* The synopsis will only be paged at the bottom, in order that each part of it as given at the end of each order, may be separated from the rest, and bound in a distinct volume if desired.

Order 1.—BIMANA.

Teeth of three sorts. The posterior extremities proper for walking, the anterior furnished with hands. Nails all flat ; teats two, pectoral ; body vertical ; stomach simple ; intestines furnished with a small cæcum ; orbital and temporal fossæ distinct.

Eats both animal and vegetable matter.

Inhabits almost the whole of the earth's surface.

Genus I. Homo.

Cutting teeth, $\frac{4}{4}$. Canine, $\frac{1}{1}-\frac{1}{1}$. Cheek teeth $\frac{5}{5}-\frac{5}{5}=32$.

I. Species 1. *Homo Sapiens,* " knowing himself." *Lin.*

Var. a. *Caucasian variety.*—Face oval ; facial angle 85°; forehead high expanding ; cheeks coloured red ; hair long and thick.

Inhabits Europe, (excepting Lapland and Finland), Western and Northern part of Asia.

b. *Mongolian variety.*—Face broad and flat, olivaceous ; facial angle 75°; eyes narrow and oblique ; hair hard, strait, black ; beard thin.

Inhabits Eastern Asia, Finland and Lapland in Europe, and the Esquimaux part of North America.

c. *American variety.*—Face broad, reddish copper-colour ; cheek-bones very prominent ; forehead short ; nose flattish : hair black ; beard weak.

Inhabits America (except the Esquimaux).

d. *Negro variety.*—Face black, projecting forward below ; facial angle 70° ; forehead narrow, slanting, arched ; lips large ; nose thick ; hair crisp.

Inhabits all the middle parts of Africa.

e. *Malay variety.*—Face brownish, rather projecting for-

3

ward ; nose rather full and broad, apex thickened ; mouth large ; hair thick, black, and curled.

Inhabits near the Ganges, the islands of the Indian Ocean, and the Polynesia.

Obs. These are the principal varieties, to one or other of which the minor ramifications may in general be traced.

ORDER II.—QUADRUMANA.

TEETH of three sorts. The four extremities furnished with hands ; teats two or four, pectoral ; bones of the arms and legs separate, capable of pronation and supination ; stomach simple membranaceous ; intestines short, with a small cæcum ; the orbital and temporal fossæ distinct.

Eats fruit, roots, and insects. *Lives* in trees. Intelligent, active.

Inhabits the warm parts of America, Africa, and India.

Genus I. SIMIA.

Form approaching that of man, cutting teeth $\frac{4}{4}$, canine $\frac{1\cdot1}{1\cdot1}$, grinders $\frac{5\cdot5}{5\cdot5}$, bluntly tubercular. Nostrils close, separated merely by a thin septum ; teats two, pectoral ; tail wanting or varying in length, never prehensile, with distinct cheek pouches, and often with callosities on the buttocks.

Eats fruit, insects, and sometimes birds.

Inhabits Africa and India, and its islands.

Sub-genus I. TROGLODYTES. Geoffroy, *Facial angle, 50°.; no cheek pouches, tail, nor callous buttocks ; arms short ; superciliary ridges distinct.*

2. 1. *S. T. Niger* (Chimpansé). Fur black.

Homo silvestris *Tyson Anat. of a pygmy.* Homo troglodytes, *Lin. Syst. Nat.* 33. Simia troglodytes, *Gmelin,* 26. S. Pygmea et S. Satyrus, *Schreb.* Troglodytes niger, *Geoffroy, Ann. Mus.* XIX. 87. Mimetes, *Leach, Jour. Phys.*

4

Man of the Woods. *Edwards, Glean.* Great Ape, *Pennant, Quadrupeds.* Jocko, *Buff.* xiv ; Pongo, *ib. supp.* vii.

Icon. *Tyson,* with the skeleton, viscera, &c., *Edwards, Buffon.* xiv. *Tulpius, Obs. med.—Schreber, l. c. tab.* 1, 2. *Audebert, t.* 1.—*Griff. Vert. Anim.*

Inhabits Africa, especially about Angola and Congo.

Obs. The adult state is unknown.

II. Pithecus.—Geoffroy. *Facial angle* 65° ; *no cheek-pouches, tail, nor callosities; arms very long; no superciliary ridges, at least in the young state.*

3. 1. *S. P. Satyrus,* (Orang-Outang). Fur brown.
Simia Satyrus. *Lin. Syst. nat.* 34. S. agrias. *Schreb.*
Orang-Outang, *Vosmaer.* Jocko. *Buff., supp.* viii.

Icon. *Abel's Embassy to China. Edwards, Glean. Vosmaer. Camper, Nat. Ver. t.* 4. *Schreb. t.* 2, B. *t.* 2, C. *Buff.* vii. *t.* 1. *Audebert Hist. des Singes, t.* 2. *Griff. Vert. Anim.*

Inhabits the eastern parts of Asia, near *Malacca and Borneo.*

Obs. Adult state not known. Presumed then to be the Pongo.

4. 2. *S. P. Wurmbii,* (Pongo). Fur black. The arms reach to the ancles; superciliary and sagittal ridges very distinct.

Pongo Wurmbii. *Geoff.* Cynocephalus Wurmbii. *Illiger.*
Pongo. *Wurmb. Mem. Soc. Bat.* ii. 245. Singe de Wurmb. *Audebert, Hist. des Singes. Blainville Jour. Phys.*
Icon. *Audebert Hist. des Singes ;* skeleton.

Inhabits Borneo and the Indian Archipelago.

Obs. Most probably the adult Orang Outang.

III. Hylobates. Illiger.—*Facial angle,* 60°.; *no cheek pouches nor tail; buttocks callous; arms excessively long.*

5. 1. *S. H. Lar.* (Gibbon). Fur black ; face surrounded with gray.

5

Homo Lar. *Lin. Mantissa*, II. 521. Simia Lar. *Gmelin.* S. longimana, *Schreb.* Pithecus Lar. *Geoff. Ann. Mus.* xix. 88.

Long-armed Ape, *Penn. Quad.* Gibbon, *Buff.* xiv.

Icon. *Schreb. t.* 3, *f.* 1. *Buff.* xiv., *t.* 2.

Inhabits East Indies, near Coromandel and Malacca, and the Molucca Islands.

6. 2. *S. H. variegatus* (Little Gibbon). Fur varied with gray brown, and deep gray.

Simia Lar. β. *Gmel.* S. longimana. *var. Scrheb.* S. longimana. *Mus. Leverianum.* Pithecus variegatus, *Geoff. Ann. Mus.* xix. 88.

Little Gibbon, *Penn. Quad.* Petit Gibbon, *Buff.* xiv.

Icon. *Schreb. t.* 3. *Mus. Leverianum, t.* 1., *Buffon,* xiv., *t.* 3.

Inhabits Malacca.

Obs. This is probably a mere variety of the last, it is distinguished, however, by being less in stature, and having the white hairs extending round the face, and also over the shoulders.

7. 3. *S. H. leuciscus* (Wow Wow). Fur ash gray; face black; callosities very large.

Simia lar. β. *Penn. Quad.* S. leucisca, *Schreb.* Pithecus leuciscus, *Geoff. Ann. Mus.* xix., 89

Wouwou, *Camper.* Moloch, *Audebert.* Gibbon cendré, *Cuvier, Règne Animal,* 103.

Icon. *Audebert* 1. § 1. *f.* 2.

Inhabits Malacca and Sunda Islands.

8. 4. *S. H. Syndactyla* (the Siamang). Fur black ; neck and breast naked ; the index and middle finger of the hinder extremities united to the end of the second phalanx.

Simia syndactyla, *Raffles, Trans. Lin. Soc.* xiii., 241.

Siamang, *F. Cuvier, Mamm. lithog.*

6

Icon. *Horsfield, Zool. Res. — F. Cuvier, l. c.*
Inhabits Sumatra.

9. 5. *S. H. Agilis* (the active Gibbon). Fur brown; back yellowish; forehead very low; orbital arches very prominent; face of the male blue black, of the female brown.

Hylobates agilis, *F. Cuvier, Mamm. lithog.* Pithecus agilis, *Desm. Mamm.* 532.

Gibbon ounko, M.M. *Diard and Duvancel.*

Icon. *F. Cuvier, Mamm. lithog.*

Inhabits Sumatra.

IV. PRESBYTIS.—Eschscholts. *Facial angle* 60°; *cheek-pouches none; callosities distinct; tail long; arms reaching to the knees.*

10. 1. *S. P. Mitrula* (the Capped Monkey). Fur finely curled, above of a bluish green colour, beneath grayish white; head crested, with a black line from the upper part of the ears across the head.

Presbytis mitrula, *Eschscholtii, Kotzebue, Voyage of Discovery,* II, 353.

Icon. Of the cranium and hands, *l. c.*

Inhabits Sumatra.

Obs. Length from head to rump one foot and a half; tail two feet. Called *Presbytis* on account of its resemblance to an old woman with a cap on her head.

VI. COLOBUS.—Geoffroy*. *Facial angle of* 40=45°; *muzzle short; anterior hands destitute of thumbs; tail very long and thin; cheek pouches and callosities distinct.*

11. 1. *S. C. polycomos* (the peruque or full-bottom

---

* It seems doubtful whether this genus exist distinct from the Macacus, the S. Silenus of Linnæus, but we have the authority of Geoffroy and others that it does.

7

monkey). Fur black, with a variegated mane covering the neck, and upper part of the back, and shoulders.

Simia polycomos, *Schreb.* S. comosa, *Shaw,* i. 59. Colobus polycomos, *Geoff. Ann. Mus.* xix. 92.

Guenon à camail, *Buff. supp.* vii. Full-bottom monkey, *Penn. Quad.* i. 197. King monkey, *Dealers.*

Icon. *Penn. Quad.* i. t. 25. *Schreb. t.* 10, *D. Shaw, Zool.* i. t. 24. *Buff.* vii. t. 17.

Inhabits the forest of Sierra Leone and Guinea.

12. 2. *S. C. ferruginosus* (Bay Monkey). Fur ferrugineous ; top of the head, hands, and tail, black.

Simia Ferruginosa, *Shaw, Gen. Zool.* i. 59. Colobus Ferruginosus, *Geoff. Ann. Mus.* xix. 92.

Autre Guenon, *Buff. suppl.* vii. 66. Bay Monkey, *Pennant, Quad.* 203.

Icon. ——

Inhabits ——

Obs. Considered by M Lacépède to be a variety of the former.

13. 3. *S. C. Temminckii* (Temminck's Colobus). Fur black ; neck, shoulders, and outer face of the thighs, black ; face, tail, and belly, white.

Colobus Temminckii, *Kuhl. Mss. Desm. Mamm.* 53.

Icon. ——

Inhabits ——

Mus. Bullock. Now in possession of M. Temminck.

VII. Lasiopyga.—Illiger. *Facial angle of 45°. ; head round, muzzle slightly prolonged ; hands very long, all provided with thumbs; thumbs of the anterior hands very short and thin; tail long ; cheek-pouches distinct ; buttocks not callous, fringed with hair.*

14. 1. *S. L. nemæus* (the Doue or Cochin China monkey). Fur varied with brilliant colours.

Simia nemæa, *Gmelin, Sys. Nat.* 34. Lasiopyga nemæus, *Illiger, Prod.* Pygathrix nemæus. *Geoff. Ann. Mus.* xix. 90. Le Douc, *Buffon*, xiv, 298, Cochin-China monkey. *Penn. Quad.* 211.

Icon. *Buff.* xiv. *t.* 41. *Supp.* vii. *t.* 23. *Audeb. Hist.* 4. § 1. *f.* 1. *Shaw, Zool. t.* 23.

Inhabits Cochin-China and Madagascar.

VIII. NASALIS.—Geoffroy, *Facial angle of* 40-45°; *head round, muzzle slightly prolonged, nose greatly prominent and elongated; ears small, round; body squat; hands long; anterior thumbs short; tail longer than the body; buttocks callous.*

15. 1. *S. N. Larvatus* (The Proboscis monkey). Fur reddish yellow; face black; nose very long.

Simia Nasalis, *Gmelin, pref. Shaw, Zool.* 1. 55. S. nasica, *Schreb.* Cercopithecus larvatus, *Wurmb. Mem. Soc. Batav.* Nasalis larvatus, *Geoff. Ann. Mus* xix 21.

Proboscis monkey, *Pen. Quad. App.* 322. Nasique. *Daubent. Mém. Acad. Scien.* Guenon à long nez. *Buffon supp.* vii.

Icon. *Buff. Supp.* vii. *t.* 11, 12. *Penn. Quad. t.* 104, 105. *Audebert t.* 4. § 2. *f.* 1.

Inhabits Borneo.

IX. SEMNOPITHECUS.—F. Cuvier. *Facial angle of* 45°, *head round, nose flat; ears moderate; limbs very long; thumbs of anterior hands very short and remote; cheek-pouches and callosities on the buttocks; tail very long and thin.*

16. 1. *S. S.* Maurus (The Negro Monkey). Fur black, with a white spot beneath, near the origin of the tail.

Simia maura, *Gmelin*, 35. Cercopithecus maurus, *Geoff. Ann. Mus.* xix. 92.

Middle-size black monkey; *Edwards, Glean.* Negro

9

Monkey, *Penn. Quad.* 206. Guenon Négre, *Buff. Sup.* VII. 83.

Icon. *Edwards, Glean. t.* 311. adult ; *Buff. Supp.* VII. *t.* 83. *Schreb. t.* 22. *B.* young.

Inhabit the Island of Java.

17. 2. *S. S. Melalophus* (The Simpai). Fur shining yellow, red above, whitish beneath ; forehead with a tuft of black hairs in the form of a band ; face blue.

Simia Melalophos. *Raffles, Lin. Trans.* XIII. Semnopithecus melalopus. *F. Cuvier, Mam.*

Simpai of the *Javanese.*

Icon. *F. Cuvier, Mam. Lithog.*

Inhabits Island of Sumatra, *Raffles.*

18. 3. *S. S. Pruniosus.* Fur blackish glazed with white; face brown.

Simia Villosa. *Griff. Vert. Ani.* 56. Semnopithecus pruniosus. *Desm. Mam. Supp. p.* 533.

Icon. *Griff. Vert. Anim. t.* 6.

Inhabits Isle of Sumatra, *M.M. Diard. et Duvancel.*

Obs. This animal differs from the S. Maurus in the want of the white spot near the insertion of the tail, and the fore-hands are black.

19. 4. *S. S. Comatus.* Fur above gray, beneath dirty white; upper part of the head covered with black hairs, forming a tuft towards the occiput.

Semnopithecus comatus, *Desm. Mam. sup.* 533.

Icon. ——

Inhabits Sumatra, *M.M. Diard et Duvancel.*

20. 5. *S. S. Entellus* (The Entellus). Fur yellowish white ; hands all black.

Simia Entellus. *Dufresne, Bul. Soc. Phil.—Schreb.—*Cercopithecus Entellus. *Geoff. Ann. Mus.* XIX. 95.

Entelle *Audeb. Hist.*

Icon. *Audeb. Hist.* 4. §. 2, *f.* 2. *Schreb. t.* 23. *f. B.*
Inhabits Bengal.

X. CERCOPITHECUS.—Geoffroy, *facial angle of* 50°; *head round, no superciliary ridges, edges of the orbits smooth, nose flat, nostrils open to the nasal fossæ; ears moderate; cheek-pouches and callosities on the buttocks; tail longer than the body.*

21. 1. *S. C. Auratus* (the golden guenon). Fur golden yellow, with long hair on the cheeks and forehead, and a black spot on the knee.
Cercopithecus auratus, *Geoff. Ann. Mus.* xix. 93.
Icon.———
Inhabits India and Molucca.

22. 2. *S. C. Talapoin* (the Talapoin Monkey). Fur olivaceous above, yellowish white beneath; tail ash-coloured; feet black.
Simia talapoin, *Gmel.* 35. Cercopithecus talapoin, *Geoff. Ann. Mus.* xix. 93.
Talapoin Monkey, *Penn. Quad.* 206. Talapoin, *Buff.* xix.
Icon. *Buff.* xix. *t.* 40. *Schreb. t.* 17.
Inhabits Africa.
Obs. Cuvier thinks this may be the young of the Malbrouc.

23. 3. *S. C. Latibarbatus* (The purple-faced or broad-bearded monkey). Fur black; with a very large laterally-extended beard; end of tail tufted; face violet purple.
Simia dentata, *Shaw, Zool.* 1, 24. S. Veter? *Shaw,* 1, 36.
Cercopithecus latibarbatus *Temm. Cat.; Geoff. Ann. Mus.* xix. 94. Adult.
Purple-faced monkey, *Penn. Quad.* Broad-toothed baboon, *Penn. Quad.* Guenon à face pourpré, *Buff. Supp.* vii. 80.
Icon. *Shaw, Zool.* 1. *t.* 13. *Penn. Quad. t.* 24. *Buff. Supp.* vii. *t.* 21.
Inhabits ———
11

24. 4. *S.C. Cephus* (Mustache Monkey). Fur greenish brown; the latter half of the tail bright red; nose and lips blue.

Simia cephus. *Lin. Sys. Nat.* 39. S. mona, *Schreb.* Cercopithecus cephus, *Geoff. Ann. Mus.* xix. 94.

Mustache Monkey,*Penn. Quad.* 205. Moustac, *Buff.* xiv. Icon. *Schreb. t.* 19, *t.* 15. *Buff.* xiv. *t.* 39. *Audebert.* Inhabits ——

25. 5. *S. C. Pileatus* (Bonneted Monkey). Fur, above brownish yellow, beneath white, with long hairs on the forehead.

Simia pileata. *Shaw Zool.* 1. 53. Cercopithecus pileatus, *Geoff. Ann. Mus.* xix. 94.

Bonneted Monkey, *Penn. Quad.* Guenon couronné. *Buff. Sup.* vii.

Icon. *Buff. Sup.* vii. *f.* 10.

Inhabits ——

Obs. This seems nearly allied with, if it be not actually the Chinese bonneted monkey, Simia sinicus of Gmelin.

26. 6. *S. C. Mona* (Varied Monkey). Fur chesnut colour, outer part of the extremities black; with two white spots on the buttock.

Simia mona and S. monacha. *Gm. Sys. Nat.* Cercopithecus mona. *Geoff. Ann. Mus.* xix. 95.

The varied monkey, *Penn. Quad.* 210. *Shaw's Zool.* 1. 54. La Mone,*Buff.* xiv., and La Mone. *Ib. Sup.* vii.

Icon. *Schreber, t.* 15 *f. A. Buff. t.* 36, and *Sup. t.* 19. *Audebert, t.* —— *Shaw's Zool. t.* 18.

Inhabits

27. 7. *S. C. Nictitans,* (White-nosed Monkey). Fur black, sprinkled with greenish gray; nose white and swollen; the anterior extremities above quite black.

Simia Nictitans. *Lin. Sys. Nat.* 40. Cercopithecus nictitans, *Geoff. Ann. Mus.* xix. 95.

Guenon à long nez proéminent. *Buff. Sup.* 7. Le Hocheur. *Audeb. Hist.* 4. *S.* 1. White Nose Monkey, *Pen. Quad.* 205.
Icon. *Buff. Supp.* vii. *t.* 18. *Audeb.* 4 § 1. *t.* 2.
Inhabits Guinea.

28. 8. *S. C. Petaurista* (Vaulting Monkey). Fur, red above; white beneath; extremities olivaceous above; gray beneath; lower part of the nose white.
Simia petaurista, *Gmelin, Sys. Nat.* 35.   Cercopithecus petaurista, *Geoff. Ann. Mus.* xix.
Le Blanc nez. *Buffon Supp.* vii. 67. L'Ascagne. *Audeb. f.* 4, §. 2.   The vaulting monkey. *Shaw, Zool.* 51.
Icon. *Audeb. Hist.* 4. §. 2. *f.* 14, 15.   *Schreb. t.* 19. *B.*
Inhabits Guinea.

29. 9. *S. C. ruber* (Red Monkey). Fur, red above; ash-colour beneath; with a narrow black or white band over the eyes.
Simia nigra.   *Gmelin, Sys. Nat.* 24. S. Patas, *Schreb.* 46. S. nigra. *Schreb. t.* 16. *B.* Cercopithecus niger, *Geoff. Ann. Mus.* xix. 96.
Le Patas, *Buff.* xiv. *t.* 25 *et* 26.   Red Monkey, *Penn. Quad.* 208.
Icon. *Buffon* xiv. *t.* 25, 26. *Schreb. t.* 16. 16. *B.*
Inhabits Senegal, commonly called the *Red Monkey of Senegal.*

30. 10. *S. C. Diana* (Palatine Monkey). Fur back bright chestnut, sides slate gray, with an oblique line of the same colour on the thighs.
Simia Diana, *Lin. Sys. Nat.* 38.   Roloway, *Gmelin, Sys. Nat.* 35.   *S. Diana faunus. Lin. Sys. Nat.* 38. Cercopithecus Diana. *Geoff. Ann. Mus.* xix. 96.
Exquima, *Marcgrave.* Roloway, *Buff. Supp.* vii. Palatine Monkey, *Penn. Quad.* 200.   Spotted Monkey, *Penn. Quad.* 201.   La Diane, *Audeb. Hist.* 4. § 2.
Icon. *Buff. Supp.* vii. *f.* 20. *Audeb. Hist.* 4. § 2. *f.* 6. *Schreb. t.* 25.

Inhabits Africa, especially near Congo and Guinea.

31. 11. *S. C. Albocinereus.* Fur, gray above ; deeper on the arms ; whitish below, with a line of ridged black hairs traversing the forehead; hands black ; tail brown.

Cercopithecus albocinereus, *Desm. Mamm.* 534.

Icon.

Inhabits Sumatra.

XI. CERCOCEBUS.—Geoffroy, *Facial angle 45°, head triangular ; muzzle longish; the upper orbital edge, rising again and cut internally ; nose flat and convex; the thumbs of he anterior hands thin, rather close to the fingers, those of he hinder larger and more distant; the buttock with very arge callosities; tail longer than the body ; cheek pouches distinct.*

32. 1. *S. C. Cynosurus,* (the Malbrouck). Fur, olive brown above, whitish beneath, with a whitish band over the eyes.

Simia cynosurus, ♂ *Gmelin, Sys. Nat.* 30. S. faunus *Lin. Sys. Nat.* 36. Cercocebus cynosuros. *Geoff. Ann. Mus.* XIX. 96.

Malbrouc, *Buffon,* XIV. Jeune Callitriche, *Audeb. Hist.* 4. § 2. *t.* 5. Dog-tailed Baboon. *Shaw,* 32.

Icon. *Scop. Delic. t.* 19. ♂—*Schreb. t.* 14. *B. Buff.* XIV. *t.* 29 ♀ *Audeb. Hist.* 4. § 2. *f.* 5.

Inhabits Bengal.

Obs. M. Cuvier considers the *S. talapoin (n.* 22.,) to be only a young individual of this species.

33. 2. *S. C. Sabæus* (Callitrix, or Green Monkey). Fur, olive green above; dirty white beneath ; head pyramidical ; face black; cheeks with long hairs; scrotum copper green, surrounded by yellow hairs; end of the tail, yellow.

Simia Sabæa. *Lyn. Sys. Nat.* 1. 38. Cercopithecus Sabæa, *Geoff. Ann. Mus.* XIX.

Singe Vert, *Bris. Règ. An.* 205. Callitriche, *Buff.* XIV. 272.

Green Monkey, *Penn. Quad.* 203. The St. James's Monkey, *Edw. Glean.*

Icon. *Buff.* xiv. *t.* 37. *Audeb. Hist.* 4. § 2. *f.* 4. *Cuv. Menag. Mus.* 4. *t.* 18. *F. Cuv. Mam. t.* — *Edw. Glean, t.* 215. Schreb. *t.* 18.

Inhabits the Mauritius, Senegal, and the Island of Cape Verd.

34. 3. *S. C. Griseo-viridis* (the Grivet). Fur greenish gray; scrotum copper green, surrounded with white hairs; head pyramidical; tail entirely gray.

Cercopitheucus Griseo viridis, *F. Cuvier.*

Le Grivet, *Fred. Cuvier, Mam.*

Icon. *F. Cuvier, Mam.*

Inhabits Africa.

Obs. Like the *S. Cynosurus,* (*n.* 25.) but differs in the head not being so round, the scrotum green, not bright blue, fringed with orange; and from the *C. Sabæus,* in the want of the white band over the eye, and the lightness of the colour of the whole animal.

35. 4. *S. C. Pygerythræus,* (Red-vented Monkey). Fur greenish gray above; white beneath; scrotum gray green, surrounded with white hairs; vent surrounded with deep red; end of the tail black.

Cercopithecus pygerythræus. *F. Cuvier, Mam.*

Icon. *F. Cuvier, Mam. Lithog.*

Inhabits Cape of Good Hope.

36. 5. *S. C. Fuliginosus* (the Mangabey, or white eye-lid monkey). Gray, slate brown, without any spot on its head or neck; upper eye-lids white.

Simia Æthiops. *Lin., Gmelin, Sys. Nat.* 33. Cercocebus Fuliginosus, *Geoff. Ann. Mus.* xix. 97.

White eye-lid Monkey. *Penn. Quad.* 204. *Shaw,* 43. Mangabey. *Buffon,* xiv. *t.* 344.

Icon. *Schreb. t.* 20. *Buffon,* xiv. *t.* 33. *Audeb.* 4. § 2. *f.* 10.
Ency. Méth. t. 13. *f.* 3. *Shaw,* 1. *t.* 20. *F. Cuv. Mam. t.* —
Inhabits Ethiopia.

37. 6. *S. C. Æthiops* (the Collared Mangabey). Fur
wine brown, top of the head red; upper eye-lid white, and
with a white band parting the eyes and proceeding to each
side of the back of the neck.

Simia Æthiops. Var. *Gmelin,* Cercocebus Æthiops. *Geoff.
Ann. Mus.* xix. 97.

Mangabey à collier blanc. *Buff.* xiv.

Icon. *Buff.* xiv. *t.* 33. *Var.* 11; *Audeb. Hist.* 4. § 2.
f. 10. *Schreb. t.* 21.

Inhabits Ethiopia.

38. 7. *S. C. Atys* (the Atys). Fur entirely white; face
and hands flesh-coloured; ears nearly square.

Simia Atys. *Audeb. Hist.* 4. § 2. *Schreb.* Cercocebus
Atys, *Geoff. Ann. Mus.* xix. 99.

Icon. *Audeb. Hist.* 4. § 2. *f.* 8. *Schreb. t.* 14. *B.*

Inhabits ?

Obs. Most probably an Albino of some other species,
but if so the type is unknown.

XII. MACACUS.—Lacépède. *Facial angle of* 40. 45°; *superciliary and occipital ridges very distant; tail shortish; cheek-
pouches, and callosities distinct; ears angular.*

39. 1. *S. M. Silenus* (the Ouanderou). Fur black; belly
gray; mane and beard large; tail gray, largish, ending in a
tuft of hairs.

Simia Silenus, of *Lin. Sys. Nat.,* 36. 1. S. Leonina, *Gmelin
Sys. Nat.* S. ferox, *Shaw, Mus. Leverianum,* 69. Papio silenus. *Geoff. Ann. Mus.* xix. 102.

Ouanderou, *Knox's Ceylon. Buff.* xiv. *t.* 18. Macaque
à crinière. *Cuv. Règ. Anim.* 1. 108. Lion-tailed baboon,

16

Pennant. Quad. 198. Nil-bundar of *Hindus.*

Icon. *Buff.* xiv. *t.* 18. *Audeb. Hist.* 2. § 1. *f.* 3. *Pennant, Quad. t.* 44. *Shaw, Zool.* 1. *t.* 16. *Schreb. t.* 11.

Inhabits Ceylon.

40. 2. *S. M. sinicus.* Fur chesnut brown, the hairs on the crown of the head diverging from the centre to the circumference and placed in the shape of a cap.

Simia sinicus, *Gmelin, Sys. Nat.* 34. Cercocebus sinicus, *Geoff. Ann. Mus.* xix. 98. Macacus sinicus *Desm. Mam.* 64.

Bonnet-chinois, *Buffon,* xiv. 224. Chinese monkey, *Pennant, Quad.* 209.

Icon. *Buff.* xiv. *t.* 30. *Audeb.* 4. *f.* 2. 11. *Shaw. Zool.* 1. *t.* 20. *Schreb. t.* 23.

Inhabits Ceylon.

Obs. This species is very closely allied to the cercopitheci both in characters and habits, and Cuvier considers it to be the full grown *S. C. Pileatus,* (n. 25.)

41. 3. *S. M. radiatus.* Greenish brown above, clear ash colour beneath, hairs at the crown of the head diverging and forming a cap.

Cercocebus radiatus, *Geoff. Ann. Mus.* xix. 984.

Macaque foqué, *Desm. Mamm.* 64.

Icon.

Inhabits India.

Obs. M. Geoffroy considers it doubtful whether this species belongs to the *Macaci,* or to the *Cercocebi.*

42. 4. *S. M. cynomolgus.* (Hare-lipped monkey) Fur, greenish brown or olivaceous above and grayish white beneath; margin of the orbits of the males very prominent; middle of the forehead of the female ornamented with an elevated tuft of hairs.

17

Simia cynomolgos, *Lin. Sys. Nat.* 1. 38. S. cynocephalus, *Lin. Sys. Nat.* 1. 38.? Cercocebus cynomolgos and C. cynocephalus, *Geoff. Ann. Mus.* xix. 99. Macacus cyno-molgus, *Desm. Mam.* 65.

Hare-lipped monkey, *Pennant Quad.* 200. Macaque, *Buff.* xiv. 190. Aigrette. *Buff.* xiv. 190. Macaque ordinaire, *Desm. sp.* 34.

Icon. *Buff.* xiv. *t.* 21. *Ency. Méthod. t.* 14. *f.* 1.

F. Cuvier, Mam. Shaw. Zool. t. 16.

Inhabits Guinea and the interior of Africa, often brought to Europe.

XIII. PITHECUS, Gray. *Facial angle of* 40. 45°; *superciliary and occipital ridges very distinct; tail very short or only a small tubercle; cheek pouches and callosities distinct, ears angular.*

43. 1. *S. P. rhesus* (the pig-tailed baboon). Fur, greenish gray above;—tail short and wrinkled at its base; but-tocks golden yellow; extremities gray.

Simia erythrea, *Schreb.*—*S. Monachus, Schreb. t.* 15 *B.?* Pithecus rhesus, *Geoff. Ann. Mus.* xix. 101. S. Nemes-trina, *Shaw, Zool.* 1. 25.

Patas à queue courte, *Buff. suppl.* vii. Macaque à queue courte, *Buff. Suppl.* vii. Rhesus. *Audeb. Hist.* 2.§ 1. *t.* 3. Pig-tail Baboon. *Pennant. Quad.* 14. *t.* 19. Macaque maimon, *Desm. sp.* 35.

Icon. *Buff.* xiv. *t.* 19, bad. *Suppl.* vii, *t.* 14. 13. (1.) *F. Cuvier, Mam. t.* — *Ency. Méth. t.* 7, *f.* 2. *Audeb. Hist.* 2. § 1. *t.* 4. and 3. *Schreb. t.* 8, D. (1.) *Shaw. Zool.* 1. *t.* 41. (1)

Inhabits West Indies, and on the banks of the Ganges.

44. 2. *S. P. nemestrinus,* (Brown Baboon.) Fur, deep brown above with the middle of the head and back; tail

small, thin, half the length of the thighs; head and extremities yellowish.

Simia nemestrina, *Lin. Sys. Nat.* 35. S. platypygos, *Schreb. t.* 5. B. S. fusca, *Shaw, Zool.* 1. 24.

Pithecus, *Geoff. Ann. Mus.* xix. 101.

Babouin à longues jambes, *Buff. Supp.* vii. Macaque, *Cuvier, Dict. Sci. Nat.* Brown Baboon, *Pennant, Quad.* 192. Pig-tailed Monkey, *Edwards, Glean. f.* 214. Macaque Orion, *Desm. sp.* 36.

Icon. *Buff. Supp.* vii. *t.* 8. *Ency. Méth. t.* 10. *f.* 20, *Schreb. t.* 9. *Edw. Glean. t.* 214. *Shaw, Zool.* 1, *f.* 13. *Griffith, Vert. Anim. t.* 25.

Inhabits Java, Sumatra.

45. 3. *S. P. inuus,* (the Pygmy, or Barbary ape). Fur greenish gray; a small tubercle in the place of a tail.

Simia inuus, *Lin. Sys. Nat.* 35. S. silvanus, *Lin. Sys. Nat.* 1. 34 *(Junior).* S. Pithecus, *Schreb.* Macacus indicus, *Desm. Mamm.* 67.

Pithecus, *Aristotle.* Cynocephalus, *Prosper.* Simia, *Ray.* Pithèque, *Buff. Supp.* vii. Magot, *Buff.* xiv. Barbary Ape, *Pennant, Quad.* 186. Pygmy Ape, *Pennant, Quad.* 183. Alpinus Ape, *Shaw, Zool.* 1. 14. Magot or Barbary Ape, *Griff. Quad.*

Icon. *Buff.* x. iv. *t.* 8, 9, *Supp.* vii. *t.* 2, *f.* 4, 5. *Audeb.* 1, § 3, *t.* 1. *F. Cuvier, Mam. Ency. Méth. t.* 6, *f.* 3, 3 *a. et.* 1. *Shaw, Zool. t.* 7, 8. *Spec. Lin.* 1. *t.* 1. *Schreb. t.* 4, 5, *et* 4 B. *Griffith. Quad. t.* 23.

Inhabits Barbary, Egypt, and the rock about Gibraltar in Spain.

XIV. Cynocephalus.—Brisson. *Facial angle of* 30=35° ; *superciliary and occipital ridges very prominent; muzzle long and truncated at the end; canine teeth strong; cheek-pouches and callosities distinct; tail as long or longer than the body.*

19

46, 1. *S. C. Babouin*, (Little Baboon). Fur greenish yellow; face livid, flesh colour; cartilage of the nose not exceeded by the bones of the upper jaw.

Simia cynocephalus, *Lin. Sys. Nat.* 38. Papio cynocephalus, *Geoff. Ann. Mus.* XIX. 102. Cynocephalus babouin, *Desm. Mam.* 63. Cynocephalus, *Plinii.* Cercopithecus, *Johnst. Quad.*

Petit papion, *Buff.* XIV. Baboin, *F. Cuvier, Mam.*

Icon. *John. Quad. t.* 59, *last fig. Buff.* XIV. *t.* 14. *F. Cuv. Mamm. Lithog. Ency. Méthod. t.* 9, *f.* 1, 2.

Inhabits Northern Africa.

47. 2. *S. C. papio* (Guinea Baboon). Fur yellowish brown; face entirely black; cartilage of the nose exceeding the jaws at their upper extremity; upper eye-lids white.

Simia cynocephalus, *Brongn. Jour. Hist. Nat.* Cynocephalus Papio, *Desm. Mam.* 69.

Papion, *Buff.* XIV.

Icon. *Brongn. l. c. Schreb. t.* 13, *B.* copied *Brong. Buff.* XIV, *t.* 13. *Audeb. Hist.* 3, § 1. *f.* 1, *Ency. Méthod. t.* 6, *f.* 4.

Inhabits shores of Guinea.

48. 3. *S. C. porcarius* (Pig-faced Baboon). Fur greenish black above, with a mane of large hairs on the neck; face violet black; paler under the eyes, and white on the upper eyelid.

Simia porcaria, *Lin. Gmelin, Sys. Nat.* S. ursina, *Pennant.* ? S. sphingiola, *Herman.* ? Papio comatus, et P. porcarius, *Geoff. Ann. Mus.* XIX. 102, and 103. Cynocephalus porcarius, *Desm. Mam.* 69.

Guenon à face alongée, *Buffon.* Singe noir, *Vaillant.* Choak kama, *Kolbe.* Chacma, *Fred. Cuv. Mam.* Papion noir, *Cuv. Règ. An.* 110.

Icon. *Boddært, Nat. Forst.* XII. *t.* 1, 2; *Ency. Méth. t.* 8, *f.* 4 ? *F. Cuvier, Mam.*

20

Inhabits the Cape of Good Hope.

49. 4. *S. C. Hamadryas* (Gray Baboon). Fur ash-coloured, beard and mane very long, face flesh-coloured; hands black.

Simia hamadryas, *Lin. Sys. Nat.* 36. Cynocephalus hamadryas, *Desm. Mam.* 70.

Cynocephalus, *Gesner*, 862 *Clus. Exot.* 370. Tartarin *Bellon*, 101. Babouin à museau de chien, *Buff. Supp.* vii. 47. Singe de Moco, *Buff.* xix. Papion â Perruque, *Cuv. Rêg. An.* 110. Dog-faced Ape, *Penn. Quad.* 194. Dog-faced Baboon, *Shaw, Zool.* 1. 28. Gray Baboon, *Shaw Spec. Lin.* 1. *t.* 3.

Icon. *Bellon Dis.* 101. *Prop. Al. t.* 17, 19. *Edw. Glean. t.—Schreb. t.* 10. *Shaw, Zool.* i *t.* 15; *Spec. Lin. t.* 32. *F. Cuvier, Mam. t.* — *Griff. Quad. t* 28.

Inhabits Moco, Persian Gulf, and Arabia.

XXV. PAPIO.—Brisson. *Facial angle, of* 30. 35°, *superciliary and occipital ridges very prominent, muzzle long, and truncated at the end, canine teeth strong, cheek-pouches and callosities very distinct, tail very short, and perpendicular to the orsal spine.*

50. 1. *S. P. Mormon* (the Mandril). Fur olivaceous gray, brown above; white beneath, beard yellow; face blue, and the nose red, in the adult males.

Simia sphinx, *Lin. Sys. Nat.* 35. S. Mormon, *Gmel. Sys. Nat.* Cynocephalus mormon, *Desm. Mam.* 70.

Papio sus, Baboon, *Gesner, Quad.* 252. Choras, *Buff. Supp.* vii. Mantegar, *Phil. Trans. n.* 220, *Bradley, Nat.* 117 Mandril, *G. Cuvier, Mênag. Mus.* Great Baboon, *Penn. Quad.* 188. Variegated Baboon, *Shaw, Zool.* i. 17. Boggo, *Travel.*

Icon. *Gesner, t.* 253. *Bradley, Nat. t.* 15. *f.* 1. *Buffon, Supp.* vii. *t.* 9. *Penn. Quad. t.* 40, 41. *Mus. Lever.* 35. *t.* 9. Shaw, Zool. 1. *t.* 10, *F. Cuv. Mam. t.*

Junior Simia Maimon, Lin. Sys. Nat. 35. —

Le Mandril, *Buff.* xiv. Maimon, Shaw, Zool. i. 20. Ribbed nose Baboon, *Penn. Quad.* 190.

Icon. *Schreb.* i. *t.* 7. *Buff.* xiv. *t.* 16. 17. *Audeb. Hist.* 2. § 2. *f.* 1. *Shaw, Spec. Lin.* 1. *t.* 2. i. *Zool.* 1. *t.* 11.

Inhabits Africa, on Gold and Guinea Coasts.

51. 2. S. *P. leucophæus* (the Dril). Fur, brownish green, gray above; white beneath; face (in both sexes and at all ages), uniform deep black.

Simia Leucophæa. *F. Cuvier, Ann. Mus.* ix. Cynocephalus Leucophæus. *Desm. Mam.* 71.

Dril. *F. Cuvier, Mamm.* Wood Baboon. *Griff. Quad. t.* 20?

Icon. *Ann. Mus.* ix. *t.* 37. *F. Cuv. Mam. t.* — *Griffith, Quad. t.* 29 ?

Inhabits Africa?

The three following species of Pennant have great affinity to this species *if really distinct.*

52. 3. † S. *P. Sylvicola* (Wood Baboon). Fur ferruginous brown; face and hands black.

Simia Sylvicola. *Mus. Lever.* 201.

Le Babouin des Bois. *Buff. Sup.* vii. 39. Wood Baboon. *Penn. Quad.* 191.

Icon. *Penn. Quad. t.* 42. *Shaw, Zool.* 1. *t.* 12 ; *Mus. Lev. t.* 1. *Buff. Sup.* vii. *t.* 7.

Inhabits Africa, near Guinea.

53. 4. † S. *P. Sublutea* (Yellow Baboon). Fur yellow, freckled with black; face black, naked; hands above hairy.

S. Sublutea. *Shaw, Zool.* 1. 23.

Yellow Baboon. *Pen. Quad.* 191.

Inhabits Africa.

54. 5. † S. *P. Cinerea* (Cinereous Baboon). Fur

cinereous, crown of the head mottled with yellow ; face brown, beard pale.

S. Cinerea. *Shaw Zool.* 1. 23.

Cinereous Baboon. *Pen. Quad.* 87.

Inhabits Africa.

55. 6. † *S. P. Pennantii* (Pennant's Baboon). Fur black, and ash-colour, varied with reddish ; face blue ; beard pale brown, with a tuft of hairs over each eye, and on each ear.

Baboon. *Pennant Quad.*

56. 7. *S. P. Apedia* (Thumbless Baboon). Fur greenish brown ; the thumb of the anterior extremities close to the palm ; claws depressed ; other claws oblong compressed.

Simia Apedia. *Lin. Sys. Nat.* 1. 35.

Inhabits India, size of a Squirrel.

Obs. This species is only known by Linnæus' specific characters.

57. 8. *S. P. Niger* (Black Baboon). Fur entirely black ; hair woolly, except on the crown of the head, where they are long, and form a tuft on the occiput. Tail none ?

Cercopithecus Niger. *Desm. Mamm.* 534.

Icon.

Inhabits one of the Islands of the Indian Archipelago.

Obs. If this species, which is described from a mutilated specimen in the French museum, really have no tail, it will form a new section*.

---

* Cuvier placed after this Genus the Pongo, which he at the same time observed might be the young of the Orang-Outang, next to which we have placed it.

AMERICAN MONKEYS.

FORMING THE SECOND PRINCIPAL SUB-GENUS OF CUVIER.

FORM approaching that of man. Cutting teeth $\frac{4}{4}$, cheek teeth $\frac{6 \cdot 6}{6 \cdot 6}$, bluntly tubercular, or $\frac{5 \cdot 5}{5 \cdot 5}$ acutely tubercular ; nostrils separated by a broad septum and opening laterally ; teats two-pectoral ; tail long, usually prehensile, cheek, pouches or callosities none.

I. ATELES, Geoffroy. *Facial angle of* 60°, *head round, limbs very thin, anterior hands destitute of thumbs, tail very long, powerfully prehensile, having the lower part of its extremity naked.*

58. 1. *S. A. Paniscus* (the Coaita). Black; face nearly naked, copper coloured; no thumb to the anterior hands.

Simia paniscus. *Lin. Sys. Nat.* 37. Ateles paniscus. *Geoff. Ann. Mus.* vii. 269. xvi. 105.

Quatto, *Vosm.* 1768. Coaita, *Buff.* xv. 16. Four-fingered monkey, *Penn. Quad.* 216.

Icon. *Buff.* xv. *t.* 1. *Audeb. Hist.* 5. § 1. *f.* 2. *F. Cuv. Man.—Ency. Méthod. p.* 16. *f.* 1. *Schreb. t.* 26. *Shaw, Zool.* 1. *t.* 28.

Inhabits Guyana and Brazil.

59. 2. *S. A. niger*, (the Black Coaita). Fur black ; face hairy, black ; no thumb on the anterior hands.

Ateles niger, *F. Cuvier, Mamm.*

Atèle coaita de Cayène, *Geoff. Ann. Mus.* xiii. 97.

Inhabits Guyana.

Icon. *F. Cuvier, Mamm. Lithog. n.* 39. *t.* 1

Inhabits Guyana.

60. 3. *S. A. Belzebuth* (The Marimonda). Fur black, belly dirty white or yellowish in the male, and white in the young or females.

Ateles Belzebuth, *Geoff. Ann. Mus.* vii. 271. xix. 106.

Le Belzebuth, *Brisson, Règ. Anim.* 1. 211. Marimonda,

Humb. Obs. Zool. 325. Coaita à Ventre blanc, *Cuvier. Règ. Anim.* 1. 113.

Icon. *Geoff. Ann. Mus.* vii. *p.* 16.

Inhabits the banks of the Orinoco.

61. 4. *S. A. Marginatus* (The Chuva). Black, with a white ruff round the face.

Ateles Marginatus, *Geoff. Ann. Mus.* xiii. 9. xix. 106.

Chuva, *Humb. Zool. Obs.* 340.

Icon. *Geoff.* xiii. *p.* 9.

Inhabits the banks of the Santiago and Amazon.

62. 5. *S. A. Arachnoïdes* (the spider monkey). Yellow-gray, fur soft, eye-brows black, long; without any thumb on the anterior hands.

Ateles arachnoides, Spider Monkey, *Edwards, Brown, Geoff, Ann. Mus.* xiii. 90. xix. 109.

Icon. *Geoff. Ann. Mus.* xiii. *t.* 9.

Inhabits Brazil?

63. 6. *S. A. Melanochir* (Black-handed Coaita). Gray; back of the head, the extremities of the limbs, and an oblique spot, on the outside of each knee of a brown, black, or gray brown.

Ateles melanochir, *Desm. Mamm.* 76.

Inhabits —— French Museum.

64. 7. *S. A. Hypoxanthus* (The Miriki). Yellowish gray, face flesh colour, spotted with gray; base of the tail and buttocks sometimes yellow, ferruginous, the thumb of the anterior hands merely rudimentary and without a nail.

Ateles hypoxanthus, *Kuhl. MSS. Desm. Mamm.* 72. Mi, Riki, and Mono, *Brazilians.*

Inhabits Brazil. Prince Maximilian's Museum.

65. 8. *S. A. Subpentadactylus* (the Chameck). Black; the thumb of the anterior hands very small, and without a nail.

Ateles pentadactylus, *Geoff. Ann. Mus.* vii. 267, and xix. 105. Ateles subpentadactylus, *Desm. Ram.* 73.

Chameck, *Buff.* xv. 21. *Humboldt, Zool. Obs.*

II. Lagothrix. Humboldt. *Facial angle of* 50° ; *head round, the limbs proportioned to the body, anterior hands provided with a thumb. Tail strongly prehensile, with the lower part of the extremity naked.*

66. 1. *S. L. Humboldtii* (The Capparo). Blackish ash-colour, hairs long.

Simia lagothricha, *Humb. Obs. Zool.* 32. Lagothrix Humboldtii. *Geoff. Ann. Mus.* xix. 107.

Icon. ——

Inhabits the Banks of the Rio Guariara.

67. 2. *S. L. Canus* (Silver-haired Monkey). Olivaceous gray ; head, hands, and tail red-gray ; hairs short.

Lagothrix Canus. *Geoff. Ann. Mus.* xix. 107.

Icon. ——

Inhabits Brazils.

III.—Mycetes.—Illiger. *Facial angle* 30° ; *head pyramidical ; visage oblique ; os hyoides very ventricose, outside prominent ; the anterior hands provided with a thumb ; tail very long, naked at the lower part of the extremity.*

68. 1. *S. M. Seniculus* (The Mono Colorado, or Red Howling Monkey). Upper part of the body fire-red ; head, extremities, and tail very lively deep red ; face naked, black.

Simia seniculus, *Lin. Sys. Nat.* 37. Mycetes seniculus, *Illiger*, 70. Stentor seniculus, *Geoff. Ann. Mus.* xix. 107.

Alouate, *Buff.* xv. Mono colorado, *Humb. Obs. Zool.* 342. Royal Monkey, *Penn. Quad.* 215.

Icon. *Buff.* xv. *t.* 5. *Suppl.* vii. *t.* 15. *Audeb. Hist.* 5. *f.* 1. *Schreb. t.* 25. *b. Griff. Quad. t.* 27.

Inhabits Guyana, near Carthagena ; the Banks of the river Saint Magdeleine, and Brazil.

69. 2. *S. M. Ursinus* (The Araguato). Uniform golden-red ; face partly covered with hairs.

Stentor ursinus, *Geoff. Ann. Mus.* xix. 108. Mycetes ursinus, Desm. Mamm. 78.

Araguato, *Humb. Obs. Zool.* 329.

Icon. *Humb. Obs. Zool. f.* 30.

Inhabits the Province of Venezuela, New Andalusia, New Barcelona, and the Shores of the Orinoco and Brazil.

70. 3. *S. M. Stramineus* (The Arabata). Fur straw-yellow ; hairs in the middle yellow, with the base and apex brown.

Stentor stramineus, *Geoff. Ann. Mus.* xix. 108. Mycetes stramineus, *Desm. Mamm.* 78.

Arabata, *Gumil. Oren. f.* 295.

Icon. ——

Inhabits Para.

71. 4. *S. M. Fuscus* (The Guariba). Fur chestnut-brown ; back and head becoming chesnut ; the extremities of the hairs golden.

Simia belzebuth, *Lin. Sys. Nat.* 37. Stentor fuscus, *Geoff. Ann. Mus.* xix. 108. Mycetes fuscus, *Desm.* 79.

Guariba, *Margr. Brazil.* 226. Ouarin, *Buff.* xv. 5. Preacher Monkey, *Penn. Quad.* 214.

Icon. *Marg. Braz.* 226. *Buff. Suppl.* vii. *t.* 26. *Ency. Method. t.* 15. *f.* 4.

Inhabits Brazil.

72. 5. *S. M. Flavicaudatus* (The Choro). Fur blackish-brown, darker on the back ; tail, with two yellow stripes on each side.

Stentor flavicaudatus, *Geoff. Ann. Mus.* xix. 108. Mycetes flavicaudatus, *Desm. Mamm.* 79.

Choro, *Humb. Obs. Zool.* 343.

Icon. ——

Inhabits New Grenada and the Banks of the Amazon.

27

73. 6 *S. M. Niger* (The Caraya). Fur fine black in the males ; with the sides and lower part of the body yellow on the young and females.

Stentor niger, *Geoff. Ann. Mus.* xix. 108. Mycetes niger, *Desm. Mamm.* 79.

Caraya, *Azara, Quad. Parag.* VII. 208. *Humboldt. sp.* 11.

Icon. ——

Inhabits Paraguay, Bahia, and the interior of Brazil.

74. 7. *S. M. Rufimanus* (the Red-handed Howler). Black, hands and end of the tail red, face and lower part of the body naked.

Mycetes rufimanus. *Kuhl. MSS. Desm: Mamm.* 79.

Icon.

Inhabits ——. Formerly Bullock's Museum, now in M. Temminck's.

IV. CEBUS, Erxleben. *Facial angle* 60° ; *head round, muzzle short, the os hyoides not prominent; tail prehensile, hairy at the lower part of the end* *.

75. 1. *S. C. Robustus.* Fur brown, upper part of the head, neck, and a line surrounding the face, black, arms clear yellow, lower part of the neck and belly reddish chestnut in the males, and of a pale yellowish brown in the young and females.

Cebus robustus, *Kuhl. MSS. Desm. Mamm.* 80.

Icon. ——

Inhabits Brazil.

76. 2 *S. C. Apella* (The Weeper Monkey). Fur brown,

---

* The species proper to this subdivision have but little distinctive character, and authors differ considerably as to their real number. Brisson described three, Linnæus four, Gmelin six, Buffon two, and finally the Baron Cuvier inclines to the opinion that there is but one. Pending this uncertainty, we shall, in conformity with our general plan, notice the several species as indicated by previous writers, together with their synonyms, with this general observation as to the uncertainty of their distinctiveness.

deeper above and paler beneath ; top of the head, tail and feet blackish-brown ; face brown, surrounded by blackish-brown hairs ; outer part of the arms and lower part of the neck yellowish-brown.

Simia apella, *Lin. Sys. Nat.* 42. Cebus apella, *Desm. Mam.* 71.

Sajou brun. *Buff.* xv. Sajou *Audeb. Hist.* 5. § 2.

Icon. *Buff.* xv. *t.* 4. *Audeb. Hist.* 5. § 2. *t.* 2. *Ency. Méth. f.* 2. *Schreb. t.* 28.

Inhabits Guyana and Terra Firma.

77. 3. *S. C. Griseus* (The Gray Sajou). Fur yellow-brown, variegated with grayish above, and clear yellow beneath ; head capped with black ; beard none ; face surrounded with brown-black hairs ; and sometimes white under the neck and chest.

Cebus barbatus, *Geoff. Ann. Mus.* xix. 110. C. griseus, *Desm. Mam.* 81.

Sajou gris, *Buff.* xv. Sajou, *F. Cuvier Mam.*

Icon. *Buff.* xv. *t.* 5. *Ency. Méthod. t.* 16. *f.* 3. *F. Cuv. Mam.* Inhabits——— ?

78. 4. *S. C. Barbatus* (The Bearded Sapajou). Fur reddish-gray ; belly red ; beard prolonged on the cheeks ; hairs long and soft.

Cebus barbatus, *Geoff. Ann. Mus.* xix. 112. C. albus, *Geoff. Ann. Mus.* xix. 112. (2.)

Sai var. *Audeb. Hist.*

Icon. *Audeb. Hist.* 5. § 2. *t.* 6. *Ann. Mus.* xix. *f.* 12. (2.)

Inhabits Guyana.

This species varies from gray to white, according to its age and sex, from whence the *C. albus* of Geoffroy.

79. 5. *S. C. Frontatus* (The Fearful Monkey). Fur nearly uniform brown-black, with the top of the head and the extremities of the limbs darker ; hairs of the forehead quite

straight, elevated perpendicularly; with some scattered white hairs round the mouth, and on the anterior hands.

Simia trepida, *Lin. Sys. Nat.* 39 ? (1.) Cebus trepidus, *Geoff. Ann. Mus.* xix. 110 ? (1.) C. frontatus, *Kuhl. MSS, Desm. Mam.* 82.

Tufted-tailed ape. *Edw. Glean.* Fearful monkey, *Penn. Quad.*

Icon. *Edw. Glean.* 312? (1.) copied. *Ency. Méth. t.* 17. *f.* 43.

Inhabits.—— Mus. Paris.

The specimen from whence the above description was taken by Dr. Kuhl, is very like that figured by Edwards, but the white hairs were not so abundant.

80. 6. *S. C. Niger* (the Black Sapajou). Fur deep brown; face, hands and tail black; forehead and hinder parts of the cheeks covered with yellowish hairs.

Cebus niger, *Geoff. Ann. Mus.* xix. 111. Cebus apella var. *Humboldt, Zool. Obs.* 323.

Sapajou nègre, *Buffon, supp.* vii. Sajou brun. var. *Humb.*

Icon. *Buff.* vii., *t.* 18, copied *Ency. Méthod. t.* 8. *f.* 4.

Inhabits ——

81. 7. *S. C. variegatus* (the varied Sapajou). Fur blackish, sprinkled with golden yellow; belly reddish; hairs of the back of three colours, the roots brown, then red, and the apex black; the fur very soft and formed of very long woolly hair; head round; muzzle prominent; with the space between the eyes blackish brown.

Cebus variegatus, *Geoff. Ann. Mus.* xix. 111.

Icon. ——

Inhabits ——

82. 8. *S. C. fulvus* (the yellow Sapajou). Fur entirely fulvous; hair silky, straight, not waved.

Simia flavus, *Schreb.* Cebus flavus, *Geoff. Ann. Mus.* xix. 112. C. fulvus, *Desm. Mam.* 83.

Icon. *Schreb. t.* 31, B.

30

Inhabits the Brazils.

When young the upper part of the head is red, the dorsal line, tail and limbs, chestnut red, with the rest yellow.

83. 9. *S. C. albifrons* (the Ouavapavi). Fur gray, paler beneath; top of the head black; forehead and orbits white; and extremities yellowish brown.

Cebus albifrons, *Humboldt, Obs. Zool.* 323. *Geoff. Ann. Mus.* xix. 111.

Sapajou Ouavapavi, *Humboldt, Obs. Zool.*
Icon.

Inhabits the vicinity of the Orinoco, in troops.

84. 10. *S. C. lunatus*, (the Spectacle Sapajou). Fur blackish; head, forehead, and anterior extremities black; with a white band across each cheek, joining the eye-brow with the angle of the mouth.

C. lunatus *Kuhl. MS. Desm. Mam.* 84.

Icon. ——

Inhabits ——

Museum of the Academy of Heidelberg.

85. 11. *S. C. xanthosternos* (the yellow chested sapajou). Fur chestnut; with the lower part of the neck and chest of a very pale reddish yellow.

Cebus xanthosternos, *Kuhl. MS. Desm. Mam.* 84.

Icon. ——

Inhabits Brazil, between 15°. 30. south latitude, and the river Belmont, *Prince Maximilian*.

86. 12. *S. C. fatuellus* (the Horned Monkey). Fur of the back chestnut; sides paler; belly bright red; extremities and tail black brown; with two strong brushes of hair elevated from the base of the forehead.

Simia fatuellus, *Lin. Sys. Nat.* 42. Cebus fatuellus, *Geoff. Ann. Mus.* xix. 109.

Le Sajou cornu, *Buff. Supp.* vii. 100. The Horned Monkey, *Pennant, Quad.* 221.

Icon. *Buff. Supp.* vii. *t.* 29. *Ency. Méth. t.* 17, *f.* 3. *Audeb. Hist.* 5, § 2, *f.* 3. *Schreb. t.* 27, *B. Shaw, Zool. t.* 28.

Inhabits Guyana.

87. 13. *S. C. cirrifer* (the crowned Sapajou). Fur chestnut brown; the crown of the head, extremities and tail, blackish brown; with a much elevated tuft of hairs in the shape of a horse shoe on the upper part of the forehead; head round.

Cebus cirrifer, *Geoff. Ann. Mus.* xix. 110.

Icon——.

Inhabits Brazil?

88. 14. *S. C. capucinus* (Capuchin Monkey). Fur from gray brown to olivaceous gray; crown of the head and extremities black; forehead, cheeks and shoulders, whitish gray.

Simia capucina, *Lin. Sys. Nat.* 42. Cebus capucinus, *Geoff. Ann. Mus.* xix. 111. Callithrix capucinus. *Illiger Prod.* 71.

Sai. *Buff.* xv. 51. Capucin monkey, *Pennant, Quad.* 218.

Icon. *Schreb. t.* 29. *Buff.* xv. *t.* 8. *Ency. Méthod. t.* 16. *f.* 4.

Inhabits Guyana.

89. 15. *S. C. hypoleucus* (the Cariblanco). Fur black; the forehead, sides of the head, throat, and shoulders whitish.

Simia hypoleuca, *Humb. prod.* 336? Cebus hypoleucus, *Desm. Mam.* 85.

Sai à gorge blanche, *Buff.* xv.

Icon. *Buff.* xv. *t.* 9. *Audeb. Hist.* 5, § 2, *f.* 8. *Ency. Méth. t.* 17. *f.* 1.

Inhabits Guyana.

V. CALLITHRIX, Cuvier. *Facial angle of* 60°.; *head round; muzzle short; nostrils narrower than the range of upper cutting teeth; tail not prehensile, covered with short hairs; ears very large. Living in troops, springing from branch to branch, and eating fruit, eggs, or young birds.*

90. 1. *S. C. sciureus* (the Squirrel Monkey or Caimiri). Fur olive gray; muzzle black; arms and legs bright red.

Simia sciureus, *Lin. Sys. Nat.* 43. *Callithrix sciureus, Geoff. Ann. Mus.* XIX. 113.

Saimiri, *Buff.* XV. Titi, *Humboldt. Obs. Zool.* 322. Orange Monkey, *Pennant, Quad.* 220. Squirrel Monkey, *Shaw. Zool.* I. 77 Caimeri, *South Americans.*

Icon. Buff. XV. *t.* 67. *Audeb. Hist.* 5, § 2. *f.* 7. *Ency. Méth. t.* 18. *f.* 1. *F. Cuvier, Mam. t.—Shaw, t.* 25.

Inhabits the Brazils and Guyana.

Var. b. Varied with red and black; double the size of the single coloured variety.

91. 2. *S. C. personata* (the Masked Monkey). Fur yellow gray; head and the four hands blackish; tail reddish.

Callithrix personatus, *Geoff. Ann. Mus.* XIX. 113.

The mask monkey *of Dealers.*

Icon.——

Inhabits parts of Brazil.

92. 3. *S. C. lugens* (the Widow Monkey). Fur blackish; throat and anterior hands white; tail a little longer than the body, black.

Simia lugens, *Humb. Obs. Zool.* 319. Callithrix lugens, *Geoff. Ann. Mus.* XIX. 113.

La viduita, *Humboldt.*

Icon. ——

Inhabits the forests on the banks of the Cassiquiare.

93. 4. *S. C. amictus* (the ruffed Sagoin). Fur blackish

brown; with a white half collar ; the hands of the anterior extremities dull pale yellow ; tail one-fourth longer than the body.

Simia Amicta, *Humb. Prod.* Callithrix amictus, *Geoff. Ann. Mus.* XIX. 114.

Icon.

Inhabits Brazils ?

Obs. Dr. Kuhl considers the three last to be varieties of the same species.

94. 5. *S. C. torquatus* (the Collared Callitrix). Fur chestnut brown ; yellow beneath ; with a white half collar ; tail a little longer than the body.

Callithrix torquata, *Hoffmansegg, Naturf.* 1809, x. 86. C. torquatus. *Geoff. Ann. Mus.* XIX. 114.

Icon. ——

Inhabits Brazils.

Obs. Only known by Hoffmansegg's description.

95. 6. *S. C. Moloch* (the Moloch). Fur ash-coloured, formed of annulated hairs above; temples, cheeks, and belly bright red ; end of the tail and hands grayish white.

Cebus Moloch, *Hoff. Naturf.* 1809, x. 96. Callithrix Moloch, *Geoff. Ann. Mus.* XIX. 114.

Icon. ——

Inhabits Para.

96. 7. *S. C. Melanochir* (the black handed Sagoin). Fur ash-coloured ; hinder part of the back, loins, and extremity of the tail of a reddish brown, anterior hands sooty black.

Callithrix melanochir, *Kuhl. MSS. Desm. Mamm.* 88. *C.* incanescens, *Lichtenstein, MSS.*

Icon. ——

Inhabits Brazil.

97. 8. *S. C. Infulatus* (Mitred Sagoin). Fur, above gray,

beneath yellowish red, with a large white spot, surrounded with black beneath the eyes; tail, origin reddish yellow, end black.

Callithrix infulatus, *Lichtenstein et Kuhls. MSS. Desm. Mamm.* 88.

Icon. ——

Inhabits Brasil.

VI. AOTES, Humboldt. *Facial angle——? head round and large, muzzle short; ears very short; eyes large, close together; tail long, covered with short hairs.*

98. 1. S. A. *Trivirgatus* (the Douroucouli). Fur ash-coloured; belly yellowish red; with three parallel brown lines, extended from the forehead to the occiput,

Aotus trivirgatus, *Humboldt, Zool. Obs. tab.* 806.

Cara rayada *Missionaries of Orinoco*, Douroucouli, *Humboldt. l. c.*

Icon. *Humboldt, Zool. Obs. tab.* 28, *Ency. Méthod. Supp. t.* 1. *f.* 2. *Griff. Quad. t.* 14.

Inhabits the thick forest of the Banks of the Cassiquiare, and upper Orinoco.

VII. PITHECIA, Desmarets. *Facial angle of* 60°, *head round; muzzle short; nostrils wider than the range of upper cutting teeth; ear moderately rounded; tail not prehensile covered with long hair; nocturnal.* Called Night Apes.

99. 1. S. P. *Satanas* (The Couxio). Fur, black brown on the males, red brown on the females; the hair of the head thick, covering the whole head and falling on the forehead; the beard very thick rounded; tail nearly as long as the body.

Cebus Satanas. *Hoff. Nat. Fors.* x. 93. Pithecia Satanas, *Geoff. Ann. Mus.* xix. 116. Simia Sangulata, *Trail. Wern. Trans.* vii. 167?

Couxio, *Humb. Obs. Zool.* 314.

Icon. *Humb. Zool. Obs. t.* 27. copied *Ency. Méthod. Sup. t.* 1. *f.* 4. *Werm. Trans.* VII. *t.* 9 ?

Inhabits the banks of the Orinoco.

100. 2. *S. P. Chiropotes* (the Hand-drinking Saki). Fur reddish chestnut; the hair of the head thick separated in the middle and recurved into two distinct toupees, on each side of the head; beard long and tufted.

Simia chiropotes, *Humboldt. Zool. Obs.* 113. Pithecia chiropotes, *Geoff. Ann. Mus.* XIX. 116.

Capucin de L'Orinoque, *Humboldt. l. c.*

Icon. ——

Inhabits the Desert of the Upper Orinoco and other parts of Guyana.

101. 3. *S. P. Rufiventer* (Fox-tailed monkey). Fur reddish brown; belly red; hairs brown at the origin, annulated with red and brown toward the extremity, hair of the head radiating on the top and bordering the forehead; beardless ; tail nearly as long as the body.

Simia pithecia, *Lin. Sys. Nat.* 40. Pithecia rufiventer, *Geoff. Ann. Mus.* XIX. 116.

Saki, *Buff.* XV. 90. Yarke, Singe de nuit, *Buff. Supp.* VII. 114. Fox-tailed monkey. *Penn. Quad.* 222.

Icon. *Buff.* XV. *t.* 12. *Supp.* VII. *t.* 30, 31. *Ency. Méthod. t.* 8. *f.* 3. *Mus. Leverian. t.* 5. *Shaw,* I. *t.* 25. *Griff. Quad. t.* 13.

Inhabits, French Guyana.

102. 4. *S. P. Miriquouina* (the Miriquouina). Fur, gray-brown above, cinnamon beneath; hairs of the back annulated black and white at each end; two white spots over each eye ; beard none ; tail a little longer than the body.

Pithecia miriquouina, *Geoff. Ann. Mus.* XIX. 117.

Miriquouina *Azara Hist. Paraguay,* II. 243.

Icon. ——

Inhabits the South Banks of the River Paraguay.

103. 5. *S. P. Rufibarba* (the Red bearded Saki). Fur,

above black-brown, beneath pale red; pale red above the eyes; tail pointed at the end.

Pithecia Rufibarba, *Kuhl. MSS. Desm. Mam.* 90.
Icon. ——
Inhabits Surinam. Mus. M. Temminck.

104. 6. *S. P. Ochrocephala* (Yellow-headed Saki). Fur, above of a clear chestnut, beneath yellowish red ash-colour; hands and feet black-brown; hairs of the forehead and circumference of the face of an ochreous yellow.

Pithecia ochrocephala, *Kuhl. MSS. Desm. Mamm.* 91.
Icon. ——
Inhabits Guyana.

105. 7. *S. P. Monachus* (the Monk). Fur varied with large spots of brown and dirty yellowish white; hairs brown at their base, red and golden near the extremities; hair of the head radiating from the occiput and bordering the vertex; tail a little longer than the body; beard none.

Pithecia monachus. *Geoff. Ann. Mus.* xix. 116.
Saki Moine. *Desm. Mam.* 91.
Icon. *Buff. sup.* vii. *t.* 30.
Inhabits Brazil.

106. 8. *S. P. Leucocephalus* (the Yarke). Fur black; head surrounded with dirty white; each hair of only one colour; tail nearly as long as the body; beard none.

Simia Pithecia. *Lin. Sys. Nat.* 40? Pithecia Leucocephala. *Geoff. Ann. Mus.* xix. 117.
Saki. *Buff.* xv. 9. part; Yarque. *Buff. sup.* vii. not the figure.
Icon. *Schreb. t,* 32. *Audeb. Hist.* 6. § 1. *f.* 2. *Buff.* xii *t.* 12.
Inhabits Guyana.

107. 9. *S. P. Melanocephala* (the Cacajao). Fur yel-

lowish brown; head black; beard none; tail one sixth shorter than the body.

Simia Melanocephala. *Humboldt. Zool. Obs.* 316. Pithecia Melanocephala. *Geoff. Ann. Mus.* xix. 117.

Cacajao, caruiri, chacuro, and mono—rabon *South Americans.*

Icon. *Humb. Obs. Zool. t.* 29.

Inhabits banks of the Cassiquira and Rio Negro.

Genus II.—OUISTITI or HAPALES.

FORM quadrupedal; cutting teeth $\frac{4}{4}$, canine $\frac{1}{1}$, grinders $\frac{6-6}{6-6}$, $=36$. Extremities pentadactylous, the thumb of the anterior hands in the same direction as the fingers, and not opposable; all the fingers furnished with claws instead of flat nails.

I. Jacchus. Geoffroy, *Facial angle of 50°; head round; muzzle short; occiput prominent; tail very long, covered with short hairs; upper intermediate cutting teeth larger than the lateral ones; the lower cutting teeth long, narrow; upper canine conical, lower very small; grinders acutely tubercular.*

108. 1. *S. J. Vulgaris* (the Stunted Monkey or Jacchus). Fur ash colour; the buttocks and tail annulated with gray, brown and ash; with a white spot on the forehead; and a tuft of very long ash-coloured hairs before and behind the ears; head, and upper part of the neck and shoulders reddish brown.

Simia Jacchus. *Lin. Sys. Nat.* 40. Jaccus vulgaris. *Geoff. Ann. Mus.* xix. 119. Hapale Jaccus, *Illiger Prod.* 72.

Ouistiti. *Buff.* xv. 96. Sagoin. *Clus. exot.* 372. Sanglin, or Cagui minor, *Edw. Glean.* Striated Monkey. *Pennant Quad.* 224.

Icon. *Buff.* xv. *t.* 14. *Ency. Méth. t.* 18. *f.* 4. *Audeb. Hist. Fan.* 6. §. 2. *t.* 4. *Schreb. t.* 33. *Fred. Cuvier Mam.*— *Edw. Glean, t.* 218. *Shaw. Zool.* 1. *t.* 25. *Griff. Quad. t.* 19.

Inhabits Guyana and Brazil.

Variety b. Fur red; buttocks annulated red and gray.

109. 2. *S. J. penicillatus* [the tufted jacchus]. Fur ash-coloured; buttocks and tail annulated brown and ash-colour, forehead with a white spot; and a tuft of black very long hairs before the ears; head and upper part of the neck black.

Jacchus penicillatus, Geoff. Ann. Mus. xix. 119.

Icon. ——

Inhabits Brazil.

110. 3. *S. J. leucocephalus* (the white-headed Jacchus). Fur red; head and chest white; upper part of the neck black; tail annulated brown and ash-colour; and a tuft of very long black hairs before and behind the ears.

Jacchus leucocephalus, *Geoff. Ann. Mus.* xix. 1. 20. Simia Geoffroyi *Humb. Zool.*

Icon. ——

Inhabits Brazil.

111. 4. *S. J. auritus* [the great eared Jacchus]. Fur black varied with brown; tail annulated black and ash-colour; with a white spot on the forehead; and a tuft of very long white hairs covering the inside of the ears.

Jacchus auritus, *Geoff. Ann. Mus.* xix. 119. Simia auritus, *Humb. Prod. Zool. Obs.*

Icon.——

Inhabits Brazil.

112. 5. *S. J. humeralifer* (the white-shouldered Jacchus). Fur chestnut brown; tail slightly annulated; ash-colour; shoulders, chest and arms white.

Jacchus humeralifer, *Geoff. Ann. Mus.* xix. 120. Simia humeralifer, *Humb. Prod.Zool. Obs.*

Icon. ——

Inhabits Brazil.

Obs. Some Naturalists regard the five last named as mere varieties. All of them have the tail annulated.

113. 6. *S. J. melanurus* (the Black Tailed Jacchus). Fur brown above; yellow beneath; tail of a uniform black.

Jacchus melanurus, *Geoff. Ann. Mus.* xix. 120.

Icon. ——

Inhabits Brazil.

Obs. Dr. Kuhl considers this species as the link between the S. Jacci, and S. midas, or the Ouistitis and the Tamarins.

114. 7: *S. J. Argentatus,* (the Mico,) or fair monkey. Fur white; face, hands and feet red ; tail black.

Simia argentata *Gmelin, Sys. Nat.* 41. Jacchus argentatus, *Geoff. Ann. Mus.* xix. 120.

Mico, *Buff.* xv. 121. Fair monkey, *Pennant Quad.* 226.

Icon. *Buff.* xv. *t.* 18. *Ency. Méthod. t.* 19. *f.* 2. *Audeb. Hist. f.* 6. § 2. *f.* 2. *Schreb. t.* 36. *Shaw. Zool.* i. 26. *Griff. Quad. t.* 21.

Inhabits Para.

Variety b. Tail white.

II. Midas, *Facial angle of* 50°, *head round, muzzle short, forehead extended, ears large, occiput prominent; tail very long, covered with short hairs; teeth, pointed ; canine teeth, conical, strong; grinders acutely tubercular.*

115. 1. *S. M. rufimanus.* (The Tamarin or great-eared monkey). Fur black ; buttocks variegated with gray ; hands and feet yellowish red.

Simia Midas, *Lin. Sys. Nat.* 42. Midas rufimanus, *Geoff. Ann. Mus.* xix. 121. Jacchus rufimanus, *Desm. Mam.* 94.

Tamarin, *Buff.* xv. 92. Little black monkey, *Edwards, Glean.* Great-eared monkey *Pennant,* 223. Temary, *Guyaness.*

Icon. *Buff.* xv. *t.* 13. *Ency. Méthod. t.* 19. *f.* 3. *Edwards Glean. t.* 196. *Schreb. t.* 37. (*from Edw.*) *Audeb. Hist.*

40

6. § 2. *f. 5. Shaw. Zool. t.* 26. *Griff. Quad. t.* 18.
Inhabits Guyana and Paragua.

116. 2. *S. M. ursulus* (the Negro Tamarin). Fur
black ; back waved with bright red ; hands black.

Simia Midas Var. *Shaw. Zool.* I. 65. Saguinus ursulus
Hoff. Natur. x. 101. Midas ursulus, *Geoff. Ann. Mus.* XIX.
121. Jacchus ursulus, *Desm. Mam.* 94.

Tamarin nègre *Buff. Supp.* VII. 116.

Icon. *Buff. Supp.* VII. *t.* 32. *Audeb. Hist.* 6. § 2. *f.* 6.
F. Cuvier, Mamm. Lithog.

Inhabits Para.

Dr. Shaw considered this a variety of the preceding.

117. 3. *S. M. labiatus* (the white lipped Tamarin). Fur
blackish, ferrugineous, red below; head black; nose and
edges of the lips white.

Midas labiatus, *Geoff. Ann. Mus.* XIX. 121. Jacchus
labiatus, *Desm. Mam.* 95.

Icon. ——

Inhabits Brazil.

118. 4. *S. M. chrysomelas* (the Yellow fronted Tamarin).
Fur black ; forehead and upper-side of the tail golden
yellow ; front arms, knees, chest, and sides of the head
chestnut brown.

Midas chrysomelas, *Kuhl. MSS.* Jacchus chrysomelas,
Desm. Mam. 95.

Icon. ——

Inhabits the large forests of Brazil and Para.

119. 5. *S. albifrons* (the white fronted Tamarin). Fur
black, slightly variegated with white; face black; forehead, sides of the neck and throat, covered with very short
white hairs; occiput and circumference of the ears gar-
nished with long straight deep black hairs; tail a little

41

longer than the body, brown varied with white ; round the arms reddish.

Jacchus albifrons. *Act. Stockholm.* 1819.

Icon. *Act. Stockholm.* 1819. *t.—*

Inhabits South America.

Obs. the distribution of the colours of this species is very similar to *Midas chrysomelas.*

120. 6. *S. M. rosalia* (the Silky Tamarin). Fur golden red ; hair of the head long.

Simia rosalia, *Lin. Sys. Nat.* 41. Midas rosalia. *Geoff. Ann. Mus.* xix. 121. Jacchus Rosalia. *Desm. Mam.* 95. Hapale rosalia.

Marikina *Buff.* xv. 108. Silky monkey, *Pennant Quad.* Lion monkey *Dealers.*

Icon. *Illiger, Prod.* 72.

Icon. *Buff.* xv. *t.* 16. *Ency. Méth. t.* 19. *t.* 1. *Schreb. t.* 35. *Audeb. Hist.* 6. § 2. *f.* 3. *Shaw. Zool.* 1. *t.* 25. *F. Cuvier, Mamm. t.—Griff. Quad. t.* 20.

Inhabits Guyana, and south part of Brazil near Rio Janeiro.

Variety *b.* The fur variegated with black and red.

Var. *c.* The fur and tail fine red.

121. 7. *S. M. leoninus* (the Leonine Tamarin). Fur olivaceous brown; hair of the head long; face black ; mouth white ; tail black above, brown beneath.

Simia leoninus, *Humbodlt. Zool. Obs.* 14. Midas Leoninus, *Geoff. Ann. Mus.* xix. 121. Jacchus leoninus. *Desm. Mam.* 95.

Leoncito *Humb. l. c.*

Icon. *Humb. l. c. t.* 5.

122. 8. *S. M. Œdipus* (the Pinche). Fur yellow-brown

42

above, white below ; beard long, silky white ; tail bare, reddish, upper part black.

Simia Œdipus *Lin. Sys. Nat.* 41. Midas Œdipus. *Geoff. Ann. Mus.* xix. 121. *t.* 1. Jacchus œdipus. *Desm. Mam.* 96.

Pinche, *Buff.* xv. 114. Titi de Carthagène. *Humboldt, Obs. Zool.* 337. Little lion monkey. *Edw. Glean.* 195. Red-tailed monkey. *Pennant Quad.* 225.

Icon. *Buff.* xv. *t.* 17. *Ency. Méthod. t.* 10. *f.* 5. *Edwards Glean. t.* 195.

Schreb. t. 34. (cop. Edw.) *Audeb. Hist. f.* 6. § 2. *f.* 1 *Shaw. Zool.* 1. *t.* 25.

Inhabits Carthagena, the mouth of the Rio Sinu, and Guyana.

Genus III.—Lemur.

Form approaching that of quadrupeds ; cutting teeth $\frac{4}{4}$ or $\frac{4}{6}$; canine $\frac{1\cdot1}{1\cdot1}$; grinders $\frac{5\cdot5}{5\cdot5}$ or $\frac{5\cdot5}{4\cdot4}$, obtusely tubercular ; head long, triangular ; nostrils terminal ; ears short, hidden ; eyes small ; tail mostly long ; fur woolly.

Eats fruit and roots.

Inhabits Madagascar.

I. Lichanotus Illiger. *Cutting teeth* $\frac{4}{4}$; *canine* $\frac{1\cdot1}{1\cdot1}$ *grinders* $\frac{5\cdot5}{5\cdot5}$. *Tail very short, or none.*

123. 1. *L. L. Niger*, (the Indri). Fur blackish.

Lemur indri, *Gmelin Sys. Nat.* 42. Indris brevicaudatus, *Geoff. Mag. Ency.* vii. 20, Lichanotus Indri *Illiger,* Prod. 72.

Indri, *Sonnini Voy.* 142. Indri Macauco, *Penn. Quad.* i. 228.

Icon. *Sonn. Voyage, t.* 8. *Audeb. Hist.—Ency. Méthod. Supp. t.* 2. *f.* 5. *Shaw. Zool. t.* 32.

Inhabits Madagascar.

II. Indris, Lacépède. *Cutting teeth* $\frac{4}{4}$; *grinders* $\frac{5\cdot5}{5\cdot5}$; *tail very long.*

43

125 1. *L. I. Laniger* (Flocky lemur). Fur yellow.

Lemur Laniger, *Gmel. Sys. Nat.* i. 44. Indris longicaudatus, *Geoff. Ann. Mus.* xix. 138. Lichanotus laniger. *Illiger, Prod.* 72.

Maki à bourré, *Sonn. Voy.* vii. 142. Maki fauve, *Buff. Supp.* vii. Flocky Lemur, *Shaw. Zool.* i. 99.

Icon. *Sonn. Voy.* ii. *t.* 89. *Buff. Supp.* vii. *t.* 35. *Shaw. Zool.* i. *t.* 34.

Inhabits Madagascar.

III. PROSIMIA, Brisson. *Cutting teeth* $\frac{4}{4}$; *lower horizontal; grinders* $\frac{6 \cdot 6}{6 \cdot 6}$; *tail very long.*

125. 1. *L. P. Macaco* (Ruffed Lemur). Fur varied with large regular patches of white and black ; tail black ; hairs of the cheeks very long.

Lemur Macaco, *Lin. Sys. Nat.* 44.

Vari. *Buff.* xiii. 174. Ruffed Lemur, *Pennant. Quad.* i. 231.

Icon. *Pet. Gaz. t.* 27. *f.* 5. *Buff.* xiii. *t.* 27. ♂. *Ency. Méth. t.* 20. *f.* 2. *Audeb. Hist. f.* 5. 6.. *Schreb. t.* 49.

Inhabits Madagascar.

Variety b. Fur white and gray brown.

126. 2. *L. P. Ruber* (Black and red Lemur). Fur of a fine reddish chestnut ; head, hands, belly, and tail black ; with a white spot on the neck.

Lemur ruber, *Perron & Lesueur. Geoff. Ann. Mus.* xix. 159.

Maki roux. *Fr. Cuvier. Mam. Lithog.*

Icon. *Fred. Cuv. Mam. Lithog. t.—Griff. Quad. t.* 33.

Inhabits Madagascar.

127. 3. *L. P. Catta* (Ring-tailed Lemur). Fur reddish ash-coloured above ; ash-coloured on the limbs, and white below ; tail annulated, black and white.

Lemur Catta, *Lin. Sys. Nat.* 45.

Mococo, *Buff.* xiii. 174. Macauco, *Edw. Glean.* Ring-tailed Lemur, *Shaw. Mus. Lever.* 43. Ring-tailed Macauco, *Pennant. Quad.* i. 130.

Icon. *Buff.* xiii. *t.* 22. *Ency. Method. t.* 20. *f.* 3. *Audeb. Hist. f.* 4. *Edw. Glean. t.* 197. *Schreb. t.* 41. *Shaw. Zool.* i. *t.* 35. *Mus. Lever. t.* 11. *F. Cuvier. Mam. Lithog. t.—Griff. Quad. t.* 31.

Inhabits Madagascar.

128. 4. *L. Niger* (Black Lemur). Fur black; with long hairs under the neck.

Lemur niger, *Geoff. Ann. Mus.* xix. 159. L. Macaco. Var. *Shaw.* 1. 98.

Black Maucauco, *Edw. Glean.* vii. 217.

Icon. *Edw. Glean. t.* 217.

Inhabits Madagascar.

129. 5. *L. P. Mongooz* (the Mongooz). Fur yellowish-gray above, white below; the circumference of the eyes and forehead black.

Lemur mongoz, *Lin. Sys. Nat.* 44.

Mongous, *Buff.* xiii. 198. Mongooz, *Edw. Glean.* iii. 216. Woolly Macauco, *Pennant. Quad.*

Icon. *Buff.* xiii. *t.* 26. *Edw. Glean.* iii. *t.* 216. *Ency. Meth. t.* 20. *f.* 1. *Schreb. t.—Shaw. Zool. t.* 33.

Inhabits Madagascar.

130. 6. *L. P. Fulvus* (Yellow Lemur). Fur brown above, and gray below; forehead elevated, and prominent.

Lemur fulvus, *Geoff. Ann. Mus.* xix. 161. L. Mongoz var. *Shaw.* 96.

Grand Mongous, *Buff. Supp.* vii. 118.

Icon. *Buff.* vii. *t.* 33. *Geoff. Menag. t.—*

Inhabits Madagascar.

131. 7. *L. P. Albimanus* (White-handed Lemur). Fur gray,—brown above; sides of the neck cinnamon-red; chest and hands white; belly reddish.

Lemur albimanus, *Geoff. Ann. Mus.* xix. 160.

Maki aux pieds blancs, *Briss. Reg. Anim.* 221. Mongous, *Audeb. Hist.*

Icon. *Audeb. Hist. f.* 1.

Inhabits Madagascar.

132. 8. *L. P. Rufus* (Red Lemur). Fur golden-red above, yellowish-white beneath; sides of the face and chin white, with a black band extended from the face to the occiput.

Lemur rufus, *Geoff. Ann. Mus.* xix. 160.

Maki roux, *Audeb. Hist.*

Icon. *Audeb. Hist. f.* 2.

Inhabits Madagascar.

133. 9. *L. P. Collaris* (Collared Lemur). Fur red-brown above, yellow beneath; mane red; face lead-coloured.

Lemur collaris, *Geoff. Ann. Mus.* xix. 161.

Maki d'Anjouan. Mongous. Var. *Fred. Cuvier. Mamm?*

Icon. ——

Inhabits Madagascar.

The top of the head gray, and the fur yellower, in the female.

Obs. M. F. Cuvier considers this a variety of the Mongous.

134. 10. *L. P. Albifrons* (White-fronted Lemur.) Fur reddish-gray above, whitish beneath; forehead of the male white, of the female deep gray; with a black longitudinal line on the upper part of the head.

Lemur. albifrons.

L. Angouan. *Geoff. Ann. Mus.* xix. 161.

Makis aux pieds fauves. *Brisson. Règ. Anim.* i. 221?

Icon. *Audeb. Hist. f.* 3. *F. Cuvier. Mam. t.* —

Inhabits Madagascar.

135. 11. *L. P. Nigrifrons* (Black-fronted Lemur.) Fur above ash-coloured before, and reddish-gray behind; with a

black band on the forehead; belly and under part of the thighs red.

Lemur nigrifrons. *Geoff. Ann. Mus.* xix. 160.

Maki v. 1. *Brisson. Regn. Anim.* 220.

Icon. *Petiver.—Schreb. t.* 42.

Inhabits Madagascar.

136. 12. *L. P. Cinereus* (Ashy Lemur.) Fur above yellowish-gray, beneath dirty white.

Lemur-cinereus, *Geoff. Mag. Encycl.*

Petit maki, *Buff. Supp.* vii. Grisset, *Audeb. Hist.*

Icon. *Buff. Supp.* vii. *t.* 84. *Audeb. Hist. f.* 7.

Inhabits Madagascar.

V. Stenops, Illiger. *Cutting teeth* $\frac{4}{6}$; *lower horizontal; canine* $\frac{1 \cdot 1}{1 \cdot 1}$; *grinders* $\frac{6 \cdot 6}{5 \cdot 5}$; *limbs thin; tail none.*

137. 1. *L. S. Gracilis* (The Slender Loris.) Fur reddish, with a white spot on the forehead.

Lemur Loris, *Shaw Zool.* i. 93. Loris gracilis, *Geoff. Ann. Mus.* xix. 161. Loris ceylonicus, *Fischer. Anat.* 28.

Loris, *Buff.* xiii. Loris Macauco, *Pennant Quad.* 228.

Icon. *Seba Mus.* i. *t.* 35. *Buff.* xiii. *t.* 30. *Audeb. Hist. t.* 2. *Ency. Méthod. t.* 19. *f.* 4. *Shaw. Zool. t.* 31. *Fischer. Anat. Makis. t.* 7, 8, 9, and 18.

Inhabits the island of Ceylon.

IV. Nycticebus, Geoffroy, *cutting teeth* $\frac{2 \text{ or } 4}{6}$; *canine* $\frac{1 \cdot 1}{1 \cdot 1}$; *grinders* $\frac{6 \cdot 6}{5 \cdot 5}$; *tail more or less long; extremities shortish—Nocturnal.*

138. 1. *L. N. Bengalensis* (The Slow Lemur.) Fur red; dorsal line brown; muzzle large; cutting teeth $\frac{4}{6}$; tail very short.

Lemur tardigradus, *Lin. Sys. Nat.* 44. Loris paresseux, *Cuv. Reg. Anim. t.* 118. Nycticebus Bengalensis, *Geoff. Ann. Mus.* xix. 164. Stenops Bengalensis, *Illiger, prod.* 73.

Paresseux pentadactyle du Bengal, *Vosmaër.* Loris du

Bengale, *Buff. Supp.* vii. 125. Slow Lemur, *Shaw. Zool.* i. 81. Slow-paced Lemur, *Shaw. Spec. Lin.*

Icon. *Vosmar. t.* 6. *Buff. Sup.* vii. *t.* 36. *Ency. Méth. Sup. t.* 2. *f.* 6. *Audeb. Hist. t.* 1. *Shaw. Spec. Lin. t.* 5 ; *Zool.* i. *t.* 29. *Griff. Quad. t.* 34.

Inhabits Bengal.

139. 2. *L. N. Javanicus* (Javanese Loris.) Fur red ; dorsal line deeper ; muzzle narrow ; cutting teeth $\frac{2}{4}$; tail short.

Nycticebus Javanicus, *Geoff. Ann. Mus.* xix. 164.

Icon. ——

Inhabits Java.

140. 3. *L. N. Ceylonicus* (Ceylon Loris.) Fur brownish-black ; back quite black ; cutting teeth ——

Nycticebus Ceylonicus, *Geoff. Ann. Mus.* xix. 1. 64.

Cercopithecus Zeylonicus, *Seba. Mus.* i. 75.

Icon. *Seba. l. c. t.* 47. *f.* 1.

Inhabits Ceylon.

Obs. These two species are only known by Geoffroy's description.

V. Galago, Geoffroy. *cutting teeth* $\frac{2\;or\;4}{6}$; *lower horizontal; canine* $\frac{1 \cdot 1}{1 \cdot 1}$; *grinders* $\frac{6 \cdot 6}{5 \cdot 5}$; *ears very large; hinder legs long ; tail very long*

141. 1. *L. G. Madagascariensis* (Little Galago.) Fur red ; ears half as long as the head ; tail much longer than the body, covered with short hairs ; cutting teeth $\frac{4}{4}$.

Lemur murinus, *Cimelia Physica.* 25. Lemur pusillus, *Audeb. Hist.* Galago Madagascariensis, *Geoff. Ann. Mus.* xix. 166.

Little Lemur, *Brown Illus Zool.* 108? Rat de Madagascar, *Buff. Suppl.* vii. 149. Murine Maucauco and Little Maucauco, *Pennant Quad.* i. 232.

Icon. *Buff. Supp.* vii. *t.* 20. *Audeb. Hist. t.* — *Brown. Illus. Zool. t.* 44. *Miller, Cim. Phys. t.* 13.

Inhabits Madagascar.

142. 2. *L. G. crassicaudatus* (Great Galago.) Fur red gray ; ears two-thirds of the length of the head ; tail very tufted ; cutting teeth $\frac{4}{4}$.

Galago crassicaudatus, *Geoff. Ann. Mus.* XIX. 166. Ortolienus Galago, *Illiger, prod.* 74.

Le Grand galago, *Cuv. Reg. Anim.* I.

Icon. *Cuvier, R. Ann.* IV. *t.* 1. *f.* 1. *Desm. Nov. Dict. Hist. Nat.* XIII. *t. E.* 31.

Inhabits ——

143. 3. *L. G. Guiniensis* (Potto.) Fur red, ash-coloured when very young ; tail half the length of the body.

Lemur Potto, *Gmelin. Sys. Nat.* 42. Nycticebus potto, *Geoff. Ann. Mus.* XIX. 165. Galago guinensis, *Desm. Mam.* 104.

Potto, *Bosman Guin.* II. 30.

Icon. *Bosman. l. c. t.* 4.

Inhabits Guinea.

Obs. Like N. Bengalensis, but it has a tail in Bosman's figure, which is the only one of the animal.

144. 4. *L. G. Demidoffii,* (Demidoff's Galago). Fur red brown ; ears shorter than the head, tail longer than the body ; reddish, end tufted ; cutting teeth $\frac{2}{6}$.

Lemur minutus, *Cuv. Tab. element* 101. Galago Demidoffii, *Fischer, Act. Moscow,* i. 24. *Geoff. Ann. Mus.* xix. 166.

Icon. ——

Inhabits Senegal ? Size of the common rat ; muzzle blackish.

145. 5. *L. G. Senegalensis,* (Senegal Galago). Fur red gray, beneath white, ears longer than the head ; tail longer than the body, red, end tufted ; cutting teeth $\frac{2}{6}$.

Lemur Galago, *Schreb.* Lemur Galago, *Shaw, Zool.* I. 108. Galago Geoffroyii *Fischer. Act. Moscou,* I. 25. G. Senegalensis. *Geoff. Mém. sur Makis.* 20.

Galago. *Adanson Senegal.* Whitish Lemur, *Shaw. l. c.*
Icon. *Geoff. Mem. sur Makis, t. 1. Ency. Meth. Suppl. t.*
2. *f.* 7. *Schreb. t.*—
Inhabits Senegal.

VI. CHEIROGALEUS, *Geoffroy. Teeth—? ears short, oval;
whiskers large; tail long, tufted, cylindrical, re-convolute;
hair short.*

146. † 1. *L. C. Major* (Large Cheirogaleus). Fur deep
brown, particularly between the eyes ; length eleven inches.

147. † 2. *L. C. Medius* (Middle Cheirogaleus.) Fur
lighter ; eyes surrounded with black rings, length eight
inches.

148. † 3. *L. C. Minor* (Small Cheirogaleus.) Very like
the former, but pale, and only seven inches long.

This sub-genus, containing three species, was established
by Geoffroy, from drawings of Commerson. Geoffroy suggests that Pennant knew the last, and confounded it with
his *Madagascar Rat, Galago Madagascariensis.*

VII. TARSIUS. (Storr). *Cutting teeth $\frac{4}{2}$, unequal, canine $\frac{1\cdot1}{1\cdot1}$,
small; grinders $\frac{6\cdot6}{6\cdot6}$; ears large, naked ; hinder legs very long;
tarsi long, tail long.*

149. 1. *L. T. Spectrum.* Fur reddish, ears half as long
as the head.
Didelphis Macrotarsus. *Gmel. Sys. Nat.* 109. Lemur
Spectrum, *Pallas Glires,* 274. Lemur Tarsier, *Shaw,
Zool.* i. 105. Tarsius Daubentonii. & T. Pallassii. *Geoff.
Mag. Ency.* T. Spectrum, *Geoff. Ann. Mus.* xix. 168.
T. Macrotarsus, *Illiger. prod.* 74.
Tarsier, *Buff.* xiii. 87. Tarsier Maucauco. *Penn. Quad.*
I. 231. Woolly Gerboa, *Penn. Quad.* 298.
Icon. *Buff.* xiii. *t.* 9. *Audeb. Hist. Makis, t.* 1. *Ency.
Method. t.* 22. *f.* 5. *Shaw, Zool.* 1. *t.* 35.

Inhabits Amboyna and East India Islands.

150. 2. *L. T. Fuscomanus* (Yellow-handed Tarsier.) Fur clear brown, beneath grayish white ; ears twice as long as the head.

Tarsius Fischerii *Desm. Dict. Hist. Nat. ed.* i.—Tarsier Fuscomanus. *Fisch. Anat. Maki. Geoff. Ann. Mus.* xix. 198.

Icon. *Fischer. l. c. t.* 3. 4. *Ency. Méthod. Suppl. t.* 2. *f.* 8.

Inhabits Madagascar.

151. 3. *L. T. Bankanus.* Fur brown, ears, rounded, horizontal, much shorter than the head ; tail very thin, cutting teeth $\frac{2}{4}$.

Tarsier. Bancanus. *Horsfield, Java. fasc.* 2.

Icon. *Horsfield, Java, t.* Teeth and skull.

Inhabits Borneo, one of the East Indian Islands.

Obs. According to some Naturalists these species are said all to be only varieties of ages.

VIII. CHEIROMYS, Cuvier. *Cutting teeth $\frac{2}{2}$, strong; canine $\frac{0.0}{0.0}$, leaving a space ; grinders $\frac{4.4}{3.3}$; fore legs short, with the middle finger very long and thin; hind leg long; tail long, tufted ; teats two, inguinal.*

152. 1. *L. C. Madagascariensis* (Aye, Aye). Fur brown, coarse ; tail black.

Lemur psylodactylus. *Schreb. Suppl.*—*Shaw, Zool.* i. 109. Sciurus Madagascarensis, *Gmelin. Sys. Nat.* Cheiromys Madagascariensis, *Geoff. Ann. Mus.* Daubentonii. *Geoff. Mem. Decad. Phil. et Litt. v.* 28.

Aye Aye. *Sonnerat Voy. aux Ind.* ii. 142. Aye Aye Squirrel, *Pennant,* ii. 142. Long-fingered Lemur, *Shaw, Zool.* i. 109.

Icon. *Sonnerat Voy.* vi. 88. *Buff. Suppl.* vii. 68. *Geof. Mem. t.* — *Schreb. t.* — *Shaw, Zool.* i. *t.* 34. —Skul. *Cuvier. Reg. Animal, t.* 3.

Inhabits Madagascar.

IX. L. ? TUPAIA. *Raffles. Cutting teeth* $\frac{2}{8}$; *canine* $\frac{1 \cdot 1}{1 \cdot 1}$; *grinders* $\frac{1 \cdot 1}{6 \cdot 6}$; *body elongate; head triangular attenuated, blunt; eyes large; ears large; tail very long; teats four ventral. Diurnal; lives in trees.*

153. 1. *L. T. Tana* (The Tapia Tana). Head long, muzzle very pointed; fur above reddish brown, speckled with black beneath, and an oblique redder line on each shoulder.

Tupai Tana, *Raffles Trans. Lin.* xiii. 257.

Tupai Tana, *Sumatresse.*

Inhabits Sumatra. Mus. Col. Surg.

154. 2. *L. T. Javanica.* Head long, muzzle slightly pointed ; tail very long, fur above brown, speckled with gray ; beneath gray, and with a grayish white oblique line on the shoulder.

Tupai Javanica. *Raffl. Lin. Trans.* XIII. 257.

Sorex Glis, *Diard Asiatic Register.* x. Glisorex. *Desm.* Bangsring *Javanesse:*

Icon. *Horsfield Java,* t.

Inhabits Java.

155. 3. *L. T. Ferruginea.* Muzzle slightly pointed, fur ferruginous.

Tupaia ferruginea. *Raff. Lin. Trans.* XIII. 277. Cladobates ferruginea, *F. Cuvier. Mam. Lithog.*

Icon. *Horsfield Java,* t. Teeth *l. c.* and *F. Cuvier, l. c.*

Inhabits Java.

Obs. This genus is added to the quadrumana with a mark of doubt.

Order III.—CARNASSIERS.

TEETH of three sorts, incisives, canines, and cheek teeth, more or less of a trenchant or carnivorous character. Articulation of the lower jaw crosswise, so as to prevent any other than a vertical motion. Orbits not separated from the temporal fossæ. Zygomatic arch wide and elevated. Thumb of the anterior extremities never opposable to the other fingers or toes. Stomach simple, membranaceous. Intestines short.

Eats more or less of animal and vegetable matter in the different species, but never grass or leaves.

Habits various. More or less savage, as their physical traits are more or less of a carnivorous character.

Inhabits nearly all the habitable parts of the globe

This order is divided into four families, *viz.*

1. CHEIROPTERA.   3. CARNIVORA.
2. INSECTIVORA.   4. MARSUPIATA.

Family I.—CHEIROPTERA.

Fingers of the anterior extremities connected by a membrane, which spreads from the anterior to the posterior extremities, and in many of the species also connects the latter to each other, forming altogether an apparatus more or less effective for flight. Incisives various in number. Canines more or less strong. Cheek-teeth, in general, having their crowns furnished with several acute points; but in the first group of the first genus a single regular furrow or indentation passes along the whole series, both sides of each tooth approaching the figure of the transverse section of a cone, a little convex, notched on the upper edge from right to left. Mammæ, in general, two, pectoral.

53   2 F

Genus I.—Vespertilionidæ *, Bats, generally.

Anterior fingers excessively elongated, and the membrane between them spread over a large surface, thereby enabling the animal to keep up a continued and rapid flight.

⁎ With frugivorous cheek-teeth.

Sub-genus I. Pteropus. Brisson. *Incisive teeth* $\frac{4}{4}$, *conical in shape; canine* $\frac{1}{1}\frac{1}{1}$; *cheek-teeth* $\frac{5}{6}\frac{5}{6}$, *presenting a surface neither flat nor aculeate, each tooth having two roof-shaped ridges, forming a longitudinal furrow between them, extending along the whole series. No membranaceous appendage to the nose. Tail short, or wanting. Interfemoral membrane sloped off. The index-finger has a third phalanx and a nail. Tongue papillary. Habits nocturnal, gregarious. Regimen frugivorous.*

§ 1. Without a tail

153. 1. *P. Edulis*, (the great Black Pteropus, or Eatable Bat.) Black, with the upper part of the neck ochreous-red; back covered with black and white hairs intermixed; length of body one foot, expanse of wings five feet.

Pteropus Edulis, *Peron* and *Lesueur. Geoff. Ann. du Mus d'Hist. Nat. t.* xv. *p.* 90. Pteropus Javanicus, *Desmarest. Ency. Méthodique, sp.* 136, and *Horsfield's Zoological Researches, No.* iv.

Kalong of the *Javanese.* Malanon Bourou of the *Malays.* Icon. *Horsfield's Zool. Researches, No.* iv.

Inhabits Java.

Var. a. With a collar of lighter brown round the neck, and a general mixture of brown hairs with the black.

Obs. The external characters of this species are considerably subject to vary, and it seems probable that it exists

---

* The Greek termination is here employed to distinguish the whole genus of Bats collectively from the sub-genus, Vespertilio.

in various parts of India, more or less diversified in its co-lour.

154. 2. *P. Edwardsii* (Rousette of Edwards.) Red; back chestnut-brown.

Vespertilio Vampyrus, *Linn.* Great Bat of Madagascar, *Edwards.*

Icon. *Edwards's Birds, f.* 180.

Inhabits Madagascar.

155. 3. *P. Vulgaris* (Common Rousette.) Underparts black, except about the pubis where the colour is red ; back inclining to red ; covered, particularly on the underparts, with very thick hair. Length of body about ten inches. Expanse of wings three feet and upwards.

Vespertilio Ingens, *Clus.Exotic. Tab. p.* 94. V. Vampyrus, *Linn.* Pteropus Vulgaris, *Geoff. Ann. du Mus. t.* xv. Chien Volant, *Daubenton, Mem. de l'Acad. Roy. des Sciences,* 1759. Rousette *Brisson, Règne Anim.* 216.

Icon. *Buff. t.* x. *pl.* 14.

Inhabits the Isle of France and Bourbon.

Var. a. *Ann. Mus. t.* vii. *p.* 227. Brightish red and yellow.

156. 4. *P. Rubicollis* (the Red-collared Rousette.) Gray-brown; red round the neck. Ears short. Length of body about eight inches; expanse of wings about two feet.

Pteropus Fuscus, *Brisson, Règne Anim. p.* 217. Pteropus Rubicollis, *Geoff. Ann. Mus.* xv. *p.* 93.

Rougette, *Buffon, t.* x.

Icon. *Buffon, t.* x. *pl.* 17.

Inhabits the Isle of France.

157. 5. *P. Griseus* (the Gray Rousette.) Head and neck

bright red, the rest reddish-gray. Length of body seven inches; expanse of wings one foot eight inches.

Pteropus Griseus, *Geoff. Ann. Mus.* xv. *p.* 94.

Icon. *Geoffroy, Ann. Mus.* xv. *pl.* 6.

Inhabits the Island of Timor.

158. 6. *P. Leschenaultii* (the Spotted Rousette.) Ashy-yellow above, varied with white beneath ; the membrane near the body spotted white, ranged on a parallel line, with similar spots between the neck and the arms.

Pteropus Leschenaultii, *Desmarest, Ency. Méthod. Mammalogie, sp.* 142.

Icon. ──

Inhabits the environs of Pondicherry.

159. 7. *P. Rostratus* (the Lowo Assu, or Dog Bat of Java.) Body of an uniform grayish-Isabella colour, deeper on the top of the head ; muzzle elongated. Length of body three inches and a half; expanse of wings about eleven inches.

Pteropus Rostratus, *Horsfield, Zoological Researches No.* iii. Macroglossus, *F. Cuvier, Dents des Mammifères, p.* 40?

Icon. *Ib.*

Inhabits Java.

Obs. M. F. Cuvier *(Dents des Mammifères)* says, that the Pteropus Rostratus of Horsfield, which he identifies with Pteropus minimus of Geoffroy, has the cheek-teeth $\frac{4}{4}$, and he therefore makes a distinct genus of it, under the name of Macroglossus. We cannot but conclude that Dr. Horsfield has correctly stated the teeth of his Rostratus, and therefore that the Macroglossus of M. F. Cuvier, from a head brought to France by M. Duvaucel, is distinct from the former.

§ II. With a tail.

160. 8. *P. Stramineus* (Lesser Ternate Bat.) Reddish yellow ; neck red ; tail very short ; length of body upwards of five inches ; expanse of wings two feet.

Pteropus Stramineus, *Geoff. Ann. Mus.* xv. *p.* 95.

Lesser Ternate Bat, *Pennant, Synop.* Chien Volant, *Seba, Thes.* I.

Icon. *Pennant, Synop. tab.* 31, *f.* 1. *Seba, Thes. tab.* 57, *f.* 1 *and* 2.

Inhabits Timor.

Var. a. With the fur of the back erect. Inhabits Ternate.

161. 9. *P. Ægyptiacus* (Egyptian Rousette.) Head shorter and larger than the others of this division. Fur gray-brown, deepest on the back, of a soft silky texture. Body rather larger than the last, but expanse of wings one foot eight inches.

Pteropus Ægyptiacus, *Geoff. Mém. de l'Institut. d'Egypte,* and *Ann. Mus. tom.* xv. *p.* 96.

Icon.

Inhabits Egypt.

Obs. This species is found suspended to the ancient buildings of Egypt, in the manner of the Common Bat.

162. 10. *P. Amplexicaudatus* (Long-tailed Rousette., Reddish gray. Tail longer than others of this division, extending half beyond the interfemoral membrane. Length of body between four and five inches. Expanse of wings about one foot six inches.

Rousette Amplexicaude, *Geoff. Ann. du Mus. t.* xv.

Icon. *Geoffroy, Ann. Mus. pl.* 4.

Inhabits Timor.

163. 11. *P. Marginatus* (the Bordered Rousette.) Olive-

brown, with a white border round the ears. Body about three inches long ; expanse of wings about a foot.

Cynopterus Marginatus, *F. Cuvier, Dents des Mammifères, p.* 39.

Rousette à oreilles bordées, *Geoff. Ann. du Mus. t.* xxv. *p.* 97.

Icon. *Ibid. pl. 5.*

Inhabits Bengal.

Obs. M. F. Cuvier *(Dents des Mammifères)* edits a new genus under the name Cynoptères, from a head imported by M. Duvaucel, with the cheek teeth $\frac{44}{33}$. M. Cuvier identifies this with P. Marginatus, which M. Geoffroy had treated, from the number and character of the teeth to be a Pteropus.

164. 12. *P. Minimus* (the Kiodote.) Fur bright red, and woolly. Tongue extensible, two inches in length, thick and covered with horny papillæ, and the point turned backwards. About the size of P. Marginatus.

Pteropus, *Desmarest, Ency. Method. Mammalogie, sp.* 147

Rousette Kiodote, *Geoffroy, Ann. du Mus. tom.* xv. *p.* 97.

Icon. ——

Inhabits Java

Obs. We have already noticed that this is identified by M. F. Cuvier with P. Rostratus, and is treated by him as generically distinct.

165. 13. *P. Palliatus* (the Mantled Rousette.) Covered with silky straw-coloured hair. The membrane of the wings attached to the dorsal line, and having the appearance of a mantle. Length of body about four inches ; expanse of wings one foot fourteen inches. Length of tail seven inches.

Pteropus palliatus, *Geoff. Ann. du Mus.* xv. *p.* 99.

Icon. ——

Inhabits India.

Obs. This species, says Desmarest, when better known. will probably form a new genus, intermediate between Pteropus and Cephalotes.

II. CEPHALOTES. Geoffroy. Dentition as stated by M. Geoffroy: *Incisive teeth* $\frac{4}{6}$, *in the upper jaw perfectly insulated and distant from each other, in the lower almost close; canine* $\frac{1}{1}\frac{1}{1}$; *cheek-teeth* $\frac{4}{4}\frac{5}{4}$, *in general worn down, the posterior with large upper surface, without tubercles or ridges.* According to M. F. Cuvier, the teeth are: *Incisors* $\frac{4}{2}$; *canines* $\frac{1}{1}\frac{1}{1}$; *cheek-teeth* $\frac{4}{5}\frac{4}{5}$. *No membranaceous appendage to the nose. Index-finger of one known species with a nail, of the other without. Tail very short. Interfemoral membrane sloped off; membrane of the wings attached to the dorsal line.*

166. 1. *C. Peronii* (Peron's Cephalote.) Fur in some brown, in others red. Wanting a nail on the index finger. Body six inches long; expanse of wings about two feet two inches. Tail nearly an inch long.

Cephalotes Peronii, *Geoff. Ann. du Mus. t.* xv. *p.*
Icon. *Geoffroy, Ann. Mus.* xv.
Inhabits the Isle of Timor.

167. 2. *C. Pallasii* (Pallas's Cephalote). Fur cinereous-gray above, pale white beneath, and undulated on the belly. Nostrils prolonged into a tube, very distant and open. The index-finger provided with a nail. Body about four inches long; wings one foot four inches wide. Tail less than an inch long.

Cephalotes Pallasii, *Geoff. Ann. du Mus. t.* xv. Vespertilio Cephalotes, *Pallas, Spic. Zool. fasc.* iii. Harpyia, *Illiger, Prodromus Anim.* Cephalote, *Buffon, Supp. tom.* iii.
Icon. *Pallas, l. c. tab.* 1 and 2. *Buffon, do. tab.* 52.
Inhabits the Moluccas.

Obs. Pallas states that his individual had but two upper

incisors ; and none in the lower jaw whence Illiger has treated it as a genus, under the name *Harpyia*. Geoffroy inclines to the opinion that the individual in question had them originally, but that they were lost.

*** Bats, properly speaking, with insectivorous cheek-teeth.

† Middle finger with three bony articulations, the other fingers with two.

III. MoLossus. Geoffroy. *Incisive teeth $\frac{2}{2}$, in the upper jaw bifid, converging, and separated from the canine teeth ; in the lower very small, and crowded together, each having two small points ; canine $\frac{1\cdot1}{1\cdot1}$; cheek-teeth $\frac{4\cdot4}{4\cdot4}$, large, furnished with several sharp points. Head and muzzle very large. Nostrils open. Ears large, united at their base, and provided with a smaller secondary tragus. No membranaceous appendage to the nose. Interfemoral membrane narrow, and cut rectangular. Tail long.*

168. 1. *M. Longicaudatus* (Bulldog Bat.) Fur ashy yellow. A sort of band or rising of the skin passes from the end of the muzzle to the forehead. Length of body under two inches. Tail nearly as long.

Molossus Longicaudatus, *Geoff. Ann. du Mus. tom.* vi. *p.* 155. Vespertilio Molossus, *Gm.*

Mulot Volant, *Daub. Buffon, tom.* x. Bulldog Bat, *Pen. Quad.* ii. *p.* 13.

Icon. *Buffon, t.* x. *tab.* 19, *f.* 2. *Schreber, tab.* 59.

Inhabits Martinique.

Obs. Desmarest is of opinion that Geoffroy is wrong in identifying this M. Longicaudatus with the Mulot Volant of Daubenton.

169. 2. *M. Rufus* (the Red Molossus.) Fur deep red colour above, lighter underneath ; muzzle very thick and

short. Length of body above three inches; expanse of wings eighteen. Length of tail under two.

Molossus Rufus, *Geoff. Ann. du Mus. tom.* vi. *p. 155.*

Icon. ——

Habitat unknown.

170. 3. *M. Ater* (the Black Molossus.) Black, with a silvery tinge on the back. Rather less than M. Rufus.

Molossus Ater, *Geoff. Ann. du Mus. tom.* vi. 155.

Icon. ——

Habitat unknown.

171. 4. *M. Obscurus* (the Brown Black Bat of d'Azara.) Blackish brown above, dark beneath ; each hair white at its root. Length of body between three and four inches; expanse of wings about a foot. Tail an inch and a half long.

Molossus Obscurus, *Geoff. Ann. du Mus. tom.* vi. 155. The dark Bat or Ninth Bat of *d'Azara, Quad. du Paraguay, tom.* ii. 288.

Icon. ——

Inhabits Paraguay.

Obs. Desmarest doubts the identity of the species of Geoffroy with that of d'Azara.

172. 5. *M. Fusciventer* (the Brown Belly Bat.) Cinereous brown above, cinereous beneath, except the belly, which is brown in the middle of it. Body about two inches long.

Vespertilio molossus, *var. β, Gm.* Molossus Fusciventer, *Geoff. Ann. Mus.* vi. 155. Second Mulot Volant of *Daubenton's Buffon, t.* x.

Icon. *Buff. t.* x. 19.

Habitat unknown.

Obs. Very similar to M. Longicaudatus, but distinguished by the brown mark on the belly.

3.73. 6. *M. Castaneus* (Chestnut-coloured Molossus Bat.)
Chestnut above, whitish beneath ; a band from the nose to
the forehead as in M. Longicaudatus. Length of body
about five inches; expanse of wings about one foot one
inch. Tail two inches.
Molossus Castaneus, *Geoff. Ann. Mus. tom.* vi. 155.
Chestnut Bat, or Sixth Bat of *d'Azara, Quad. du Paraguay,*
tom. ii. 282.
Icon. ——
Inhabits Paraguay.

174. 7. *M. Laticaudatus* (Broad-tailed Molossus Bat.)
Dark brown above, lighter beneath ; the tail surrounded
by an extension of the interfemoral membrane ; upper lip
marked with vertical ridges ; tongue appearing double ;
ears joined at their base. About the size of M. Castaneus.
Molossus Laticaudatus, *Geoff. Ann. Mus. tom.* vi. 156.
Dark Bat, or Eighth Bat of *d'Azara, Quad. du Paraguay.*
Icon. ——
Inhabits Paraguay.

175. 8. *M. Crassicaudatus* (Great-tailed Molossus Bat.)
Cinnamon-colour, lighter beneath ; interfemoral membrane
enveloping half the tail. Body between three and four
inches long: expanse of wings eleven inches. Length of
tail above an inch.
Molossus Crassicaudatus, *Geoff. Ann. du Mus. tom.* vi. 156.
Cinnamon Bat, or tenth Bat of *d'Azara, Quad. du Pa-*
raguay.
Icon. ——
Inhabits Paraguay.

176. 9. *M. Amplexicaudatus* (Guyane Molossus Bat.)
Blackish, but lighter underneath ; interfemoral membrane
larger than in the preceding species, and entirely embracing
the tail.

Molossus Amplexicaudatus, *Geoff. Ann. Mus. tom.* vi. 156. *Buffon, Sup. tom.* vii. 294.
Icon. *Buffon, Sup.* vii. *pl.* 75.
Inhabits Cayenne.

177. 10. *M. Acuticaudatus* (Sharp-tailed Molossus.) Black-brown ; interfemoral membrane large, enveloping the tail except just the end. Length of body under two inches. Tail about the same length.
Molossus Acuticaudatus, *Desmarest, Ency. Method. Art. Mammalogie, sp.* 160.
Icon. ——
Inhabits Brazil.

178. 11. *M. Ursinus* (Ursine Bulldog Bat.) Black body, and jaws robust. Ears falling over the forehead.
Molussus ursinus, *Spix, Sim. Braz.* 59.
Icon. *Spix, Sim. Braz. t.* 35. *f.* 4.
Inhabits Para, Brazil.

179. 12. *M. Nasutus* (Proboscis Bulldog Bat.) Nose lengthened. Ears distant over the forehead. Body above brown-black, below brown. Tail nearly half free beyond the membrane.
Molossus Nasutus, *Spix, Sim. Braz.* 59.
Icon. *Spix, Sim. Braz. t.* 35, *f.* 7.
Inhabits sides of the river St. Francis, Brazil.
Probably a new subgenus.

180. 13. *M. Fumarius* (Smoky Bulldog Bat.) Body black-ish-brown ; face, ears, and wings very black.
Molossus Fumarius, *Spix, Sim. Braz.* 59.
Icon. *Spix, Sim. Braz. t.* 35, *f.* 5, 6.
Inhabits Brazil.

IV. Nyctinomus. Geoffroy. *Incisive teeth* $\frac{2}{4}$, *conical and contiguous in the upper jaw, small in the lower; canine* $\frac{1-1}{1-1}$; *cheek teeth* $\frac{4-4}{5-5}$, *furnished with sharp tubercles* = 28. *Nose flat, and on a level with the lips, which are deeply cleft or wrinkled. Ears large, and united with exterior tragus. Tail long, extending in part beyond the interfemoral membrane. No appendage to the nose. Wings very large. Hind feet covered with long hair.*

181. 1. *N.Ægyptiacus* (Egyptian Nyctinome.) Red above, brown beneath; upper lips much wrinkled. Interfemoral membrane enveloping half the tail, and destitute of muscular bands. Length of body about three inches.

Nyctinome d'Egypte, *Geoff. Mem. de l'Institut de l'Egypte, Hist. Nat. tom.* ii. 28.

Icon. *Geoff. Egypt. pl. 2, No. 2.*

Inhabits Egypt.

182. 2. *N. Bengalensis* (Bengal Nyctinome.) Tail thick. Upper lips having several folds. Interfemoral membrane with muscular bands *. About the size of N. Ægyptiacus.

Vespertilio Plicatus, *Buchanan, Voyage to India. Transactions of the Lin. Soc.* v.263. Nyctinomus Bengalensis, *Geoff. Ægypt. His. Nat. tom.* ii. 130.

Icon. *Lin. Soc. Transactions, vol.* v. *t.* 13.

Inhabits Bengal.

Obs. Buchanan describes but two incisives in each jaw.

183. 3. *N. Acetabulosus* (Port Louis Nyctinome.) Brown-

---

* Mr. Gray observes, from a specimen in spirits in the British Museum, which agrees with Buchanan's description, that the interfemoral membrane is destitute of muscular band, but that it is plaited on each side of the tail, which gives it the appearance represented in Buchanan's plate, which has been mistaken for muscular bands by Geoffroy; consequently that the first describer's name is most characteristic.

black. Interfemoral membrane enveloping two-thirds of the tail. Smaller than the other two species.

Vespertilio Acetabulosus, *Herman, Obs. Zool. p.* 19. Nyctinomus Acetabulosus, *Ency. Méthod. Art. Mammalogie, sp.* 263. Nyctinomus Mauritianus, *Geoff. Egypt. Hist. Nat.* ii. 130. *Horsfield's Java, No.* 5.

Icon. ——

Inhabits the environs of Port Louis, in the Island of Mascareigne.

184. 4. *N. Dilatus* (Dilated Nyctinome Bat.) Blackish brown, paler underneath. Wings dilated. Tail slender, attached halfway down to the interfemoral membrane, which is furnished with a few muscular bands.

Nyctinomus Dilatus, *Horsfield's Java, No.* 5.

Icon. ——

Horsfield's Java.

185. 5. *N. Tenuis* (Lowo-churut of Java.) Blackish-brown. Wings of great length, and very narrow. Tail slender, the latter half free beyond the interfemoral membrane ; edge of the interfemoral membrane folded, and furnished with muscular fibres.

Nyctinomus Tenuis, *Horsfield, Zool. Researches in Java, No.* 5.

Icon. *Horsfield's Java.*

Inhabits Java.

186. 6. *N. Braziliensis* (Brazilian Nyctinome.) Generally of a cinereous brown colour, lighter on the lower parts by varying also from yellow hair to black hair, in different individuals. Upper lip not so deeply notched as in the Egyptian species. Ears with folds or wrinkles.

Nyctinomus Braziliensis, *Isidore Geoffroy St. Hilaire,*

Annales des Sciences Naturelles for *April,* 1824, *and Zoological Journal, No.* iii. *p.* 233.

Icon. *Ann. des Sci.* 1824. *Zool. Journal, No.* iii. *pl.* 11.

Habitat Brazil.

187. 7. *N.?* *Murinus* (Murine Nyctinome.) Body blackish above, brown underneath ; wings, ears, and head black. Interfemoral membrane destitute of muscular bands. Tail about two-thirds exserted. Length of body two inches and a half; of tail an inch ; expanse of wings eight inches.

Nyctinomus Murinus, *Gray, MSS.* from a specimen in the British Museum, the teeth of which cannot be examined.

Icon. ——

Inhabits Jamaica ? according to Redman.

Obs. If the habitat is correctly stated, this is the second species found out of the old continent.

V. Cheiromeles. Horsfield. *Incisive teeth* $\frac{2}{2}$; *canine* $\frac{1 \cdot 1}{1 \cdot 1}$; *cheek-teeth* $\frac{4 \cdot 4}{5 \cdot 5}$. *Face conical. Ears distant and spreading; operculum short, semicordate blunt. Interfemoral membrane short. Tail exserted. Thumb distinct; claw flat, fringed on the edge with a series of bristles.*

188. 1. *C. Torquatus* (Collared Cheiromeles.) Neck covered with longish hairs ; back naked and dotted.

Cheiromeles Torquatus, *Horsfield, Zool. of Java, No.* vii.

Icon. *Horsfield Java, No.* vii., and dissection of head.

Inhabits Indian Archipelago.

VI. Stenoderma. Geoffroy. *Incisive teeth* $\frac{4}{4}$, *according to Geoffroy,* $\frac{2}{4}$ *according to Cuvier; Canine* $\frac{1}{1}$; *cheek-teeth* $\frac{4 \cdot 4}{4 \cdot 4}$ $= 28$. *Ears moderate, lateral and distinct. Interfemoral membrane merely rudimentary. Nose simple. Tail none.*

189. 1. *S. Rufa* (the Red Stenoderme.) Bright chestnut colour. Ears small, lateral, and isolated without oreillon. Tail none. Length of body about three inches; expanse of wings under a foot.

Stenoderme roux, *Geoff. Mém. de l'Institut de l'Egypte, Hist. Nat. tom.* ii.

Icon. ——

Inhabits.

VII. Noctilio. Geoffroy. *Incisive teeth* $\frac{4}{4}$, *the two upper intermediate teeth larger than the others, the lower incisors placed before the canine teeth; canine* $\frac{1.1}{1.1}$, *very strong; cheek-teeth* $\frac{4.4}{4.4}=26$, *furnished with sharp tubercles. Ears small, lateral and insulated. Interfemoral membrane large. Tail extending a little beyond the membrane. Muzzle short, thick, cleft, and furnished with warts or fleshy tubercles. Nose without appendage. Claws of hind feet very large.*

190. 1. *N. Leporinus* (the Peruvian or Hare-lipped Noctilio.) Fur of an uniform reddish yellow; as big as a Rat.

Vespertilio Leporinus, *Gm.* i. 47. Noctilio Americana, *Linn. S. No.* i. 88. *Geoff.* Noctilio Unicolor, *Collect. du Mus. d'Hist. Nat.*

Peruvian Bat, *Pennant.* Chauve-Souris de la Vallée d'Ylo, *Feuillée Obs.* i. 623. Reddish Bat of *d'Azara Quad. du Paraguay, tom.* ii. 280.

Icon. *Shaw's Zoology, t.* i. *p.* 1. *pl.* 41. *Schreb. f.* 60.

Inhabits Brazil, Paraguay, and Peru?

Var. β. With a whitish band down the back. Body above four inches long; expanse of wings eighteen inches. Noctilio Dorsatus, *Geoff.* Pteropus Leporinus, *Erxleben.*

Var. γ. Back reddish, belly white Peruvian Bat, var. β, *Pennant.* Noctilio à ventre blanc, *Geoff. Collect. du Mus. d'Hist. Nat. de Paris.*

191. 2. *N. Rufus* (Red Bulldog Bat.)   Body above and below red; the four legs and ears nearly naked, reddish.

Noctilio Rufus, *Spix. Sim. Braz.* 57.

Icon. *Spix, Sim. Braz. f.* 35, *f.* 1.

Inhabits Brazil.

192. 3. *N. Albiventer* (White-bellied Bulldog Bat.) Body above fuscous-brown, beneath whitish; with a whitish line down the centre of the back.

Noctilio Albiventer, *Spix, Sim. Braz.* 58.   Not the Albiventer of Geoffroy.

Icon. *Spix, Sim. Braz. f.* 35, *f.* 2, 3.

Inhabits the banks of the River St. Francis, Brazil.

VIII. PHYLLOSTOMA. Geoffroy.   *Incisive teeth* $\frac{4}{4}$, *pressed close between the canine teeth, the intermediate being the largest; canine teeth* $\frac{1}{1}\frac{1}{1}$; *cheek-teeth* $\frac{5}{5}\frac{5}{5} = 32$.   *The nose supporting two membraneous crests, one like a leaf, and the other like a horseshoe.   Ears large, naked, not united.   Oreillon internal.   Tail and interfemoral membrane varying in the several species.   Tongue furnished with sharp horny prickles.*

*** Tail distinct, shorter than the extent of the inter-femoral membrane.

193. 1. *P. Crenulatum* (Indented Phyllostome Bat.) The foliaceous nasal appendage forming a long triangle with the edges jagged or indented, appended to the horseshoe membrane; under-lip furnished with warts. End of the tail freed from the surrounding membrane. Length of body between two and three inches; expanse of wings about fourteen inches.

Phyllostoma Crenulatum, *Geoff. Ann. Mus. tom.* **xv.** *p.* 183.

Icon. *Ann. Mus.* **xv.** *pl.* 10.

Habitat unknown.

194. 2. *P. Elongatum* (the Long-leafed Phyllostome Bat) Nasal leaf not jagged at the edge, but smooth, larger, and longer than in the other species. Extremity of tail free from the membrane. About the size of the last.

P. Elongatum, *Geoff. Ann. du Mus. tom.* xv. *p.* 182.

Icon. *Ann. Mus.* xv. *pl.* 9.

Habitat unknown

195. 3. *P. Hastatum* (Spear-leaf Phyllostome, or Javelin Bat,) Brownish-red colour above, yellowish-brown on the belly. Nasal leaf like a spear-head, small at the bottom and top, and swelled out in the middle; horseshoe appendage very large. A range of warts in the form of the letter V on the under lip. Tail short, altogether enclosed in the membrane, which is large. Length of body four inches; expanse of wings about one foot nine inches.

Vespertilio Hastatus, *Gmel.* i. 47. Vespertilio Perspicillatus, *Schreb.* Phyllostoma Hastatum, *Geoff. Ann. Mus.* xv. 177

Fer-de-lance, *Buffon, tom.* 13.

Icon. *Schreb. pl.* 46, A. *Buffon,* xiii. *pl.* 33. *Ann. Mus. d'Hist. Nat. tom.* xv. *pl.* 11.

Inhabits Guyana.

196. 4. *P. Planirostra* (Flat-nosed Phyllostome Bat.) Head thick, depressed above; side of the nose tubercular; front of the nasal leaf free, pendulous; lips crenulated on the edges; chin short, flattish. Length of the body three inches and three-quarters.

Phyllostoma Planirostra, *Spix, Sim. Braz.* 66.

Icon. *Spix, Sim. Braz. f.* 36, *f.* 1.

Inhab. Bahiæ, Brazil.

⁎ Without a tail.

197. 5. *P. Perspicillatum* (Spectacle Bat.) Nasal leaf

short, sloped near its termination. Two white streaks from the nostrils to the ears; blackish-brown above, clear brown beneath. About the size of the last.

Vespertilio Perspicillatus, *Lin.* V. Americanus Vulgaris, *Seba, Thes.* i. Phyllostoma perspicillatum, *Geoff. Ann. Mus. t.* xv. *p.* 176.

Le Grand Fer-de-lance, *Buff. Sup. t.* vii.

Icon. *Seba, Thes.* i. *pl.* 55. *Buffon, Sup.* vii. *pl.* 74.

Habitat Guyana, Paraguay ?

Obs. The first Bat of d'Azara is probably a variety of this.

198. 6. *P. Lineatum* (Streaked Phyllostome Bat.) Brown, lighter underneath, with one white streak from the occiput to the os coccygis, one from each nostril to the ear, and one also from each corner of the mouth to the ear. Length of body about three inches, expanse of wings fifteen inches.

Phyllostoma Lineatum, *Geoff. Ann. Mus. t.* xv. 180.

Second Bat of *d'Azara, Quadrupeds of Paraguay,* ii. 271.

Icon. ——

Habitat Paraguay.

Obs. M. d'Azara's enumeration of the teeth does not accord precisely with those proper to this sub-division.

199. 7. *P. Rotundum* (Round-leaved Phyllostome Bat.) Reddish-brown, with the nasal leaf circular at its extremity. Size of the last.

Phyllostoma Rotundum, *Geoff. Ann. Mus. tom.* xv. 181.

The third Bat of *d'Azara, Quadrupeds of Paraguay,* ii. 273.

Icon. ——

Inhabits Paraguay.

Obs. This species runs on the ground with more ease than its congeners, and is said, like the Glossophagi, to suck blood.

200. 8. *P. Lilium* (the Lily-leafed Phyllostome Bat.) Reddish-brown; lighter underneath; the nasal appendage large, narrow at the base, and erect, in form of a lily-leaf. About the size of P. Lineatum.

Phyllostoma Lilium, *Geoff. Ann. Mus. tom.* xv.

Fourth Bat of *d'Azara, Quadrupeds of Paraguay.*

Inhabits Paraguay.

201. 9. *P. Spectrum* (Spectre or true Vampyre Bat.) Brownish-red colour above; reddish-yellow underneath ; nasal membrane long and high; jaws elongated. Length of body about six inches.

Vespertilio Spectrum, *Linn.* Canis Volans maxima aurita, *Seba, Thes.* 1. Andira Guaca, seu Vespertilio Cornutum, *Piso.* Phyllostoma Spectrum, *Geoff. Ann. Mus. tom.* xv. 174. Vampyrus spectrum, *Leach, Lin. Trans.* XIII. 80.

Spectre Bat, *Pennant's Quad.* II. 308.

Icon. *Ann. Mus. tom.* xv. *pl.* 11. *Piso Braz.* 230. *Schreb. f.* 45. *Seba, Thes.* 1. *pl.* 56. *Nouveau Dict. d'Hist. Nat. pl.* M. 28. *f.* 3.

Inhabits New Spain.

M. Auguste St. Hilaire briefly refers to three other species of this division in the French Museum.

IX. VAMPYRUS *. Spix. *Incisive teeth* $\frac{4}{4}$, *conical, the two*

---

* Vampyrus, it is understood, was long ago appropriated by M. Geoffroy (in a MS. communication to Dr. Leach) as a generic name to V. Spectrum of Linnæus ; but Spix, in his splendid work on the animals of Brazil, now publishing, has adopted it for three species there described, the Cirrhosus, Soricinus, and Bidens. These, it will be observed, differ in the character of their dentition, as V. Spectrum, though differing in the number of the cheek-teeth from the Phyllostomata in general, has been commonly arranged in that genus. Mr. Gray proposes to treat V. Spectrum of Linnæus as generically distinct from Phyllostoma, under the name of Vampyrus, as originally applied to it by Geoffroy, and to divide the three species of Spix's genus Vampyrus above-mentioned into two genera, the one under the name Istiophorus, including Cirrhosus and Soricinus, and the other under that of Tonatia including Bidens only.

intermediate in the upper jaw being largest; canine teeth $\frac{1}{1}$; *cheek-teeth* $\frac{4 4}{4 4}$, *the first with one tubercle, and the remainder with three. Mouth rather obtuse; lower jaw verrucose. Tail short, involved in the membrane, except just at the apex.*

202. 1. *V. Cirrhosus* (Bearded Vampyre of Spix.) Head oblong; nasal leaf pendulous; lips and chin bearded. Expanse of wings four inches and a half.
Vampyrus Cirrhosus, *Spix, Sim. Braz.* 64
Icon. *Spix, Sim. Brazil. f.* 36.
Inhabits Brazil.

203. 2. *V. Soricinus* (Soricine Vampyre of Spix.) Body less robust than the last; mouse-colour on the back, brownish-gray underneath; chin smooth.
Vampyrus Soricinus, *Spix, Brazil.* 65.
Icon. *Spix, Brazil. f.* 36, *f.* 2, 6.
Inhabits Brazil.

204. 3. *V. Bidens* (Two-toothed Vampyre of Spix.) But two incisors in the lower jaw; blackish-brown above, mouse-coloured underneath.
Vampyrus Bidens, *Spix, Sim. Brazil.* 65.

X. GLOSSOPHAGA. Geoffroy. *Incisive teeth* $\frac{4}{4}$, *rangea regularly; canine,* $\frac{1}{1}$; *cheek-teeth,* $\frac{3 3}{3 3}$. *The tongue very long and extensible, acting as an organ of suction. Nose carrying a small crest, in shape like a lance-head. Interfemoral membrane and tail little or none. Sanguisugous by means of the tongue.*

205. 1. *G. Soricina* (Leaf Bat of Pennant.) Ashy-brown above, bright-brown underneath; muzzle long; nose surmounted with a small spear-shaped appendage; no tail; body about two inches long; expanse of wings about ten inches.

Vespertilio Soricinus *Pallas, Spic. Zool.* Phyllostoma Soricinum, *Geoff. Ann. Mus.* xv. and Glossophaga Soricina, *ejusd Mém du Mus. d'Hist. Nat. tom.* iv.

La Feuille, *Vicq. d'Azyr Syst. des Anim. tom.* iii. Leaf Bat, *Penn. Quad.* ii. *p.* 309. Jamaica Bat, *Edwards.*

Icon. *Pallas, Spic. Zool. fasc.* iii. *pl.* 3 and 4. *Schreb. tab.* 7. *Edwards, pl.* 201. *Geoff. Ann. Mus. tom.* xv. *pl.* 11.

Inhabits Surinam, Cayenne, &c.

206. 2. *G. Amplexicaudata* (Knobbed-tail Glossophag Bat.) Blackish-brown ; interfemoral membrane large ; a short tail, terminated with a nodosity.

Glossophaga Amplexicaudata, *Geoff. Mém. du Mus. d'Hist. Nat. tom.* iv.

Icon. *Geoff. Mém. du Mus.* iv. *f.* 18, A.

Inhabits Brazil, the neighbourhood of Rio de Janeiro, &c.

207. 3. *G. Caudifer* (Tailed Glossophag Bat.) Blackish-brown ; interfemoral membrane short, with a tail extending beyond it.

Glossophaga Caudifer, *Geoff. Mém. du Mus. tom.* iv.

Icon. *Geoff. pl.* 17.

Inhabits Brazil.

208. 4. *G. Ecaudata* (Tailless Glossophag Bat.) Dark brown ; interfemoral membrane short ; no tail.

Glossophaga Ecaudata, *Geoff. Ann. du Mus. d'Hist. Nat. tom.* iv. 418.

Icon. *Geoff. Ann. du Mus. pl.* 18 B.

Inhabits as the last.

XI. MORMOOPS. Leach. *Incisive teeth* $\frac{4}{}$, *the two intermediate in the upper jaw largest ; canines* $\frac{1.1}{1.1}$; *cheek-teeth* $\frac{5.5}{5.5}$. *Ears large, close, furnished with auricles. Nasal appendage one, erect, confluent with the ears. Index-finger two joints, middle finger four, the rest three.   Tail enveloped in mem-brane, except the last joint.*

209. 1. *M. Blainvillii* (Blainville's Mormoops Bat.) Na-sal leaf plaited; ears above bilobed; labial processes di-vided.

Mormoops Blainvillii, *Leach, Lin. Trans.* xiii. 76.

Icon. *Lin. Trans.* xiii. *f.* 7, *from Museum of Mr. Brooks.*

Inhabits Jamaica.

XII. MEDATEUS. Leach.   *Incisive teeth* $\frac{4}{4}$, *the two inter-mediate in the upper jaw longest; canine* $\frac{1}{1}$; *cheek-teeth* $\frac{4}{3}$. *Nasal appendages two, one vertical, the other lunate and ho-rizontal.   Tail none.   Lips furnished with a series of warts.*

210. 1. *M. Lewisii* (Lewis's Medateus Bat.)   Blackish; nasal leaf vertical, spear-shaped ; ears rounded.   Expanse of wings seventeen inches.

Medateus Lewisii, *Leach, Linn. Trans.* xiii. 81.

Icon. ——

Inhabits. ——

†† Index-finger with one bony articulation, the other fingers with two each.

XIII. MEGADERMA. Geoffroy.   *Incisive teeth* ♀; *canine teeth* $\frac{1}{1}$; *triangular in the upper jaw, inclining backward in the lower cheek-teeth* $\frac{4}{5}=26$. *Ears very large, and united; interior ears much developed.   Three appendages to the nose, one erect, one foliaceous or horizontal, and the third like a horse shoe.   Tail none.   Interfemoral membrane square.   Third finger of the hand without the first phalanx.*

211. 1. *M. Spasma* (Cordated Bat of Pennant.) Reddish, brighter on the head ; the erect appendage to the nose mo-derate in size, and heart-shaped, the foliaceous appendage

74

of the like shape, but very large. Tragus semicordate. Body about four inches long.

Vespertilio Spasma, *Lin.* i. 47. Glis Volans Ternatanus, *Seba, Thes. tom.* i. Megaderma Spasma, *Geoff. Ann. Mus. tom.* xv. 195.

Icon. *Seba, Thes. pl.* 56. *f.* 1. *Geoff. Ann. du Mus. pl.* 12. *Schreber, f.* 48. *Shaw, Gen. Zool.* i. *f.* 42.

Inhabits the Isle Ternate.

212. 2. *M. Lyra* (Lyre Leaf Megaderme Bat.) Red on the back, yellow underneath; with the nasal appendages so disposed as to assume the shape of a lyre. Body three inches long; expanse of wings fourteen inches.

Megaderma Lyra, *Geoff. Ann. Mus. tom.* xv. 190-198.

Icon. *Ann. Mus.* xv. *pl.* 12.

Habitat unknown.

213. 3. *M. Frons* (Foliaceous Megaderme Bat of Daubenton.) Cinereous, with a slight tinge of yellow; foliaceous appendages two, one horizontal, and the other vertical, resembling a leaf. Body nearly three inches long.

Megaderma Frons, *Geoff. Ann. Mus. tom.* xv.

La Feuille, *Daub. Mém. de l'Académie des Sciences, An.* 1759.

Inhabits Senegal.

214. 4. *M. Trifolium* (the Loro of Java.) Mouse-colour, with the inner auricle trifoliated. Body three inches long; expanse of wings, about a foot.

Megaderma Trifolium, *Geoff. Ann. Mus. tom.* xv.

Icon. *Geoff. Ann. du Mus. pl.* 12.

Inhabits Java.

X. RHINOLPHUS. Geoffroy. *Incisive teeth* $\frac{2}{4}$, *the upper incisors very small, and not permanent; canine,* $\frac{1-1}{1-1}$; *cheek-*

teeth, $\frac{44}{44}$, *furnished with sharp points,* $= 30.$ *Nose furnished with a crest, shaped like a horseshoe, and surmounted with a leaf. Ears distinct. Interfemoral membrane large. Two pectoral mammæ, and two warts on the pubes, having the appearance of mammæ, but destitute of lactiferous glands. Tail long, free.*

215. 1. *R. Ferrum Equinum* · (Horseshoe Rhinolphus Bat.) Ash colour, mixed with red above; yellowish-gray beneath; membrane black; ears long and pointed; the anterior nasal membrane like a horseshoe, the posterior assimilated to a lance head; the length of body about three inches; expanse of wings sixteen inches.

Vespertilio Ferrum Equinum, var A. *Lin.* Noctilio Ferrum Equinum, *Kuhl, Deutsch. Fledermaus.* Rhinolophus Major, and R. Unihastatus, *Geoff. Collection du Mus. &c. Ann. Mus. tom.* xx. Vespertilio Hippocrepis. *Herman,* 257. *Obs. Zool.*

Grand Fer-à-Cheval, *Daubenton, Mém. de l'Acad.* 1759.

Icon. *Buffon, tom.* VIII. *pl.* 20. *fig.* 1 and 2.   Head, *Horsfield's* Java, No. 6.

Inhabits Europe.

Obs. This and the succeeding, have generally been treated, especially lately, by Dr. Kuhl, as mere varieties. Geoffroy St. Hilaire, however, considers them as distinct, both by the nasal membrane, and the form of the ears.

216. 2. *R. Ferrum Equinum Minor* (the Lesser Horseshoe Bat.)   Similar to the last, but the appendage forming a double spear head, and less in dimensions.

Vespertilio Ferrum Equinum, *var.* B. *Lin.*   V. Hipposideros, Bechstein, *Leach, Zoological Miscellany.* V. Minutus, *Montagu, Linnæan Transactions,* IX. 163. Rhinolophus Bihastatus, *Geoff. Ann. Mus.* xx. 295.

Icon. *Leach, Zool. Misc. tom.* III. *p.* 121. *Buffon, tom.* VIII. *pl.* 17. *f.* 2. *Geoffroy, Ann. Mus. tom.* xx. *pl.* 5.
Inhabits Europe, including England.

217. 3. *R. Tridens* (Trident Rhinoloph Bat.) Nasal appendage simple, erect, and tridented; body about two inches long; expanse of wings, ten inches.

Rhinolophus Tridens, *Geoff. Disc. de l'Egypte, tom.* II. and *Ann. Mus. tom.* xx. 260.

Icon. *Geoffroy's Egypt, tom.* II. *pl.* 2.
Inhabits Egypt.

218. 4. *R. Speoris* (Pitnosed Rhinoloph Bat.) Reddish-gray; nasal leaf simple, rounded; with a purse or cavity on the forehead.

Vespertilio Speoris, *Schneider*. Rhinolophus Marsupialis, *Geoff. Cour. Public.* 1805.

Rhinolophe Cruminifere, *Peron and Lesueur, Voyage to Australasia*. Pitnosed Bat, *Shaw, Zool.* I.

Icon. *Peron and Lesueur, Voy. Aust. Atlas, pl.* 35.
Inhabits the Isle of Timor.

219. 5. *R. Diadema* (Diadem Rhinoloph Bat.) Brighter red than the other species; nasal appendage disposed like a diadem; tail as long as the thighs; no frontal cavity; body about four inches long.

Rhinolophus Diadema, *Geoff. Ann. Mus. tom.* xx.

Icon. *Ann. Mus. pl.* 5, *the head; pl.* 6, *the animal entire.*
Inhabits the Isle of Timor.

Obs. The R. Commersonii, described by M. Geoffroy, in the Ann. Mus. tom. xx. differs from R. Diadema principally in having the tail a third shorter, and the foliaceous appendage about a third less; the interfemoral membrane terminates by a re-entering angle. It has been seen and noticed only by Commerson, and its specific distinctness seems doubtful

220. 6. *R. Affinis.* Yellowish-brown above, yellow underneath, deeper on the throat and breast; tail shorter than the legs; cartilaginous septum of the nose crooked; ears large, bent at the outer side, with a large accessory lobe at their base, size of Horseshoe Bat.

Rhinolophus Affinis, *Horsfield's Java.*

Icon. ——

Inhabits Java.

221 7. *R. Minor.* Lead colour, or silvery above, gray underneath; septum, tail, and ears, like the last; expanse of wings nine inches.

Rhinolophus Minor, *Horsfield's Java.*

Icon. ——

Inhabits Java.

222. 8. *R. Nobilis.* Pure brown above, varied with gray underneath; the nasal membrane extended across the nose, in the form of a shelf; tail as long as the legs; expanse of wings nineteen inches and a half.

Rhinolophus Nobilis, *Horsfield's Java*

Icon. ——

Inhabits Java.

223. 9. *R. Larvatus.* Deep brown above, with a golden lustre, more intense posteriorly; membrane blackish brown, with a yellowish tint, varying according to the disposition of the light; expanse of wings twelve inches and a half.

Rhinolophus Larvatus, *Horsfield's Java.*

Icon. ——

Inhabits Java.

224. 10. *R. Vulgaris.* Brown above, uniform gray beneath; tail a little longer than the feet; upper nasal membrane stretched transversely; ears patulous, with a hairy

lobule at the base ; expanse of wings twelve inches and a half.

Rhinolophus Vulgaris, *Horsf. Zool. Java, No.* vi

Icon. ——

Inhabits Java.

225. 11. *R. Deformis.* Brown above, gray underneath ; skull elongated, and compressed ; upper nasal membrane transverse ; large, erect, approximated ears ; expanse of wings twelve inches.

Rhinolophus Deformis, *Horsf. Zool. Java, No.* vi.

Icon. ——

Inhabits Java.

226. 12. *R. Insignis.* Dark-brown above ; tail a little longer than the feet, with an elongated frontal sinus between the skin and the skull ; mouth contracted, ascending transversely ; upper nasal membrane transverse, and partially concave ; ears large and patulous, with the extremity nearly circular.

Rhinolophus Insignis, *Horsf. Zool. Java, No.* vi.

Icon. ——

Inhabits Java.

XIV. Nycteris. Geoffroy. *Incisive teeth $\frac{4}{6}$, lobed; canine $\frac{1}{1}\frac{1}{1}$; cheek-teeth $\frac{44}{44}$, with sharp tubercles, = 30. Nostrils covered with a cartilaginous and moveable opercule. Forehead with deep longitudinal groove. Interfemoral membrane larger than the body, comprehending the tail, which is terminated in the form of the letter* T ; *with a pouch on each side of the mouth, communicating to a large membranaceous sac, formed by the skin of the body, according to M. Geoffroy.*

227. 1. *N. Hispidus* (the Rough-haired Nycteris Bat.) Reddish-brown above ; under parts, head, except the crown,

throat, breast, and belly, yellowish-white; ears large, tragus simple; length of body, an inch and a half; expanse of wings, eight or nine inches.

Vespertilio Hispidus, *Gm. Syst. Nat.* Nycteris Daubentonii, *Geoff. Egypt.* ii. 387.

Compagnol Volant, *Daub. Mém. de l'Acad. des Sciences,* An. 1759. Autre Chauve-souris, *Buffon,* x.

Icon. *Buff. tom.* x. *pl.* 20. *f.* 1 & 2.

Inhabits ——.

228. 2. *N. Geoffroyii*, (Geoffroy's Nycteris.) Gray-brown above, lighter underneath; ears very large, tragus spiral; lower lip having a large wart at its extremity, situated between two lengthened furrows, in form of the letter V.

Nycteris Geoffroyii, *Desm. Mam.* 127. Nycteris Thebaida, *Geoff. Egypt, tom.* ii.

Icon. *Geoff. Egypt.* ii. *pl.* 1. *fig.* 2. *and Skull, f.* 4. *f.* 111.

Inhabits Egypt, and probably Senegal.

229. 3. *N. Javanica* (Nycteris Bat, of Java.) Bright-red on the upper part of the body, reddish-ash colour underneath.

Nycteris Javanica, *Geoff. Egypt, Hist. Nat. tom.* ii.

Icon. ——

Inhabits Java.

XV. RHINOPOMA. Geoffroy. *Incisive teeth $\frac{4}{4}$, the upper incisors separated from each other; canine teeth $\frac{1.1}{1.1}$; cheek-teeth $\frac{4.4}{4.4}$; nose long, truncated, and surmounted with a small leaf, nostril operculated; ears large, united, and hanging over the face; inner ears; forehead large, concave; inter-femoral membrane narrow, and cut square; tail long, extending beyond the membrane.*

230. 1. *R. Microphylla* (Small-leaved Rhinopome Bat.)

Ash-coloured; tail very long and thin; nostrils capable of being closed and opened at the will of the animal, as in the Seals.

Vespertilio Microphyllus, *Brunnich, Description of the Copenhagen Museum.* Rhinopoma Microphylla, *Desm. Mam.* 129.

Chauve-Souris d'Egypte, *Bélon, de la Nature des Oiseau, book* II. *ch.* 19.

Icon. *Brunnich, l. c.* VI. *p.* 50. *f.* 1, 2, 3, and 4.

Inhabits the Pyramids of Egypt.

231. 2. *R. Caroliniensis* (Rhinopome Bat of Carolina.) Brown, with a tail long, but thicker than in the preceding species.

Rhinopoma Caroliniensis, *Geoff. Col. Mus. Desmarest, Nouveau Dict. d'Hist. Nat. tom.* XXIX. *p.* 258.

Inhabits Carolina, according to M. Brongniart, the possessor of the individual described by Desmarest.

Obs. M. Geoffroy is of opinion that this is not a true *Rhinopoma.*

XVI. Taphozous. Geoffroy. *Incisive teeth* $\frac{0}{4}$; *canine* $\frac{1\cdot1}{1\cdot1}$; *cheek-teeth* $\frac{4\cdot4}{5\cdot5}$; *a furrow on the nose, as in the two preceding divisions, but not furnished with a laminous appendage; ears moderate, separated from each other; no external lesser ears; interfemoral membrane large, tail not so long as the membrane, and exserted on its upper side.*

232. 1. *T. Senegalensis* (the Taphozous Bat of Senegal.) Brown above, mixed with ash-colour on the under parts.

Taphozous Senegalensis, *Geoff. Descrip. Egypt. Hist. Nat.* II. 127.

Loret Volant, *Daubenton, Mém. de l'Acad. des Sci. Année* 1759.

Icon. ——
Inhabits Senegal.

233. 2. *T. Mauritianus* (Taphozous Bat of the Mauritius.)
Brownish red colour above, inclining to red underneath;
nose more pointed than in the preceding; tail shorter than
the thighs; inner ears, with a sinewy edge.

Taphozous Mauritianus, *Geoff. Descrip. Egypt.* ii. 127.
Icon. ——
Inhabits the Isle of France.

234. 3. *T. Perforatus* (the Perforated Taphozous Bat.)
Red-gray above, cinereous beneath, but the lower part of
each hair white; inner ears in form of a hatchet, and terminated by a rounded edge.

Taphozous Perforatus, *Geoff. Descrip. Egypt.* ii. 127.
Icon. *Geoffroy, l. c. pl.* 3. *n.* 1. *Skeleton and head, f.* 4.
f. 4. 4. 4.
Inhabits the ancient buildings of Egypt.
Obs. M. Desmarest thinks it probable that this and the
T. Senegalensis are the same.

235. 4. *T. Lepturus* (Slender-tailed Taphozous Bat.) Gray,
paler underneath; membrane folding so as to form a sort
of pocket.

Taphozous Lepturus, *Geoff. Descrip. Egypt.* ii. 126.
Vespertilio Lepturus, *Schreb.* i. 173. V. Marsupialis,
Muller. Naturfoscher, Supp. 19. Saccopteryx Lepturus,
Illiger, Prodromus.

Pouched Bat and Slender-tailed Bat, *Pennant, Quad.* 312
and 315.

Icon. *Schreb. Saught.* 1. *tab.* 57.

Inhabits Surinam.

Obs. M. Geoffroy thinks this species is indigenous in
India, and not at Surinam.

XVII. Myopteris. Geoffroy. *Incisive teeth* $\frac{2}{2}$, *those below bilobed; canine* $\frac{1}{1}\frac{1}{1}$; *cheek-teeth* $\frac{4}{3}\frac{4}{3}$; *nose without leaf, membrane, or furrow; muzzle short and thick.*

236. 1. *M. Daubentonii* (Daubenton's Myopteris Bat.) Top of the head and back brown; the under parts pale white, with a slight tinge of yellow.

Myopteris Daubentonii, *Geoff. Descrip. Egypt.* ii. 113.
Rat Volant, *Daubenton, Mém. de l'Acad. des Science, Ann.* 1759.
Icon. ——
Habitat unknown.

XVIII. Celæno. Leach. *Incisive teeth* $\frac{2}{2}$, *the upper acuminated and simple, the lower formed, as it were, of four columns; cheek-teeth* $\frac{6}{8}$; *the anterior teeth, in both jaws, acuminated, the three posterior acutely tuberculated.*

237. 1. *Celæno Brooksiana* (Brooks's Celæno Bat.) Back ferruginous; belly and shoulders yellowish; membrane black; ears acuminated, distinct, the anterior margin rounded, the posterior straight; oreillon very small; tail doubtful.

Celæno Brooksiana, *Leach, Lin. Trans. tom.* xiii. *p.* 70.
Icon. ——
Habitat unknown. Mus. Mr. Brooks.

XIX. Aello. Leach. *Incisors* $\frac{2}{4}$; *cheek-teeth* $\frac{8}{12}$, *the two upper anterior acuminated, the third bifid, and the fourth with three edges; in the lower jaw, the three anterior acuminated, the three posterior bifid.*

238. 1. *A. Cuvieri* (Cuvier's Aëllo Bat.) Isabella ferruginous-colour; wings dark-brown; ears short, approximated, broad; no oreillon; tail not reaching beyond the interfemoral membrane.

Aëllo Cuvieri, *Leach, Lin. Trans. tom.* XIII. *p.* 71.
Icon. ——
Habitat unknown. Museum of Mr. Brooks.

XX. Scotophilus. Leach. *Incisive teeth* $\frac{4}{6}$; *in the upper jaw the two lateral teeth shorter; cheek-teeth* $\frac{8}{8}$, *furnished with acuminated processes.*

239. 1. *S. Kuhlii* (the Scotophilus Bat of Kuhl.) Ferruginous, with the ears, nose, and wings, brown; ears distinct; oreillon small; tail reaching to the end of the membrane.
Scotophilus Kuhlii, *Leach, Lin. Trans. tom* XIII. *pl.* 1. *p.* 72.
Icon. ——
Habitat ——? Mus. of Mr. Brooks, and British Museum

XXI. Artibeus. Leach. *Incisive teeth* $\frac{4}{4}$, *the two intermediate in the upper jaw the largest; in the lower jaw, truncated, the two intermediate the largest, reeded in front; cheek-teeth* $\frac{8}{10}$; *the hinder teeth small.*

240. 1. *A. Jamaicensis* (Jamaica Artibeus Bat.) Dark-brown above, mouse-coloured underneath; the ears, nasal appendages, and membranes, dark-brown, with two nasal appendages, one horizontal, the other vertical and acuminated, marked with a streak anteriorly; no tail.
Artibeus Jamaicensis, *Leach, Lin. Trans. tom.* XIII. *pl.* 1. *p.* 75.
Icon. ——
Habitat Jamaica. British Museum

XXII. Diphylla. Spix. *Incisive teeth* $\frac{4}{4}$, *the upper middle largest, apex six pointed; canine teeth* $\frac{1}{1}\frac{1}{1}$, *scarcely exserted; cheek-teeth* $\frac{44}{44}$? *or* $\frac{88}{88}$? *short apex crenulated; lips*

84

smooth; nose with two short, erect, truncated leaves, placed close together, and not elongated on the sides; the hinder legs nearly as long as the arms; tail and interfemoral membran deficient.

241. 1. *D. Ecaudata* (Tail-less Diphylla Bat.) Body hairy-woolly ; back fuscous-brown ; head and abdomen beneath, brownish-gray ; wings blackish, nearly naked ; face near the ears nearly naked.

Diphylla Ecaudata, *Spix, Sim. Braz.* 68.

Icon. *Spix, Sim. Braz.* 136. *f.* 7.

Inhab. Brazils.

Obs. This sub-genus is very peculiar for its two leaves, from whence its name. Dr. Spix describes the cheek-teeth eight above, and eight below, but he does not state whether he means eight or four on each side ; in some places, where he describes them in the same way, he evidently means the former, as his context illustrates, and in others, the latter.

XXIII. Monophyllus, Leach. *Incisive teeth $\frac{4}{6}$, the two intermediate the largest; canine teeth $\frac{1\cdot1}{1\cdot1}$; cheek-teeth $\frac{5\cdot5}{6\cdot6}$, the two first, in the upper jaw, distant, the rest tuberculated on both edges; the second and third, in the lower jaw, with a space between them.*

242. 1. *M. Redmani* (Redman's Monophyllus Bat.) Brown above, mouse-colour beneath ; all the membranes, ears, and nasal appendage brown. But one nasal appendage erect, and acute ; ears round ; beard elongated.

Monophyllus Redmani, *Leach, Lin. Trans. tom.* XIII *pl.* 1. *p.* 76.

Icon. ——

Habitat. Jamaica

XXIV. Dysopes. F. Cuv. *Incisive teeth $\frac{2}{4}$, upper close elongate, elliptical; canine $\frac{1-1}{1-1}$; cheek-teeth $\frac{4-4}{5-5}$.*

243. 1. *D. Mops.*
Dysopes Mops, *F. Cuv. Dents de Mamm.* 49.
Icon. ——
Inhab. India.
M. F. Cuvier has given no further characters of this sub-genus.

XXV. Nyctophilus. Leach. *Incisive teeth $\frac{2}{6}$, the upper elongated, conical, and short, the under equal, with three cutting-edge, jagged, canine teeth $\frac{1-1}{1-1}$; cheek-teeth $\frac{4-4}{4-4}$; the first, in the upper jaw, acute, and with one tubercle; the second and third with four tubercles, and the fourth with three. In the lower jaw, the first is acute and conical, the other three tuberculated.*

244. 1. *N. Geoffroyi* (the Nyctophilus Bat of Geoffroy.)
Back dirty-brown, under parts whitish; ears broad; membrane blackish; tail as long as the interfemoral membrane; two nasal appendages, erect, the posterior the longest.
Nyctophilus Geoffroyi, *Leach, Lin. Trans. tom.* XIII. *pl.* 1. *p.* 78.
Icon. ——
Inhabits

XXVI. Thyroptera. Spix. *Teeth ——? body slender, small; nose simple; wings very narrow, running down to the tarsus; thumb of the hand armed below with a rather concave patella; interfemoral membrane expanded, not extending beyond the feet; tail long, exserted beyond the membrane.*

245. 1. *T. Tricolor* (Three-coloured Thyroptera.) Body

above fuscous brown, beneath pure white; wings and legs pure black.

Thyroptera Tricolor, *Spix, Sim. Braz.* 61.

Icon. *Spix, Sim. Braz. t.* 36. *f.* 69.

Inhab. Shores of the Amazon, Brazils.

XXVII. PROBOSCIDEA. Spix. *Incisive teeth* $\frac{2}{8}$, *upper very small, distant, diverging, lower lobed, placed in a semicircle; canine teeth* $\frac{1}{1}\frac{1}{1}$; *cheek* $\frac{5}{5}\frac{5}{5}$, *the front one small, and the rest with many tubercles; wings narrow; tail long, half involved in the interfemoral membrane.*

246. 1. *P. Saxatilis* (Rock Proboscidea.) Body, above, variegated with gray and brown; beneath, ash mouse-colour; wings and feet fuscous brown.

Proboscidea Saxatilis, *Spix, Sim. Braz.* 62.

Icon. *Spix, Sim. Braz. t.* 35. *f.* 8.

Inhab. rocky places on the shores of St. Francis, in Brazil.

247. 2. *P. Rivalis* (River Proboscidea.) Body smaller; above fuscous brown, beneath pale brown.

Proboscidea Rivalis, *Spix, Sim. Braz.* 62.

Icon. ——

Inhab. Shores of the Amazon, Brazils.

XXVIII. VESPERTILIO. Lin. *Incisive teeth* $\frac{4}{6}$, *the upper teeth separated in pairs, cylindrical, and pointed, the lower very close with two cutting lobes directed forward; canine teeth* $\frac{1}{1}\frac{1}{1}$; *cheek teeth* $\frac{4}{3}\frac{4}{3}$, $\frac{6}{6}\frac{6}{6}$, $\frac{5}{5}\frac{5}{5}$, *or* $\frac{5}{6}\frac{5}{6}$; *the anterior cheek teeth simply conical, the posterior having several sharp points or prominences. The nose simple, without membranaceous appendage, ridge, or furrow. Ears lateral and distinct, internal ears visible. Index finger with but one phalanx, the middle with three, the annular and little finger with two.*

2 H 2

Tail not exceeding the interfemoral membrane. Sebaceous glands under the skin of the face, assuming different forms and dimensions in the different species.

248. 1. *V. Murinus* (the Common Bat.) Reddish brown above, deeper according to the age of the individual, grayish beneath. Face nearly naked; forehead very hairy; nose prominent, exceeding the under lip; nostrils opening laterally; eyes large; ears naked, inclined backwards, separate, with the points turning forward. Length of body about four inches; expanse of wings nearly eighteen inches.

Vespertilio murinus, *Linnæus, Sys. Nat.* I. 47. V. Myotis, *Bechstein and Kuhl, Deut. Flederm. sp.* 4, from an aged individual. V. Major Vulgaris, *Klein, Quad.* 61. La Chauve-souris, *Buff. t.* VIII.

Common Bat, *Pennant, Quad.* II. 119. Short-eared English Bat, *Edwards's Birds*, 201.

Icon. *Buff.* VIII. *pl.* 20. *Ann. Mus. t.* VIII. *pl.* 47 and 48. *Schreb. tab.* 51. *Ency. Méthod. f.* 33. *f.* 2. *Edwards's Birds, f.* 201. *f.* 2. *Pennant's Brit. Zool.* I.

Inhabits Europe, and probably the eastern parts of Asia.

249. 2. *V. Serotinus* (the Serotine Bat.) Back red-brown colour, brighter in the females. Membrane black. Ears oval, but approaching a triangle; inner ears pointed. Length of body under three inches; expanse of wings about fourteen inches.

Vespertilio Serotinus, *Gmel. Sys. Nat.* I. 41. Vespertilio Noctula, *Geoff. Ann. Mus.* VIII. 193. Blasse Fledermaus, Speck-fledermaus, and Spatling *of the Germans.* La Serotine, *Daub. Buff. t.* VIII.

Icon. *Buff. l. c. pl.* 18. *Schreb. tab.* 53. *Geoff. Ann. du Mus.* VIII. *pl.* 47 and 48. *Daub. Mém. Acad. Sci.* 1759, *f.* 2. *f.* 1.

Habitat. Europe and Great Britain

250. 3. *V. Noctula* (the Noctule Bat.) Body yellow, with the membranes brown-black. Ears like those of the Serotine, but rather less in proportion to the head. Length of body about three inches ; expanse of wings about sixteen inches.

Vespertilio Noctula, *Herman's Obs. Zool.* 17. *Gmelin, Sys. Nat.* i. 48. V. Lasiopterus, *Schreb.* V. Proterus, *Kuhl, Deutch Flederm.* 33. V. Serotinus, *Geoff. Ann. Mus.* viii.

Noctule Bat, *Pennant's Quad.* 369. La Noctule, *Daubenton, Mém. Acad. Sci.* 1759, 380. Great Bat, *Pennant's Brit. Zool.* i.

Icon. *Schreb. Saugth, tab.* 58. *Daub. Mém. de l'Acad. &c.* 1759, *tab.* 15. *f.* 1. Young, *F. Cuv. Mam. No.* 38, *t.* 3.

Inhabits the whole of Europe, but especially Germany.

Obs. The Noctule and the Serotine have been very much confounded. Dr. Kuhl has given the distinctive characters of the two at different ages

251. 4. *V. Pipistrellus* (the Pipistrelle.) Back blackish brown, under parts inclining to yellow. Ears shaped like those of the preceding; inner ears rounded at their termination. Length of body little more than an inch ; expanse of wings about seven inches.

Vespertilio Pipistrellus, *Lin. Schreb. Geoff. Kuhl.*

La Pipistrelle, *Daub. Mém. de l'Acad. &c.* 1759. The Pipistrelle, *Pen. Quad.* ii. 318.

Icon. *Daubenton, l. c. fig.* 3. *Buff. t.* viii. *pl.* 18. *f.* 2. *Schreb. tab.* 54. *Geoff. Ann. Mus. t.* viii. *pl.* 47 and 48.

Inhabits various parts of Europe.

Var. a. With the points of the hairs ash-coloured; found in Egypt by M. Geoffroy.

Icon. *Geoff. Descrip. d'Egypte, f.* 1. *f.* 3. *Skull, f.* 4. *f.* 585.

252. 5. *V. Pictus* (the Kirivoula, or Striped Bat.) Back of a bright yellowish red colour, yellow underneath; fingers along the wings bright yellow; membrane brown red Length of body about two inches; expanse of wings about eight inches.

Vespertilio Ternatanus, *Seba, Thes.* V. Pictus, *Gm. Pallas, Geoff.* V. Kirivoula, *Boddaert, Elench. Anim.*

Striped Bat. *Pennant.* Muscardin Volant, *Daub. Mem. de l'Acad.* 1759.

Icon. *Ann. du Mus. t.* viii. *pl.* 48. *Buff. t.* x. *pl.* 20. *Seba, Thes. tab.* 56, *fig.* 23.

Inhabits the East Indies, especially Ceylon, where it is called Kirivoula.

253. 6. *V. Lasiurus* (Rough-tailed Bat.) Colour varied between yellowish gray and bright red; tail thick; ears oval and short. Length of body about two inches; expanse of wings about ten.

Vespertilio Lasiurus, *Gm.*

Rough-tailed Bat, *Pennant, Shaw.*

Icon. *Ann. Mus. t.* viii. *pl.* 47. *Schreb. tab.* 62, B.

Inhabits Cayenne.

254. 7. *V. Nigrita* (Senegal Bat.) Yellow-brown above, ashy brown underneath. Ears smaller than in most of the Vespertiliones. Length of body about three inches; expanse of wings about fifteen.

Vespertilio Nigrita, *Gmel.*

Marmotte Volante, *Daub. Mém. de l'Acad.* 1759. Chauve-souris Etrangère, *Buff. t.* x. Senegal Bat, *Pennant, Quad.* 281.

Icon. *Buff. l. c. pl.* 18. *Schreb. Saugt. tab.* 58. *Ann. Mus. t.* viii. *pl.* 47.

Inhabits Senegal.

90

255. 8. *V. Nasutus* (Great Serotine Bat.) Red-brown colour on the back, bright yellow on the flanks, pale yellow on the belly. Muzzle long and pointed.

Vespertilio Nasutus, *Shaw. Gen. Zool. t.* i. 142.

Grande Serotine de la Guyane, *Buff. Supp. t.* vii. Great Serotine, *Pennant, Quad. t.* ii. 318.

Icon. *Buff. l. c. pl.* 73.

Inhabits Guyane.

256. 9. *V. Pygmæus* (Pigmy Vespertilio Bat.) Brown, deeper on the back and head than on the under part ; muzzle short and obtuse ; ears shorter than the head, broad at the base, rounded, tragus linear. Expanse of wings about five inches.

Vespertilio Pygmæa, *Leach, Zool. Journal,* iv. 589.

Icon. *Zool. Journal,* iv. *f.* 22.

Inhabits Devonshire.

Obs. Nearly allied to V. Pipistrellus.

257. 10. *V. Braziliensis* (Brazil Vespertilio Bat of Spix.) Black wings.

Vespertilio Braziliensis, *Spix, Brazil,* 63

Icon. *Spix, Brazil, t.* 36. *f.* 8.

Inhabits Brazil.

258. 11. *V. Hilarii* (Vespertilio Bat of Isidore St. Hilaire.) Ears small and triangular ; tail as long as fore-arm ; interfemoral membrane naked.

Vespertilio Hilarii, *Isidore Geoffroy St. Hilaire, Ann. des Sciences Nat.* iii. 440

Icon. ——

Inhabits Brazil.

259. 12. *V. Polythrix* (Indented Vespertilio.) Ears small, notched at the external margin ; tail as long as fore-arm ;

interfemoral membrane with scattered hairs on the upper side; face hairy.

Vespertilio Polythrix, *Isidore Geoffroy St. Hilaire, Ann. des Sciences Nat.* III. 440.

Icon. ——

Inhabits Brazil.

260. 13. *V. Lævis* (Smooth Vespertilio.) Ears long; tail as long as the body; face partly naked.

Vespertilio Lævis, *Isidore Geoff. St. Hilaire, Ann. Sci. Nat.* III. 445.

Icon. ——

Inhabits Brazil.

261. 14. *V. Temminckii* (Temminck's Vespertilio Bat.) Head cuneate, top and sides flat; ears shorter than the head, oblong, rounded; tragus elongate, falcate. Fur silky, hair very short, olive-brown, beneath dirty yellow, sides pale rufous. Incisive teeth $\frac{4}{4}$.

Vespertilio Temminckii, *Horsf. Java, No.* 8.

Icon. *Horsfield's Java, No.* 8.

Inhabits Java.

262. 15. *V. Adversus.* Head wedge-shaped, high behind; muzzle, snout, broad; ears erect, tragus linear. Fur rather woolly, hairs long, above shining gray-brown, underneath whitish ash-coloured. Incisive teeth $\frac{4}{6}$; canine $\frac{1}{1}$; cheek teeth $\frac{44}{55}$.

Vespertilio adversus, *Horsfield's Java, No.* 8.

Icon. ——

Inhabits Java.

263. 16. *V. Hardwickii* (Hardwicke's Bat.) Head globose, tumid; muzzle short, depressed, lower cutting teeth simple. Ears very broad, lobe round, produced concave;

tragus linear, lanceolate, erect. Fur woolly, very soft, hairs very long, basis woolly; above brown ash-coloured, underneath dirty gray. Incisive teeth $\frac{4}{6}$; canine $\frac{1\cdot1}{1\cdot1}$; cheek teeth $\frac{6\cdot6}{6\cdot6}$.

Vespertilio Hardwickii, *Horsfield's Java*, No. 8.

Icon. ——

Inhabits Java.

264. 17. *V. Tralatitius* (Transposed Vespertilio.) Head wedge-shaped, above broad, face bristly. Ears large, flat, broad; tragus linear, erect, blunt. Fore-arm long. Fur soft, sooty black. Incisive teeth $\frac{4}{6}$; canine $\frac{1\cdot1}{1\cdot1}$; cheek teeth $\frac{5\cdot5}{6\cdot6}$. Body three inches; expanse of wings ten inches.

Vespertilio Tralatitius, *Horsf. Java*, No. 8.

Lowo-Manir, *Javaneesse.*

Icon. ——

Inhabits Java.

265. 18. *V. Imbricatus* (Tiled Bat.) Head and snout short, broad. Ears broad, obtuse; tragus short, semilunar Fur shining fulvous brown; eyes and upper side of ears covered with thick fur. Incisive teeth $\frac{4}{6}$; canine teeth $\frac{1\cdot1}{1\cdot1}$; cheek teeth $\frac{4\cdot4}{3\cdot3}$.

Vespertilio Imbricatus, *Horsfield's Java*, No. 8.

Icon. ——

Inhabits Java.

266. 19. *V. Carolinensis* (Carolina Vespertilio of Geoffroy.) Brown-red above, yellow underneath. Ears oblong, tragus semicordate.

Geoff. Ann. Mus. t. VIII. *pl.* 47

Icon. *Geoff. l. c.* VIII. 47.

Inhabits South Carolina.

267. 20. *V. Discolor* (Dingy Vespertilio.)   Hairs of the back brown, with the points white ; under parts pale white,
Vespertilio discolor, *Natterer.*
Icon. *Kuhl, Fledermaus,* 43.
Inhabits Europe.

268. 21. *V. Emarginatus* (Bordered-eared Vespertilio.) Reddish gray above, ash-coloured underneath.
Vespertilio Emarginatus, *Geoff. Ann. Mus. t.* VIII. *pl.* 46 and 48.
Icon. ——
Inhabits Great Britain and the north of France.

269. 22. *V. Mysticinus* (Red-brown Vespertilio of Leisler.)   Hairs of the back brown, tipped with brown-red ; some hairs on the upper lip in form of whiskers.
Vespertilio Mysticinus, *Leisler, Kuhl, Deut. Flederm.* 58.
Icon. ——
Inhabits Europe

270. 23. *V. Borbonicus* (Bourbon Vespertilio.)   Red above, white underneath.
Vespertilio Borbonicus, *Geoff. Ann. Mus. t.* VIII. *pl.* 46.
Icon. ——
Inhabits Isle of France.

271. 24. *V. Brasiliensis* (Brazilian Vespertilio of Desmarest.)   Dark brown, each hair tipped with chestnut ; membrane chestnut and black.
Vespertilio Braziliensis, *Desmarest, Nouveau Dict. d'Hist. Nat. t.* XXXV.
Icon. ——
Inhabits Brazil.

94

272. 25. *V. Bechsteinii* (Bechstein's Vespertilio.) Red-gray above, white underneath.
Vespertilio Bechsteinii, *Kuhl, Deutsch Flederm.*
Icon. *Kuhl, Deutsch Flederm. pl. 22.*
Inhabits

273. 26. *V. Nattererii* (Natterer's Vespertilio.) Yellow-gray above, white underneath. Membrane dark gray, interfemoral membrane in festoons.
Vespertilio Natererii, *Kuhl, Deut. Flederm. pl. 25.*
Icon. *l. c. f. 23.*
Inhabits Europe, England.

274. 27. *V. Leisleri* (Leisler's Vespertilio.) Hair long, deep brown, but tipped with red-brown colour ; lower sides of the membranes along the arms very hairy.
Vespertilio Leisleri, *Kuhl, D   Flederm.* V. Dasicarpus, *Leisler.*
Icon. ——
Inhabits Europe.

275. 28. *V. Schreibersii* (Schreiber's Vespertilio.) Ashy gray above, paler beneath, sometimes mixed with yellowish white.
Vespertilio Schreibersii, *Kuhl, Deutsch Flederm*
Icon. ——
Inhabits Europe.

276. 29. *V. Kuhlii* (Kuhl's Vespertilio.) Bright brown-red above, yellow beneath.
Vespertilio Kuhlii, *Kuhl, Deutsch Flederm.* Discovered by Natterer at Trieste.
Icon. ——
Inhabits Europe.

277. 30. *V. Daubentonii* (Daubenton's Vespertilio.) Gray-red above, white beneath.

Vespertilio Daubentonii, *Kuhl, Deutsch Flederm.* 51.

Icon. *Kuhl, l. c. t.* 25, *f.* 2.

Inhabits Europe.

The three following American species from d'Azara are referred by Geoffroy to the Vespertiliones. Desmarest inclines to the opinion that they belong rather to the divisions established by M. Rafinesque, which he has named Hypercodon and Nycticeius, but without sufficiently pointing out their distinctive characters.

278. 31. *V. Villosissimus* (Shaggy Vespertilio.) Pale brown; ears like those of a Rat. The seventh Bat of *Azara's Quad. of Paraguay, Geoff. Ann. du Mus. t.* VIII.

279. 32. *V. Ruber* (Red Vespertilio.) Red on the upper parts, yellow underneath; ears like the preceding. Eleventh Bat of *Azara's Quad. of Paraguay. Geoff. Ann. Mus. t.* VIII.

280. 33. *V. Albescens* (Silvery or Black Vespertilio.) Nearly black, with white points on the back; ears like the preceding. Twelfth Bat of *Azara, Quad. of Paraguay. Geoff. Ann. Mus. t.* VIII. *pl.* 18

Azara describes a variety of this with more white about the lower part.

XXV. PLECOTUS. Geoff. *Incisives* $\frac{4}{?}$; *canines* $\frac{1-1}{1-1}$; *cheek teeth* $\frac{5-5}{6-6}$. *Ears larger than the head, and united at their base. In other respects, agrees with Vespertilio.*

281. 1. *P. Auritus* (the Long-eared Bat.) Gray, darker above than underneath. Length of body nearly two inches; expanse of wings eleven or twelve inches.

Vespertilio Auritus, *Lin*

L'Oreillard, *Daub. Mem. de l'Acad. des Sciences,* 1759.
Long-eared Bat of *English naturalists.*

Icon. *Daubenton, l. c. tab.* 1, *f.* 2. *Buffon, t.* VIII. *pl.* 17.
Schreiber, tab. 50. *Edward's Birds, t.* 201, *f.* 3. *Pennant. Shaw.*

Inhabits Europe, common in England.

Var. a. Of Egypt; less than the common species; rather redder; last vertebra of the tail detached from the membrane.

Icon. *Geoff. Desc. d'Egypte, t.* 2. *f.* 3.

Var. b. Of Austria; bigger than our variety, colour deeper.

282. 2. *P. Barbastellus* (the Barbastel.) Deep brown, with the point of each hair yellow; membrane black. About as big as the Auritus.

Vespertilio Barbastellus, *Gm.* 1, 40.

La Barbastelle, *Daub. Mem. de l'Acad.* 1759. Pennant's Quadrupeds, II. 319.

Icon. *Daub. l. c. pl.* 2. *f.* 3. *Buffon, t.* VIII. *f.* 19. *Schreb. tab.* 55. *Geoff. Ann. du Mus. t.* VIII. *pl.* 46.

Inhabits France and Germany, but rarely met with, particularly in the former country.

283. 3. *P. Maugei* (Porto Rico Bat.) Blackish brown above, clear brown beneath, posterior part of the body white, membranes gray. Rather larger than the Barbastel.

Vespertilio Maugei, *Desmarest, Nouveau Dict. d'Hist. Nat. t.* XXXV.

Inhabits the island of Porto Rico; discovered there by Maugé

284. 4. *P. Timoriensis* (Timor Bat.) Blackish brown above, ashy brown underneath. About the size of the last.

Vespertilio Timoriensis, *Geoff. Ann. Mus. t.* VIII.
Icon. *Geoff. Ann. Mus. pl.* 47.
Inhabits the Island of Timor.

285. 5. *P. Velatus* (Veiled-eared Bat.) Chestnut above; grayish-brown beneath; tail as long as the body, entirely involved; ears long, with two longitudinal plaits, hanging over the face; auricule, elongate, naked; face, partly naked; expanse of wings thirteen inches and a half.

Plecotus Velatus, *Isid. Geoff. Ann. Sci. Nat.* III. 446.
Icon. ——
Inhabits

M. Rafinesque has proposed three more sub-genera of Bats; but the French naturalists, not having examined the species, do not as yet admit them. His sub-genera are named: 1. ATALAPHA, without incisive teeth*, including the V. NOVABORASCENSIS of Gm. and Pennant, and a second species he names A. Sicilienne.

His second sub-genera, HYPEXODON, has no incisors in the upper jaw, and six in the lower. It includes but one species, the Mustache Hypexanthus.

His third subgenus, NYCTICEIUS, has two incisors above, separated by a great interval, and six below. It includes the Black Shoulder Bat, and the netted Bat.

Rafinesque also describes the Blue Wing Bat, the Black Back Bat, the Sparred Bat, the Monk Bat, the Black-faced Bat, and the Big-eared Bat, but without placing them decidedly in either subdivision of the genus †.

---

* It is known that these teeth occasionally fall out in the Vespertilionidæ, which renders this character more doubtful.

† We cannot conclude this long list of species of the Vespertilionidæ without, in a more particular manner, reminding the reader that we by no means vouch for the propriety of the specific distinctness of each. This, in most cases, must be left entirely to their original describers.

Genus II. GALEOPITHECUS, PALLAS.

Incisive teeth $\frac{4}{6}$, the two intermediate, in the upper jaw smaller than the others; the edges of the lower incisors in dented. Canine teeth $\frac{1}{1}\frac{1}{1}$, small, but very sharp at the point; cheek teeth $\frac{5}{5}\frac{5}{5}$, the anterior similar to the canine teeth, the posterior furnished with several points; the membrane envelops the neck, extremities, fingers, and the tail; fingers of the hands not longer than those of the feet; nails slender and semicircular; mammæ two, pectoral.

286. 1. *G. Rufus* (the Colugo.) Fur red, lighter on the belly and internal sides of the limbs; no spots; length of body about a foot.

Lemur Volans, *Linn. Sys. Nat.* i. 45. Galeopithecus, *Pallas, Act. Acad. Sc. Peters.* 1780. 280.

Flying Macauco, *Penn.* 1, 234. Flying Colugo, *Shaw,* 1, 116. Galeopithèque Roux, *Geoff. Mag. Encycl. Audebert, Hist. Nat. des Singes.*

Icon. *Audebert, Hist. des Singes. Pallas, Petersburg Transactions,* 1780, *copied in Shaw's Zool.* i. 38. *Pennant, Quad.* i. 50. *copied in Shaw, Zool.* i. 35.

Inhabits the Molucca, Philippine, and Pelew Islands.

287. 2. *G. Variegatus* (Varied Colugo.) Brown-red, varied on the back, and the sides spotted with white; length of body, about six inches.

Galeopithecus Variegatus, *Cuv. Tabl. Elem. des Anim. p.* 107. *Audebert, Hist. des Singes,* &c.

Icon. *Audebert, Hist. des Singes,* &c. *pl.* 2, *of the Galeopitheci.*

Obs. Probably either a young individual of the last preceding, or a variety of it.

288. 3. *G. Ternatensis* (the Ternate Colugo.) Grayish-red; tail slightly spotted, smaller than the preceding.

Felis Volans Ternatea, *Seba, Mus.* 1. *tab.* 58. Galeopithecus Ternatensis, *Geoff. Mag. Ency.*

Obs. Known only by Seba's short description, and is probably a variety of G. Rufus.

FAMILY II.—INSECTIVORA.

Cheek-teeth furnished with various sharp points; canines,
in some species, very long, in others short; in the latter
case, called lateral incisors or false cheek-teeth; incisors
also, varying both in number and length; teats ventral, or
ventral and pectoral; legs short; mode of locomotion always plantigrade; all the feet pentadactylous, except in
one species *.

∗ First tribe, with two long incisors in front, followed
by other lateral incisors or false canines, not longer than
the cheek-teeth.

Genus I. ERINACEUS.

Incisive teeth $\frac{6}{6}$; the intermediate teeth, in the upper
jaw, very long, separated from each other, cylindrical, and
directed forward; canine teeth $\frac{1}{1}\frac{1}{1}$, shorter than the cheek-
teeth; cheek-teeth $\frac{5}{4}\frac{5}{4}$; body capable of a spherical shape
at the will of the animal, protected, as to the upper parts,
by prickles, the under parts furnished with coarse hair;
nails constructed for digging; tail short or wanting; teats
ten, six pectoral, and four ventral.

289. 1. *E. Europæus* (the Common Hedgehog.) Ears
short; prickles very sharp, about an inch long, set in clusters, diverging in their directions, and crossing each other,
with the points white; the hair of the under part of a dirty-
white colour.

Erinaceus Europæus, *Linn.*

∗ The Chrysoclore, which has but three toes on the anterior feet.

Le Herisson, *Buff. tom.* viii. Common Hedgehog, *Pennant, Quad.* 316.

Icon. *Schreb. tab.* 162. *Buffon, t.* 8. *pl.* 6. *var.* A. *Pennant, Quad. tab.* 28. *f.* 3.

Inhabits all the temperate parts of Europe.

Obs. Var. A., with shorter nose; spines less extended; hair of a deep-red. The E. Sibiricus of Erxleben appears also to be a variety of this.

290. 2. *E. Auritus* (Long-eared Hedgehog.) Ears two-thirds the length of the head; a little less than the common species.

Erinaceus Auritus, *Pallas, Nov. Com. Petrop. tom.* xiv. Herisson, d'Egypte. *Geoff. Egypt.*

Icon. *Pallas, l. c. tab.* 21. *fig.* 4. *Schreber, tab.* 163.

Inhabits North-western Asia and Egypt.

Obs. In Seba, Thes. tab. 51., is a figure of Porcus Aculeatus, Pendent-eared or Malacca Hedgehog; and in the same work, tab. 49, is another of the E. Inauris, Earless or American Hedgehog; but their correct classification, in this order, is doubtful.

Genus II. SOREX.

Incisive teeth $\frac{2}{2}$, in the upper jaw, indented at their base; in the lower, proceeding horizontally from their alveoli, and turned upwards towards their points, where they are sometimes of a brown colour; lateral incisors, or false canines $\frac{3 \text{ or } 4}{2}$ conical, small, shorter than the cheek-teeth; cheek-teeth $\frac{4 4}{3 3}$; muzzle and nose much elongated, the latter moveable; ears and eyes small; tail varying in length, round, compressed or four-sided; pentadactylous; nails, crooked, short, curved, and pointed; teats six or eight, both pectoral and ventral; sebaceous gland, on each flank, exuding a scented unction.

291. 1. *S. Araneus*, (Common Shrew, vulgarly called the Shrew Mouse.) Mouse-coloured, lighter underneath; tail sub-quadrated, not quite so long as the body; ears large and naked; incisors altogether white?

Sorex Araneus, *L.* VIII. La Musaraigne, *Buff. tom.* VIII. Icon. *Daubenton, Mém. de l'Acad. des Sciences,* 1756. *pl. 5. Buffon,* as above, *pl.* 10. *f.* 1. *Geoff. Ann. du Mus. t.* 17. *pl.* 2. *f.* 2. *Schreb. tab.* 160.

Inhabits Europe.

Obs. The Shrew is subject to considerable superficial variety.

292. 2. *S Fodiens* (the Water Shrew.) Black above, white beneath; ends of the incisors brown; tail square; ears capable of being hermetically closed, by means of three valves.

Sorex Fodiens, *Pallas.* Sorex Carinatus, *Hermann, Obs. Zool.* 46. Sorex Daubentonii, *Erxleb.* Musaraigne d'eau, *Daubenton, Mém. de l'Acad. des Sciences,* 1756. Le Greber, *Vic. d'Azyr. Syst. Anatom. tom.* III. 35.

Icon. *Schreber. Daubenton, l. c. pl.* 5. *fig.* 2. *Buffon, tom.* VIII. *pl.* 10.

Inhabits Europe.

293. 3. *S. Tetragonurus* (Square-tailed Shrew.) Ears short; blackish above, ashy-brown underneath; tail long, and perfectly square.

Sorex Tetragonurus, *Herman. Obs. Zool. Geoff. Ann. Mus. t.* 17. *pl.* 2. *f.* 3.

Icon. ——

Inhabits France.

294. 4. *S. Constrictus* (Flat-tailed Shrew.) Dark ash-colour; tail flat at its insertion, and at its extremity, but round in the middle; ears hid in the fur.

Sorex Constrictus, *Herman. Obs. Zool. Geoff. Ann. Mus.
t.* 17. S. Cunicularius, *Bechstein*, Musaraigne Plaron. *Vicq. d'Azyr.*

Icon. *Ann. Mus.* xvii. *f.* 3.

Inhabits

295. 5. *S. Lineatus* (Streaked Shrew.) Brownish-black, lighter underneath; throat ash-coloured; a white streak from the forehead to the nostril, and a spot on each ear; tail round.

Sorex Lineatus, *Geoff. Ann. Mus. t.* 17.

Icon. ——

Inhabits France.

296. 6. *S. Remifer* (Oared Shrew.) Larger than the preceding species; blackish-brown above, lighter underneath; throat ash-coloured; tail square at its insertion, but flattened towards the point.

Sorex Remifer, *Geoff. Ann. Mus. t.* 17. *pl.* 2. *f.* 1.

Icon. *Geoff. l. c.* 12. *f.* 1.

Inhabits France.

297. 7. *S. Leucodon* (White-toothed Shrew.) Back brown; belly and flanks white; tail slightly quadrated.

Sorex Leucodon, *Hermann, Obs. Zool.* 49.

Icon. ——

Inhabits the vicinity of Strasbourg.

298. 8. *S. Indicus* (Indian Shrew.) Larger than the European species. Fur short, gray-brown, tinted red above; tail round, half the length of the body.

Sorex Indicus, *Geoff. Ann. Mus. tom.* 1. and *t.* 17. 309.

Icon. *Geoff. Ann. Mus. tom.* 1. *pl.* 15. *f.* 1. *Buffon. Sup. tom.* 7. *pl.* 71.

Inhabits India, especially Pondicherry and Tranquebar.

299. 9. *S. Capensis* (Cape Shrew.) Ash-coloured, with

a tint of yellow ; tail red, round, and half the length of the body.

Sorex Araneus Maximus, *Petiver.* Sorex Capensis, *Geoff. Ann. Mus. t.* 17. 184.

Icon. *Petiver, t.* 23. *f.* 9. *Valentin. Mus.* ii. 2.

Inhabits the Cape of Good Hope.

300. 10. *S. Myosurus* (Rat-tailed Shrew.) Entirely white; tail round, denuded ; muzzle, thick.

Sorex Myosurus, *Pallas, Act. Petrop.* 1781. *tom.* ii. Musaraigne à queue de Rat, *Desmarest, Ency. Méthod. Art. Mammalogie, sp.* 242.

Icon. *Pallas, l. c. pl.* 4. *f.* 1. *Geoff. Ann. Mus. t.* 17. *pl.* 3. *f.* 2 and 3.

Obs. Pallas describes the male of a brown-black, but M. Geoffroy refers his individual to another species ; the female, according to the former naturalist, quite white, which seems to be the result of albinism. The distinction, as a species, therefore, is uncertain.

301. 11. *S. Collaris* (the Collared Shrew.) Black, with a white collar round the neck.

Sorex Collaris, *Geoff. Mém. Mus.* 1. 309.

Icon. ——

Inhabits the Islands at the mouth of the Meuse.

302. 12. *S. Etruscus* (Tuscan Shrew.) Ashy-gray; white underneath ; ears round ; large tail, subquadrate.

Sorex Etruscus, *Savi, Nuo. Giornal.* 1. 60.

Icon. ——

Inhabits Tuscany.

Others have been also named as distinct, as the S. Murinus of Gmelin, probably the S. Indicus ; the S. Minimus of Pallas, the S. Cæcutiens of Laxman, the S. Minutus or Pygmæus of the same ; the S. Exilis, said to be the smallest of quadrupeds, and the S. Pusillus of Gmelin, probably a Desman, the distinctive pretensions of which seem very doubtful.

It may be useful to notice here that the S. Aquaticus is now arranged in a distinct subdivision, *viz.*, the Scalope; the S. Cristatus, in like manner, is now the Condylure ; the S. Brasiliensis, the Didelphis Tricolor, the S. Auratus, the Chrysoclore, and the S. Moschatus, the Desman.

Genus III. MYGALE.

Character of the teeth, according to Geoffroy, from the type of the Desman of the Pyrenees. Incisive teeth $\frac{2}{4}$, the two upper large, very strong, conical ; lower incisors like those of Sorex; false canines or lateral incisors $\frac{4}{1\frac{2}{2}}$; cheek-teeth $\frac{8}{6}$, the four posterior above, and the three underneath, bristled with points; nostrils pierced at the end of a flexible sort of proboscis ; no conque to the ears ; tail long, scaly, and laterally compressed ; toes palmated.

303. 1. *Mygale Moscovitica* (the Desman.) Brown above, white underneath.

Mygale Moscovitica, *Geoff. Ann. Mus.* XVII. 192. Mus Aquaticus Exoticus, *Clusius Exot.* Castor Moschatus, L. Sorex Moschatus, *Pallas.* Sorex Moschoviticus, *Charleton Exot.* Mus Aquatilis, *Aldrovandus.* Glis Moschiferus, *Klein.*

Desman, *Buffon,* t. 10. Musk Shrew, *Pennant, Shaw.* Icon. *Buffon,* t. 10. *pl.* 2. *Schreber, tab.* 159.
Inhabits Southern Russia.

304. 2. *M. Pyrenaïca* (Desman of the Pyrenees.) Brown above, gray underneath ; tail longer than the body, cylindrical for the greater part of its length, and laterally compressed toward its extremity.

Geoff. Ann. Mus. t. 17. 193.

105

Icon. *Geoff. Ann. Mus. pl.* 4. *f.* 1.

Inhabits the vicinity of Tarbes, at the foot of the Pyrenees.

Genus IV. TUPAIA.

Incisive teeth $\frac{2}{6}$; canine $\frac{1+1}{1+1}$; cheek-teeth $\frac{7+7}{6+6}$; body elongate; head triangular, attenuated, blunt; eyes large; ears large; tail very long; teats four, ventral. Diurnal. Inhabits trees.

305. 1. *T. Tana* (the Tupaia Tana.) Head long; muzzle pointed; fur above, reddish-brown, speckled with black beneath, and an oblique red line on each shoulder.

Tupaia Tanaia Tana, *Raffles. Lin. Trans.* XIII. 257.

Tupaia Tana, *Sumatratresse.*

Icon. *Horsfield, Zool. Java.* No. III. and head and teeth.

Inhabits Sumatra. Mus. Col. Surg.

306. 2. *T. Javanica* (the Javanese Tupaia.) Head long; muzzle slightly pointed; tail very long; fur above brown, speckled with gray, with a grayish-white oblique line on the shoulder.

Tupaia Javanica, *Raffles. Lin. Trans.* XIII. 267. Sorex Glis, *Diard. Asiatic Researches,* x. Glisorex Desmarest, *Mam. Sup.* Bangsring *Javanesse.*

Icon. *Horsfield's Java,* No. 3.

Inhabits Java.

307. 3. *T. Ferruginea* (the Ferruginous Tupaia. Muzzle slightly pointed; fur ferruginous.

Tupaia Ferruginea, *Raffles. Lin. Tr.* XIII. 277. Clodabates Ferruginea, *F. Cuvier, Mam. Lithog.*

Icon. Head and teeth, *Horsfield's Java,* No 3.

Inhabits Java.

Genus V. SCALOPS.

Upper incisors two, very long and large, followed by three lateral incisors or conical, or canine teeth, leaving a void space between them and the two incisors; cheek-teeth $\frac{3\cdot3}{3\cdot3}$, with several points; in the lower jaw, are two very small central incisors, with a larger incisor on each side; lower conical teeth increasing successively toward the cheek-teeth; muzzle elongated and cartilaginous; no external ears; anterior toes large, united as far as the third phalanx, armed with long, strong, and flat nails, constructed for digging.

308. 1. *S. Canadensis* (the Canadian Scalops, or Shrew Mole.) Fur gray-brown; eyes hidden within the fur; nose long, terminated by a button-shaped cartilage.

Sorex Aquaticus, L. Talpa Fusca, *Pennant. Quad.* 314. Musaraigne Taupe, *Cuv. Tab. Element des Anim.*

Icon. *Schreber, tab.* 158.

Inhabits from Canada to Virginia, in the United States.

VI. CHRYSOCHLORIS.

Incisors $\frac{2}{4}$ in the upper jaw, strong and sharp, the intermediate lower incisors very small; conical teeth $\frac{3}{3}$, small; cheek-teeth $\frac{6\cdot6}{5\cdot5}$, of the insectivorous character; anterior extremities with only three toes, armed with strong nails, assimilated to those of the Mole; hinder extremities with five toes; eyes very small; external ears wanting; muzzle terminated with a cartilaginous appendage.

309. 1. *C. Capensis* (the Cape Chrysoclore.) Fur brown but giving, in certain angles of light, a brilliant metallic green and copper colour; hind toes five.

Chrysocloris Capensis, *Desm. Ency. Méthod.* 156. Aspalax, *Seba.* Talpa Sibirica Aurea, *Brisson.* Talpa Asiatica, *Gm.*

Musaraigne Dorée, *Cuvier, Table Element. des Anim*
Icon. *Seba, Thes.* 1. *tab.* 52. *Schreber, tab.* 157.

Inhabits the Cape of Good Hope.

Obs. Seba states the habitat of this animal to be Siberia, but it is now known to be an African species.

The Red Mole of Seba, Thes. *t.* 52. *f.* 2. Gmelin, Pennant, &c., and the Long-tailed Mole of Penn. Arctic Zoology, are thought, by the Baron Cuvier, to be proper to this sub-division; but see his note on the Chrysoclore in the text.

*** Second Tribe of Insectivorous Mammalia, with several incisors smaller than the canine teeth, like the Quadrumana, &c.

Genus VII. TALPA.

Incisors $\frac{6}{8}$, small and vertical in the upper jaw, forming an arch, and a little inclining in the lower; canine $\frac{1}{1}$, triangular; cheek teeth $\frac{7}{6}$, the three anterior in the upper, and the two in the lower jaw smaller than the rest; head elongated, terminated with a sort of boutoir; eyes very small; no external ears; pentadactylous; fore feet very large, turned, with the lower edge trenchant; toes united to the nails, which are strong, and slightly arched; hinder feet weak.

310. 1. *T. Europæa* (the Mole.) Glossy cinereous black, like velvet; tail scaly like a rat's; limbs short.

Talpa Europæa, L. i. 81. T. Vulgaris, *Brisson.*

La Taupe, *Buffon, tom.* viii. Mole, *Pen. Brit. Zool.* 52.

Icon. *Buffon,* viii. *pl.* 12.

Inhabits nearly all Europe, but not Ireland or Greece.

There are several varieties of the Mole pointed out by naturalists, as the Spotted Mole; T. Variegata of Brisson; T. Maculata of Klein; the Yellow Mole of Pennant; and the Albinose, or White Mole.

Genus IX. CENTENES.

Incisors $\frac{6}{6}$ or $\frac{4}{8}$; canine $\frac{1}{1}$, similar to the cheek-teeth $\frac{66}{66}$, the first very small, the rest pointed; muzzle pointed; ears scarcely visible; body low, covered with spines, but not capable of being formed into a ball. Toes five; tail none.

311. 1. *C. Semispinosus* (the Tendrac.) Covered with stiff spines on the upper part of the body, with annular varieties of colour; silky hairs on the under parts; four incisors in the lower jaw only. Larger than the Common Hedgehog.

Erinaceius Ecaudatus, L. E. Ecanthurus, *Boddaert Elem. Anim.* 129. Setiger Ecaudatus, *Geoff. Col. du Mus.* Tendrac, *Buff. tom.* XII. Tenrec, *Cuvier, Table Elementaire des Animaux, et Règne Animal.*

Icon. *Buff.* as above, *pl.* 57.

Inhabits Madagascar.

312. 2. *C. Setosus* (The Tanrec.) Spines long and flexible; six incisors in each jaw. Larger than the preceding.

Erinaceus Setosus, L. E. Tanrec, *Boddaert, Elem. Anim.* 129. Tendrac Setiger Inauris, *Geoff. Coll. du Mus.* Cuv. *Tab. Elementaire des Animaux, et Règne Animal.*

Icon. *Buffon,* XII. *pl.* 56.

Inhabits Madagascar, and now the Isle of France

313. 3. *C. Semispinosus* (Radiated Tenrec*.) Covered with silky hairs and flexible spines intermixed, radiated yellow and black.

Erinaceus Ecaudatus, L. E. Tanrec, *Boddaert, Elem. Anim.* 129. E. Semi-spinosus, *Cuv. Tab. Elementaire des Animaux, et Règne Animal.* Setiger Variegatus, *Geoff. Coll. Mus. d'Hist. Nat.*

* There is some confusion as to the synonymes of these three species.

Jeune Tanrec, *Buff. Supp. tom.* III. 214. Eteocles **Semi-**spenosus, *Gray, Med. Repos* 1821.

Icon. *Buffon, pl.* 37. *Schreber. t.* 165.

Inhabits Madagascar.

Genus VII. CONDYLURA *.

Incisor teeth $\frac{4}{4}$, the two intermediate in the upper jaw larger than the rest, which are conical in shape; canine teeth $\frac{1}{1}\frac{1}{1}$; cheek teeth $\frac{6}{7}\frac{6}{7}$; muzzle very long, extremity ciliated; ears, none; external eyes small; feet pentadactylous; nails before famed for digging, those behind weak and small.

314. 1. *C. Cristata* †, (Radiated Condylure.) Nostrils surrounded by a circle of membranaceous processes, radiating from a centre; tail short.

Sorex Cristatus. *Lin.* I. Condylura Cristata, *Desmarest, Journal de Physique*, 1819.

Radiated Mole, Pen. *Quad.* 313. Taupe de Canada, *Delafaille, Buff.* VI.

Icon. *Pen. Quad. t.* 28. *f.* 4. *Delafaille, Essai sur la Taupe,* 1769. *Buff.* II. *t.* 37.

Inhabits Canada.

315. 2. *C. Longicaudata,* (Long-tailed Condylure.) No nasal processes; tail half the length of the body.

Talpa Longicaudata, *Erxleb. Sys. Anim.* I. 118. **Condy**lura? Longicaudata, *Desmorest, Mam.* 158.

Long-tailed Mole, *Pen. Quad.* 314.

Icon. *Pen. Quad. t.* 18. *f.* 2.

Inhabits North America.

---

* Delafaille's figure erroneously represents the tail as knobbed; from which character, Illiger named the genus κονδαλος nodus, and ουρη cauda.

† Its insertion in this genus is conditional.

Family III.—CARNIVORA.

Six incisors in each jaw ; cheek-teeth never furnished with sharp points, as in the family of Insectivora, but either trenchant or tuberculous, or both ; the species more or less carnivorous, in proportion to the trenchant or tuberculous character of these teeth ; canines long and strong.

Tribe I. Plantigrades.

Beasts of prey that bring the whole sole of the foot from toe to heel to the ground in walking.

Genus I. Ursus.

Cheek-teeth proper $\frac{3}{3}$, large and entirely tuberculous, with $\frac{3}{4}$ or $\frac{4}{4}$ false molars, very small, which come late, and soon fall out ; body thick, and covered with thick hair ; ears largish, slightly acuminated ; toes five, furnished with strong curved claws, fitted for digging ; tail short ; mammæ six, two pectoral and four ventral.

316. 1. *U. Arctos* (Common European Bear.) Brown ; forehead convex above the eyes ; muzzle truncated.

Ursus Arctos, L.

Ours, *Buff. tom.* viii. The Brown or Common Bear of the English.

Icon. *Buff. tom.* viii. *pl.* 31. *Perrault, tab.* 9. *F. Cuvier, Mam. Lithog. f. Baron Cuvier, Menag. Mus.*

Inhabits the highest mountains and largest forests of Europe, and the temperate and southern parts of Asia.

317. 2. *U. Niger Europæus* (European Black Bear.) Brownish-black ; nose red ; muzzle reddish yellow brown ; cranium flat.

Black Bear of Europe. *Baron Cuvier, Ossemens Fossiles,* iv. 316.

Icon. ——

Inhabits Europe.

Obs. There appear to be many varieties of the Common Bear, from white through different shades of brown to black, and such have been frequently treated as distinct.

318. 3. *U. Niger Americanus* (Black Bear of America.) Forehead and nose nearly on one inclined line; fur shining black.

Ursus Americanus, *Pallas, Spic. Zool.*

Ours d'Amérique, *Cuv. Menag. du Mus.* Black Bear, *Pen. Quad.* II.

Icon. *Cuv. l. c.*

Inhabits North America.

Obs. The Yellow or Cinnamon Bear of various degrees of intensity, proper to America, are all considered by the Baron as varieties of this species.

319. 4. *U. Maritimus* (Arctic Bear.) Head elongated; skull flat; fur long, soft and white.

Ursus Maritimus, L. Ursus Albus, *Brisson.*

Ours Blanc, *Buff. Supp. tom.* III. Polar Bear, *Pennant's Synopsis*, 192.

Icon. *Buff. Supp.* III. *pl.* 34. *Pennant, Synopsis, pt.* 20. *f.* 1. *Pallas, Spic. Zool.* XIV. *f.* 1.

Inhabits the coast of the Polar Sea, principally in America.

320. 5. *U. Candescens* (the Grisly Bear.) Fur long, cinereous gray, very thick, especially about the neck.

Ursus Ferox, *Lewis and Clark, Journey to the Missouri. Warden, Description of the United States.* Ursus Candescens, *Hamilton Smith.* Ursus Cinereus, *Desm. Mam.* 164. Ursus Horribilis, *Ord.*

Grisly Bear, *Lewis and Clark, Journey to the Missouri*.
Gray Bear, *Hearn's Voyage. Warden's United States*.

Icon. ――――

Inhabits North America, especially the vicinity of the Missouri.

321. 6. *U. Labiatus* (Long-lipped Bear.) Brown-black, long rough hair ; lips long and extensible.

Bradypus Ursinus, *Shaw, Gen. Zool. vol.* I. Ursus Labiatus, *Blainville, Nov. Bull. de la Société Philom.* 1817. Prochilus Ursinus, *Illiger*. Melursus, *Meyr*. Ursus Longirostris, *Teidman*.

Ursiform Sloth, *Pennant*.

con. *Shaw, l. c. pl.* 4. *Catton's Anim.*

Inhabits the mountainous districts of India.

322. 7. *U. Malayanus* (Malay Bear.) Black, with a large heart-shaped patch of yellowish white on the throat. Fur short and smooth.

Ursus Malayanus, *Raffles, Lin. Trans.* XII. *p.* 1. *Horsfield's Zool. Java, No.* 4.

Icon. *Horsfield, l. c.*

Inhabits India.

323. 8. *U. Thibetanus* (the Thibet Bear.) Black, under jaw white, pectoral patch forked, and continued to the middle of the belly.

Ursus Tibethianus, *Baron Cuvier and F. Cuvier, Mam. Lithog.*

Icon. *F. Cuvier, l. c.*

Inhabits Thibet.

Genus PROCYON.

Canine teeth large and compressed on each side ; cheek teeth $\frac{6.6}{6.6}$, the three first pointed, the three posterior tuberculous ; body slightish ; muzzle pointed ; ears small ; tail long and pointed; standing on the heel of the hinder legs, but walking on the toes; six teats, ventral.

324. 1. *P. Lotor* (the Racoon.) Fur grayish slate-coloured; muzzle white, with a brown streak across the eyes; tail annulated, dark-slate colour and white.

Ursus Lotor, *L.* i. 70. Vulpes Americana, *Charleton.*

Mapach *of the Americans.* Racoon, *Anglo Americans,* Agouara, Pope *d'Azara, Quad. of Paraguay.*

Icon. *Buffon, tom.* viii. *pl.* 43. *Pennant,* ii. 2. *Shaw,* i. 105.

Inhabits South America

Obs. M. Geoffroy has designated two varieties, the Yellow and the Brown-throated Racoon, and Buffon probably another, under the name of Meles Abba.

325. 2. *P. Cancrivorus* (Crab Racoon.) Clear, uniform, cinereous-brown above, yellowish-white underneath; the rings of the tail less distinct than in the other species.

Ursus Cancrivorus, *Cuv. Règne Anim.* Procyon Cancrivorus, *Geoff.*

Raton Crabier, *Buff. Sup.* 6. 236. Chien Crabier, *Laborde.*

Icon. *Buff. Sup. tom.* vi. *pl.* 32.

Inhabits South America.

Genus II. NASUA.

Teeth similar to those of the preceding sub-genus; body long, thin; nose elongated and moveable; feet semipalmate, armed with strong nails; tail long; teats six, ventral.

326. 1. *N. Rufa* (the Red Coati.) Bright-red; muzzle grayish-black, with three white spots about each eye.

Nasua rufa, *Desm. Mam.* 170.

Quachi Valmont de Bomarre, *Dict. d'Hist. Nat.* Coati Roux, *F. Cuvier, Mam. Lithog.*

Icon. *F. Cuvier, l. c.*

Inhabits South America.

327. 2. *N. Fusca* (the Brown Coati.) Brown or yellowish above, yellowish-gray underneath ; three white spots about each eye, as in the preceding species, but with a white streak down the nose in addition.

Viverra Nascica, V. Quasje, *Gm.* i. 64. Viverra Rufa, *Schreb. tab.* 118.

Coati Mondi Marcg. Brasil Coati, *Azara, Anim. de Paraguay.*

Icon. *Perrault, Hist. des Anim. tom.* ii. *pl.* 37. *Buff. tom.* viii. *pl.* 47 & 48. *Schreb. tab.* 118. *F. Cuv. Mam. Lithog.*

Inhabits South America.

Genus III. CERCOLEPTES.

POTTOS. Illiger. Cheek teeth $\frac{5}{5}$, the two first pointed in front, the three posterior tuberculous ; body thin ; head round ; muzzle not elongated ; tongue extensible ; ears oval ; large membranous pentadactylous, toes armed with strong, crooked nails ; tail long and prehensile, like that of the Sapajous.

328. 1. *Pottos Caudivolvulus* (the Potto.) Fur silky bright brownish-yellow.

Viverra Caudivolvula, *Schreb.* Potot, *Buff. sup. tom.* iii. Yellow Maucoco, *Pennant, Quad.* Kinkajou, *Desm. Mam.*

Icon. *Pennant, Quad. pl.* 16. *Schreb. tab.* 125. *Buff Sup.* iii. *pl.* 51.

Inhabits parts of South America.

Genus IV. MELES.

Cheek-teeth $\frac{3}{6}\frac{4}{6}$; the first very small, the second and

115

third pointed, the fourth trenchant on the external side, the fifth turberculous and large ; the penultimate in the lower jaw the most trenchant of that range. Body thick ; legs low ; muzzle not long ; ears short and round ; eyes small ; tail very short ; a pouch under the tail, containing a fetid secretion.

329. 1. *Meles Vulgaris* (the Common Badger.) Gray-brown above, black underneath ; a longitudinal black band on each side of the head, passing round the eye and ear.

Taxus, or Meles, *Ray*. Ursus Meles, *L.*

Blaireau *Buff. tom.* VII. Badger, *Pen. British Zool.*

Icon. *Schreb.* 142. *Buff. Supp. tom.* III. *pl.* 49.

Inhabits Europe.

Obs. The country people pretend to distinguish two varieties, under the names of the Dog-Badger and the Hog-Badger, but they are not authenticated.

330. 2. *M. Labradorica* (American Badger.) Pale yellowish-gray ; belly and throat white, with a longitudinal band on the side of the head, passing above the eye and ear.

Ursus Labradoricus *Gmelin, Syst. Nat.* 1102. Meles Labradoria, *Sabine, Ross's Voy. App.* 649.

American Badger, *Penn. Quad.* II. 15. Carcajou, *Buff. Supp.* 242.

Icon. *Shaw, Zool.* I. *t.* 106. *Buff.* II. *f.* 49.

Inhabits Hudson's Bay.

Genus V. GULO.

Cheek teeth $\frac{5}{6}$; the three first in the upper, and the four in the lower jaw, small, succeeded by a larger carnivorous or trenchant tooth, and small tuberculous tooth at the back. Body low ; head moderately elongated ; ears short and round ; tail short ; pentadactylous, toes armed with crooked nails.

331. 1. *Gulo Vulgaris* (the Glutton.) Fur deep brownish-red, darker on the back.

Ursus Gulo, Mustela Gulo, *Lin. Sys. Nat.* i. 67. Meles Gulo, *Boddaert. Gm. Sys. Nat.* i. 104.

Glouton, *Buff. Suppl. tom.* iii. 240.

Icon. *Pallas, Spic. Zool.* 14. *tab.* 2. *Schreb. tab.* 144 *Buff. l. c., pl.* 48. *Shaw's Zool.* i. *t.* 104.

Inhabits the coasts of the Arctic Sea.

332. 2. *Gulo Wolverene* (the Wolverene.) Paler than the preceding.

Ursus Luscus, *L.* Ursus Gulo, var. *Shaw's Zool.* i. 462. Hudson's Bay Bear of *Brisson.* Quick-hatch, or Wolverene, *Edwards' Birds.*

Icon. *Edwards' Birds, pl.* 108. *Pennant's Quad. pl.* 20. *fig.* 2. *Shaw's Zool.* i. *t.* 105.

Inhabits the coasts of the Arctic Sea.

The Baron Cuvier treats the above two as different species; Desmarest merely as varieties.

333. 3. *G. Vittatus* (the Grison.) Black, spotted with white; top of the head and neck gray; white band passing from the forehead to the shoulders; body elongated like the Weasels.

Viverra Vittata, *L.* Lutra Vittata, *Wern. Trans.* iii. Fuine de la Guyane et Grison, *Buff. Sup. tom.* viii. Petit Furet, *Azara, Quad. Paraguay.*

Icon. *Buffon, Sup.* 8. *pl.* 23 and 25. *F. Cuvier, Mam. Lithog.* Trail, *Wern. Trans. t.* 19. *f.* 506.

Inhabits South America.

334. 4. *Gulo Barbatus* (Galera or Taira.) Brownish-black, with a white patch covering the under part of the neck and throat; body Weasel formed.

Mustela Barbara, *L.*

Taira or Galera, *Buff. Supp. t.* 7. The Great Weasel of Azara, *Quad. of Paraguay.* Galera, *Brown's Jamaica.* Cariqueibein, *Marcgrave;* la Saricovienne, *Buff. t.* XIII. ?

Icon. *Buff. Supp. tom.* VII. *pl.* 60. *Brown, Jamaica, pl.* 49. *f.* 1.

Inhabits parts of South America.

Obs. The two last-mentioned species have the cheek teeth $\frac{4.4}{3.3}$, and may, therefore, be treated as a sub-genus.

II. RATELLUS. F. Cuvier. *Cheek teeth* $\frac{4}{3}\frac{4}{3}$; *two false in the upper jaw, and three in the lower.*

335. 1. *G. Ratel* (the Ratel.) Gray above, black underneath, with a longitudinal white line on each side, from the ears to the tail; body thick and heavy.

Viverra Mellivora, *Gm.* I. 91. V. Capensis, *Gm.* 89.

Rattel, *Sparman, Act. Stockholm,* 1777. The Fizzler Weasel, *Pen. Quad.* Blaireau Puant of *Lacaille's Travels.* Honey Weasel, *Shaw's Zool.* 395. Cape Weasel, *Id. Zool.* 396. *Hardwick, Lin. Trans.* v. IX. ?

Icon. *Schreb.* 125. *Sparman,* as above, *pl.* 4. *f.* 3.

Inhabits the Cape of Good Hope.

Obs. The Ursus Indicus of Shaw, Indian Badger of Pennant, are said to be varieties of the Ratel.

The Atok Gulo Quitensis, and the Mapurito of the Baron Humboldt, have been placed in this sub-genus, on account of their plantigrade motion; but M. Desmarest and other systematic writers, refer these two species to the subgenus of Mephitic Weasels, the species of which may be said to be semi-plantigrade. The Labrador Glutton, of Sonnini, appears to be a Badger.

336. 2. *G. Orientalis* (the Nyentek of Java.) Glossy reddish brown ; white patches about the head and throat ; and a long pyramidical white patch from the top of the

head to the middle of the spine; rather less than the pole-cat.

Gulo Orientalis, *Horsfield's Java.*

Nyentek of the Javanese.

Icon. *Horsfield l. c.*

Inhabits Java.

337. 3. *G. Larvatus* (the Masked Glutton.) Olive-brown and gray, tip of tail and feet black, white patches about the face; larger than the pole-cat.

Gulo larvatus, *Temminck's* and *Hamilton Smith's MSS.*

Icon. nobis.

Inhabits.

338. 4. *G. Ferrugineus* (Ferruginous Glutton.) Chesnut colour, tail black, and feet sepia; head broad and depressed, eyes near the nostrils, ears far back; four feet from the nose to the end of the tail.

G. Ferrugineus, *Hamilton Smith, MSS.*

Icon. nobis.

Habitat. ?

Tribe II. DIGITIGRADES.

Beasts of prey that walk on the toes only.

I. Subdivision of Digitigrades.

With one tubercular tooth behind the great carnivorous tooth in the upper jaw.

Genus MUSTELA. Incisors $\frac{6}{6}$; canines $\frac{1-1}{1-1}$; cheek teeth $\frac{4-4}{5-5}$, or $\frac{5-5}{5-5}$; head small and oval; ears short and round; body long vermiformed; legs short; toes five, armed with sharp crooked claws; no anal pouch, but with a small gland secreting a strong stinking unguent.

I. PUTORIUS. Cuvier. *Two false molars above, and three below; the great carnivorous tooth below without an internal tubercle; muzzle short; fœtid.*

2 K 2

339. 1. *P. Vulgaris* (the Polecat.) Fur from the root pale yellow, toward the extremity bright brown; small white spots on the head and muzzle. Length of body one foot five inches; tail six inches.

Mustela Putorius, *Lin. Sys. Nat.*

Putois, *Buff. t.* vii. Polecat, *Pennant's Quad.* i. 213.

Icon. *Schreb. tab.* 131. *Buff.* vii. *pl.* 24. *Ency. Method. t.* 82. *f.* 2. *Penn. Brit. Zool.*

Inhabits the temperate parts of Europe.

Var. a. *Furo* (the Ferret.) The fur yellow; eyes red.

Mustela Furo, *Lin. Sys. Nat.* i.

Le Furet, *Buff.* vii. The Ferret, *Penn. Quad.* 214.

Icon. *Schreb. t.* 133. *Buff.* vii. *f.* 26. *Ency. Method. t.* 82. *f.* 2.

Inhabits Spain and Africa.

Obs. It is sometimes variegated with black and brown.

340. 2. *P. Alpinus* (the Alpine Polecat.) The fur sulphur yellow, above brownish; chin white; the canine teeth without any internal tubercle. Length of body and head a foot; tail five inches.

Mustela Alpina (vel. Putorius Alpinus.) *F. Gebler, Mem. Soc. Imp. des Nat. de Moscou,* vi. 213. (1824.)

Icon. ——

Inhabits the Altaica Mountains, near Reddersk.

Obs. Shaped like the common Polecat, but smaller, and the head more elongated and acute. The fur is not used in commerce, the hairs being too short.

341. 3. *P.?* *Altaica* (the Altaican Weasel.) The tail twice as long as the head, of the same colour.

Mustela Altaica, *Pallas Zool. Ross's Ascat.* i. 98. *F. Gebler, Mem. Nat. Mosc.* vi. 213.

Icon. ——

Inhabits Altaica, near Reddersk.

342. 4. *P. Nudipes* (Java Ferret.) Fur brilliant golden yellow; forehead and tip of the tail yellow white; soles of the feet quite naked. Length of the body and head ten inches and a half, tail six inches.

Mustela Nudipes, *F. Cuv. Mam. Lithog. N. 32, t. 3.*
Icon. *F. Cuv. Mam. Lithog. N. 32. t. 3.*
Inhabits Java.

343. 5. *P. Sarmatica* (the Perouasca or Sarmatian Weasel.) Fur rich brown, spotted with yellow; throat and belly black.

Mustela Sarmatica, *Pallas, Spic. Zool.* 14.
Tiger Iltis, *Pallas, Itiner. j.* 175 and 454. Sarmatian Weasel, *Shaw.* Paræiasta, *Russians.*
Icon. *Pallas, Spic. Zool. tab.* 4. *f.* 1. *Nov. Act. Petrop.* xiv. *f.* 10. *Schreber, tab.* 132. *Ency. Method. t.* 82, *f.* 4.
Inhabits Poland and Russia Proper.

344. 6. *P. Vulgaris* (the Common Weasel.) Reddish-brown above, white underneath. Length of the body and head six inches; tail an inch and a half.

M. Vulgaris, *Lin. Sys. Nat.* i.
Belette, *Buff.* vii. Weasel, *Penn. Brit. Zool.* 39.
Icon. *Schreber, tab.* 137. A. *Buff.* vii. *pl.* 29, *f.* 1. *Pennant's Brit. Zool. fig.* *Ency. Méthod. t.* 84. *f.* 1. *Shaw, Zool. F. Cuv. Mam. Lithog. f.* 2.
Inhabits most of the temperate and the northern parts of the old world, and North America.

Var. a. *Nivalis.* White, with tip of the tail black.
Mustela Nivalis, *Lin. Faun. Suec.* 7. M. Vulgaris, B. *Gmelin, Sys. Nat.* M. Erminea, B. *Bodd.*
Belette des neiges, *Ency.*
Icon. *Ency. Méthod. t.* 83. *f.* 4.

Var. b.? *Africana.* Body above reddish brown, beneath pale yellow, with a narrow central longitudinal brown band. Length of the body and head ten inches, tail seven.

Mustela Africana, *Desm. Nouv. Dict. Hist. Nat.* xix. 376
Icon. ——
Inhabits Africa.
Obs. M. Desmarest treats this as distinct.

345. 7. *P. Erminea* (the Ermine in Winter, the Stoat in Summer.) Yellow-white beneath, tip of the tail black; back yellow-white in winter, pale chestnut brown in summer. About the size of the Weasel.

Mustela Erminea, *Lin. Sys. Nat.* i. M. Candida, *Ray. Syn.* M. Armelina, *Klein. Quad.*

In winter, Ermine, *Pennant. Quad.* L'Hermine, *Buff.* vii.
Icon. ——
Inhabits. ——

346. 8. *P. Sibirica* (Siberian Weasel.) Pale fulvous, especially on the lower parts; muzzle brown, round the nose white. Same size as the Polecat.

Mustela Sibirica, *Pallas, Spic. Zool.* 14.

Chorock, *Sonnini's Buffon*, xxxv. 19.

Icon. *Pallas, Spic. Zool. pl.* 4. *f.* 2. *Schreber, pl.* 135. B.
Inhabits Siberia.

347. 9. *P. Lutreola* (the Mink.) Blackish brown; upper lip, chin, and under the neck white. Feet semipalmate.

M. Lutreola, *Pallas, Spic. Zool.* 14. Lutra Minor, *Exr-leben, Mém. de Stockholm*, 1739.

Tuhcuri, *Finlanders.* Nærs of the *Prussians.* Mink of the *Furriers.*

Icon. *Pallas, Spic. Zool. pl.* 31. *Erxleb. Mem. Stock. tab.* 11. *Ency. Méthod. t.* 80. *f.* 1.

Inhabits Finland very generally, and is found also in all the north-eastern parts of Europe. Erxleben refers its habitat also to North America, in which, says Desmarest, other similar species are found; but the name Minx, as used by the Americans, has relation to a species of the Vison.

Var. a. With four longitudinal bands, a spot on each cheek, and the ends of the ears white.

348. 10. *P. Zorilla* (Zorillo.) Fur irregularly variegated with longitudinal black and white bands.

Viverra Zorilla, *Gmelin, Sys. Nat.* i.

Blaireau du Cap, *Kolbe, Description of the Cape*, i. 86. Putois du Cap ou Zoreille, *Buff.* xiii.

Icon. *Buff.* xiii. *pl.* 41. *Schreber, tab.* 123. *Ency. Méthod. t.* 86. *f.* 4. *Shaw's Zool.*

Inhabits the Cape of Good Hope.

Obs. Buffon and his followers have confounded this species with the American Mephitic Weasels, to which it is much assimilated in appearance.

II. MARTES. *One more false molar in each jaw than in the Putorii, and the lower large carnivorous tooth with a tubercle on the inner side.*

349. 1. *M. Vulgaris* (the Pine Marten.) Brown, with a clear yellow patch under the throat.

M. Abietinum, *Ray, Syn. Quad.* Mustela Martes, *Lin. Sys. Nat.*

La Marte, *Buff.* vii. Pine Martin, *Ray, Syn. Quad. Penn. Brit. Zool.*

Icon. *Buffon,* vii. *pl.* 22. *Schreb. tab.* 130. *Ency. Méthod. t.* 81. *f.* 4. *Pennant, Brit. Zool. Shaw, Zool.*

Inhabits the northern parts of Europe and Great Britain.

Obs. Buffon states, but erroneously, that this species is not British.

350. 2. *M. Foina* (the Fouin or Beech Marten.) Pale brown, with the under part of the throat and neck whitish.

123

Mustela Foina, *Lin. Sys. Nat.*

La Fouine, *Buff. t.* VII. The Marten, *Ray.*

Icon. *Schreb. tab.* 129. *Buff.* VII. *pl.* 18. *Ency. Méthod. t.* 81, *f.* 1. *Shaw. Zool.*

Inhabits Europe and Western Asia.

351. 3. *M. Zibellina* (the Sable.) Fur shining, blackish brown, the head and throat whitish. Feet covered with fur to the ends of the toes.

Mustela Zibellina, *Lin. Syst. Nat. Pallas, Spic. Zool. fasc.* 14.

Sable, *Heralds* and *Furriers.* Sobol, *Russians.* Sabbel, *Swedes.* Sable Weasel, *Shaw.*

Icon. *Schreb. tab.* 136. *Pallas, Spic. Zool.* 14. *tab.* 3. *f.* 2.

Inhabits Northern Asia.

352. 4. *M. Vison* (the Vison.) Brown, with the point of the lower jaw white, and the tail brown-black. Length of the head and body fifteen inches.

Mustela Vison, *Gm. Syst. Nat.*

Le Vison, *Buff. t.* XIII. Minx of the *Americans?*

Icon. *Buff.* XIII. *pl.* 43. *Schreber, tab.* 127.

Obs. According to Gmelin and Warden, the feet are semipalmate, but Cuvier says they are not so. It has considerable affinity in colouring and size to *M. Lutreola ;* but Cuvier locates the two species in two distinct sub-genera. Desmarest confounds this with the *M. Martes* in his notes.

353. 5. *M. Canadensis* (Pekan.) Head, shoulders, and upper part of the back mixed gray and brown; nose, crupper, tail, and limbs blackish brown ; Frequently, but not always, with a white patch on the throat. Length of the body and head eighteen inches, tail ten inches or a foot.

Mustela Canadensis, *Gmelin, Syst. Nat*

Pekan, *Buff. t.* XIII. Pekan Weasel, *Penn. Quad.* 331, and 204.

Icon. *Buff.* XIII. *pl.* 42. *Schreb. tab.* 134.
Inhabits Canada and the United States.

354. 6. *M. Pennanti* (the Fisher Weasel.) Fur dark at the base, yellow above, and tipped with black, becoming chestnut instead of yellow on the back ; tail black, shining ; throat brown, with a few white-tipped hairs ; belly and legs dark brown ; ears short, lighter at the tips. Length of head and body thirty inches, tail fifteen inches.

Mustela Pennanti, *Erxleben, Syst. Mam. sp.* 10. Mustela Melanorhyncha, *Bodd. Elench. Anim. sp.* 13.

Fisher Weasel, *Pennant's Quad. No.* 202.

Icon. ——

Inhabits North America.

Obs. This species, it is said, by Captain *Sabine,* does not feed on fish, but takes its food like the Pine Marten.

355. 7. *M. Rufa* (Chestnut Weasel.) Brownish red colour, deeper above, each hair annulated, brown red, and yellow ; tail brown at the extremity.

Marte Marron, *Geoff. Collect. du Mus. d'Hist. Nat.* Mustela Rufa, *Desm. Mam.* 184.

Icon. ——

Habitat unknown.

Obs. Probably a variety of the Pekan???

356. 8. *M. Sinuensis* (the Zorra.) Uniform blackish gray ; belly and interior of the ears white. Body less vermiformed than in the other species of this subgenus, and more like that of the Kinkajou.

Mustela Sinuensis, *Humb. Voy. dans l'Amérique, Mérid. partie Zoologique.* Marte Zorra, *Desmarest Ency. Méthòd, sp.* 286.

Icon. ——?

Inhabits the warm part of New Grenada.

357. 9. *M. Leucotis* (White-eared Weasel.) Sepia brown colour; inside of ears white. Twenty inches long.

Mustela Leucotis, *Temminck and Hamilton Smith, MSS.*

Icon. —— nobis.

Inhabits.

The *Mustela Cuja* and the *Mustela Quiqui* have also been named by Molina (Chili, 272 and 258), but their specific distinctness seems uncertain, as is also that of the white cheek Weasel of *Penn.* (M. Flavigula, *Bod.* and M. Quadricolor of *Shaw*.)*

III. MEPHITIS. Cuvier. *Cheek teeth $\frac{4}{4}$, two false or small anterior cheek teeth above, and three below; the great carnivorous tooth provided with two tubercles on the inner side, the posterior tooth tuberculous and very long and large. Anterior toes furnished with long digging nails; heel very little raised in walking; the palm and heel hairy.*

Obs. Following the nomenclature and description of many travellers, zoologists, and systematic describers, there would be nineteen species of the Mephitis proper to America to be described, differing principally in colour. The Baron Cuvier *(Ossemens Fossiles, IV.)* inclines, however, to the opinion that all these are but varieties of one species, which varieties, however, are very local, and seem to present a character of permanency.

The following nomenclature, therefore, taken from the Baron's researches and from the Encyclopedie Methodique of Desmarest, treats all the American Mephites merely as varieties The insertion of them, however, here as such,

---

* The following species of Shaw are referred to other genera·
The *Gray-headed Weasel*, the *Guiana Weasel*, and the *Galera;* all appear to be varieties of *Gulo Barbatus;* the South American Weasel is the Gulo Vittatus; the Woolly Weasel (the *M. Guyanensis* of Lacep.) is said by Desmarest to be a young *Coati.* The *Musky Weasel* and the *Slender-toed Weasel,* described from drawings, are very obscure.

must be considered conditional only, as several of them are still thought by able observers to be distinct.

It must be premised, however, that the Zorille of Buffon (Viverra Zorilla, Gm.) belongs to the division Martes, and that the Coasse of the same writer, at least in the Baron Cuvier's opinion, is established only from an imperfect skin of the Coati.

a. M. Americana (American Mephitic Weasel.) Fur soft and shining, marked by white longitudinal bands upon a blackish brown ground ; tail long and furry.

358. 1. Mustela Americana, *Desm. Mam.* 186. Viverra Striata, *Shaw's Zool.* I. 387. V. Putorius, *Gmelin, Syst. Nat.* I. 87. V. Conepatl, *Gmelin*, I. 88. V. Mephitis, *Gmelin*, 1. 88. V. Chinge, *Shaw's Zool.* I. 390. V. Putorius, *Mutis, Act. Holm.* V. Marputio, *Gmelin*, I. 88. Mephitis Chilensis, *Geoff.* Gulo Quitensis, *Humboldt, Rec. Obs. Zool.* M. Interupta, *Raff. Ann. Nat.* 3. Gulo Marpurito, *Humboldt, Obs. Zool.* Striated Weasel, *Pennant, Quad.* II. 64. Conepate, *Buff.* XIII. 288. Conepatl, *Hernand. Mex.* 232. La Chinche, *Buff.* XIII. 294. Skunk, *Americans.*

Var. a. *Yagouare* of d'Azara, Quad. du Paraguay. Blackbrown, brightening with the increased age of the individual, with two white stripes stretching to the tail. Some individuals are without the white stripes ; others have them very obscurely indicated, and others again have the stripes extending along the sides of the tail.

Var b. Polecat of Kalm, *Skunk* of the Americans. Brown-black, with a white stripe down the dorsal line and another on each side of it.

Var. c. The *Zorille* of Gemelli Carreri Voyag. Described only as being black and white, with a very fine tail.

Var. d. The *Mapurita* of Gumilla, Natural History of the Orenoco, Mafutiliqui of the Indians. Spotted black and white.

Var. e. The *Puant* of Lepage, Dupratz, Hist. de la

Louisiane. Male black; female black, bordered with white.

Var. f. The *Orthula* of Mexico, Fernandez, Hist. Nouv. Hisp. Black and white, with yellow in some parts.

Var. g. The *Tepemaxtla* of the same. Without any yellow ; tail annulated black and white.

Var. h. The *Atok* or *Zorra* of Quito, *Gulo Quitensis*, Humboldt. Body marked with two white stripes ; tail mixed black and white ; tongue aculeated. *Annals of Nature.*

Var. i. The *Ysquiepatl* of Hernandez. Marked with several white stripes.

Var. k. The *Polecat* of Catesby's Carolina, tab. 62. Nine white stripes.

Var. l. *Conepate* of Buffon, t. 13. pl. 40. Six white stripes.

The Baron Cuvier thinks that this figure is made up from that of Catesby.

Var. m. The *Conepatl* of Hernandez. Two white stripes on the tail only.

Var. n. The *Mapurito* of Mutis, Act. Holm. 1769, the *Viverra Mapurito* of Gmelin, and *Glouton Mapurito* of Humboldt, Observ. Zool. One white stripe commencing on the forehead, and terminating half down the back. Tail white at its extremity.

Var. o. *Moufette de Chili*, Buff. Sup. t. 7. pl. 57, *M. Chiliensis* Geoff. Dict. des Sciences Nat. pl. 19. fig. 1. Fur brown red, with two white stripes on the sides of the body uniting in the form of a crescent behind the head; tail white and brown.

Var. p. The *Chinche* of Buff. t. 13. pl. 39, *Viverra Mephitis*, Gm. Two white stripes very wide, and large toward their posterior termination ; tail with long white hairs, mixed with a few black; forehead with a longitudinal white band, joined to that of the back; rest of the body brown, more or less deep, with two small white spots on the shoulders and belly.

Var. q. The *Chinche* of Feuillée, Journal du P. Feuillée, 1714. With two white stripes, terminating on the sides.

Var. r. *Mephitis interrupta, Rafinesque.* Brown, with two short white parallel rays on the head, and eight upon the back, of which four are equal and parallel, and four rectangular, and placed in opposite directions.

Var. s. *Chinga, Molina.* With a band of round white spots on the back.

Var. t. *Mephitic Weasel of Bengal.* Shaw's Gen. Zool. 1st part. Two spots on the head, four white dorsal stripes, and tail furry.

359. 2. *M. Meliceps* (Telagon.) Deep brown, especially on the upper part; forehead marked with a white spot, which is extended into a dorsal line ; length of the body and head sixteen inches; tail one inch.

Mephitis Javanensis, *Desm. Mam.* 187. Myadeus Meliceps, *Horsf. Zool. Resch.* vi. *Raffles.* Telagon, *F. Cuvier, Mam. Lithog.* 27.

Telagon, *Raffles. Lin. Trans.* xiii. Telagon, *F. Cuv. Mam. Lithog.* xxvii.

Icon. *Horsf. Zool. Resch.* vi., and head. *F. Cuvier Mam. Lithog.* xxviii.

Inhabits Java.

Obs. Dr. Horsfield makes a distinct genus of this, under the name Myadeus.

III. Lutra. Ray. *Head large, flattish. Ears short. False grinders,* $\frac{3.4}{3.3}$; *the lower great carnivorous tooth, with two points on its outer side ; toes webbed, nails crooked. Tail slightly flattened horizontally.*

360. 1. *L. Vulgaris* (the Common Otter.) Brown above;

whitish underneath; tail more than half as long as the body.

Lutra Vulgaris, *Erxleb.* Mustela Lutra, *Lin. Sys. Nat.* i. 66.

La Lutre, *Buff.* vii. Common Otter, *Shaw. Zool.* 437. Greater Otter, *Pennant, Quad.* ii. 77. Otter, *Penn. Brit. Zool.*

Icon. *Schreb. Saugth. tab.* 126, A. *Buff.* vii. pl. 11. *Pennant, British Zoology.*

Inhabits Europe.

Var. a. *Maculata, the Spotted Otter,* with a great number of little round white spots on the flanks : found near Paris.

361. 2. *L. Brasiliensis* (the Brazilian Otter.)

Brown or yellow, with the throat white or yellowish.

Lutra Brasiliensis, *Ray, Syn. Quad. Desm. Mamm.* 188. Mustela Lutra Brasiliensis, *Gmelin, Sys. Nat.* Lutra Brasiliana, *Shaw. Zool.* i. 446.

Saricovienne de la Guyane, *Buff. Sup. tom.* vi. Brasilian Otter, *Pennant, Quad.* ii. 79.

Icon. *Ency. Méthod. Supp. t.* 5. *f.* 3. *Cuv. Reg. Anim.* iv. *t.* 4. *f.* 3.

Inhabits the rivers of both Americas, especially of Guyana.

362. 3. *L. Canadensis,* (Canadian Otter.) Glossy brown ; chin and throat dusky white ; neck and head long, and the ears closer together than in the L. Vulgaris ; legs short ; tail pointed, and as long as the body.

Lutra Canadensis, *Sabine, Franklin, Voy. Ap.* 653. Mustela Hudsonica, *Lacépède.*

American Otter, *Sabine.* Loutre de Canada, *Lacépède.*

Inhabit Copper Mine River.

The Baron Cuvier unites these two varieties, but they are separated by M. F. Cuvier.

363. 4. *L. Insularis.* (Trinity Otter.) Hairs scattered, polished; body clear chestnut brown; throat and chin yellowish white.

Lutra Insularis. *F. Cuv. Dict. Sci. Nat.* xxvii.

Icon.

Inhab. Isle of Trinity.

364. 5. *L. Lataxina* (Carolina Otter.) Hairs long, twisted, wool abundant; above deep brown black; and the cheeks, chin, and throat, pale bluish gray; frontal region of the skull rather concave.

Lutra Lataxina, *F. Cuvier, Dict. Sci. Nat.* xxvii.

Icon. ——

Inhab. Carolina.

365. 6. *L. Enudris.* Above clear bay, paler beneath; throat and sides of the face nearly white; the curve of the profile of the top of the head slightly but regularly arched, from the occiput to the end of the nose.

Lutra Enudris. *F. Cuv. Dict. Sci. Nat.* xxvii.

Icon.

Inhab.

366. 7. *L. Nair* (Pondicherry Otter.) Hair long, deep chestnut; lower part of the neck, throat, and belly, clear reddish white; the cheeks marked under each eye with a reddish brown band.

Lutra Nair. *F. Cuv. Dict. Sci. Nat.* xxvii.

Icon.

Inhab. Pondicherry.

367. 8. *L. Inunguis* (Clawless Otter.) Body elevated on its legs; toes large, shortly palmated, clawless; fur soft, chestnut brown, deeper on the buttock, tail, and legs; head and shoulders brownish gray; lips, chin, and lower part of the neck and chest white.

131

Lutra Inunguis, *F. Cuv. Dict. Sci. Nat.* xxvii.

Icon.

Inhab. Cape of Good Hope.

Obs. Deference for the highly respectable zoologist, who has treated these several otters as distinct, obliges us to separate them specifically; otherwise we should have been inclined to say merely, and judging from description alone, that America produces several varieties of the otter.

368. 9. *L. Leptonyx* (Javanese Otter.) Shining fulvous brown; throat dull yellow; tail less than half the body; claws short, blunt, nearly laminar.

Lutra Leptonyx, *Horsfield, Java,* vii. L. Barang. *F. Cuv. Dict. Sci. Nat.* xxvii.? Mustela Lutra, *Marsden, Sumatra,* 113?

Welingsang, or Wargul, *Javanesse.* Angingayer, *Malay.* Simung, *Sumatresse, Raffles. Lin. Trans* xiii. Anzing Ager, *Marsden, Sumatra,* 113. Gryze Otter, *Wurm. Batav. Soc. Trans.* ii. 457.

Icon. *Horsf. Java, t. Marden, Sumatra, t.* 12, 13?

Inhabits Java, Sumatra?

IV. Enhydra. Fleming. *Cutting teeth,* $\frac{6}{4}$; *cheek teeth,* $\frac{44}{55}$; *false molares,* $\frac{22}{33}$. *Body very long. Hind legs and tail short.*

This subgenus, of Fleming, is intermediate between the otter and the seals, both in the number of its teeth and form of its body and feet. The character of the cutting teeth is from the specimen in the British Museum.

369. 1. *E. Marina* (Sea Otter.) Chestnut brown.

Mustela Lutris, *Lin. Sys. Nat.* i. 66. Lutra Marina, *Erxleb.* Enhydra Marina, *Fleming, Phil. Zool.*

Sea Otter, *Cook's Voyages. Menzies and Home, Phil. Trans.* 1796. Lutra Marina, or Sea Otter, *Steller. Nov.*

Comm. Petrop. ii. 367. Loutre du Kamtchatka. *Geoff. Col. Mus.*

Icon. *Cook's Voy. t.* 43. *Schreb. t.* 128. *Steller. l. c.* ii. *t.* 26. *Shaw. Zool. t.* 101. *Ency. Method, t.* 79. *f.* 3.

Inhab. Bering's Strait and Kamtchatka.

II. SUBDIVISION of the Digtigirades.

With two tubercular teeth behind the great carnivorous tooth in the upper jaw.

Genus CANIS. Lin. Incisors $\frac{6}{8}$; canine $\frac{1}{1}\frac{1}{1}$; cheek-teeth $\frac{6}{7}\frac{6}{7}$; the three first in the upper jaw; and the four in the lower trenchant, but small, and called also false molars. The great carnivorous tooth, above bicuspid, with a small tubercle on the inner side, that below with the posterior lobe altogether tubercular, and two tuberculous teeth behind each of the great carnivorous teeth. Muzzle elongated, (sometimes rather short in the tame varieties;) tongue soft; ears erect, (sometimes pendent in the domestic varieties.) Fore feet pentadactylous; hind feet tetradactylous. Teats both inguinal and ventral.

Dogs, properly speaking. Pupil of the eyes circular.

370. 1. *Canis Familiaris* (the Dog.) Tail recurved into an arch; muzzle more or less lengthened. Fur varying in the nature of its hair. Tail generally tipped with white.

Canis Familiaris, *Lyn. Sys. Nat.* i. 56.

Canis, *Gesner. Quad.* 91.

Le Chein, *Buffon, Hist. Nat.* v. The Dog, *Pennant. Quad.* Icon. *Lin. Amœn. Acad.* iv. 43. *t.* 1. *j.* 1.

This species is exceedingly subject to vary in the form, colour, and quality of the fur. In arranging the varieties, we have followed the method adopted by M. F. Cuvier.

SECT. I. Head more or less elongated, parietals shelving in an insensible manner towards each other, condyles of the lower jaw on the same line with the upper molar teeth.

Var. a. *C. F. Australasiæ* (the Dingo.) Size and form of the shepherd's dog, with the head resembling that of the Fox. Fur thick, formed of two kinds of hair; the woolly, gray; the silky, yellow or white; upper part of the body, and head and tail deep yellow : lower parts paler; muzzle and inner side of the thighs white; tail with 18 vertebræ.

C. F. Australasiæ, *Desm. Mam.* 191; *Zool.* 277.

Dingo, or Australasian Dog, *Shaw.*

Var. b. *C. F. Sumatrensis.* Countenance of a fox, nose pointed; eyes oblique; ears rounded, very hairy; muzzle foxy brown, much mixed with black ; tail pendulous, bushy, reaching nearly to the heel.

Canis Familiaris Sumatrensis, *Hardwicke, Lin. Trans.* XIII. 235.

Icon. *Hardwicke, Lin. Trans.* XIII. *t.* 23. *Shaw, Zool.* I. *t.* 78.

Inhab. Sumatra.

Very lively, running with its tail extended horizontally, the head high, and the ears straight, courageous and voracious, but very volatile, and scarcely to be rendered tractable.

Var. c. *C. F. Laniarius* (the Mâtin). Head elongated; forehead flat; ears erect at their base, and half drooping; form long and strong without being thick ; legs long, nervous, and very strong; tail recurved; hair on the upper part of the body short, and the lower part and tail longer; yellow fox colour, with obliquely disposed parallel interrupted bands on the flanks, sometimes white, gray, brown, or black : length of the body and head thirty-five inches.

C. F. Laniarius, *Gmelin, Sys. Nat.*

Mâtin. Buff. v.

134

Icon. *Buffon*, v. *t.* 25.   *Ency. Méthod.* t 103. *f.* 2. *Bewick. Quad.*

Very courageous and intelligent. Buffon thinks that this variety, proper to temperate climates, becomes the great Danish dog when transported to the North, and the greyhound when bred in the South. Crossed with the bulldog, the offspring is the mastiff. *C. F. Anglicus* of *F. Cuv.*

Var. d. *C. F. Danicus* (the Great Danish Dog.) Head like that of the mâtin; body generally white, marked with numerous small round black spots, but it is sometimes gray, or brown. This variety is remarkable for its acquired attachment to horses.

C F. Danicus, *Desm. Manm.* 191.

Danois, *Buffon*, v. The Dalmatian or Spotted Dog. *Shaw, Zool.* i. 282. The Danish Dog, *Pennant.* Coach Dog.

Icon. *Buffon*, v. *t.* 26.   *Bewick. Quad.*

Var. e. *C. F. Grajus* (the Greyhound.) The greyhound, properly speaking, has several other varieties assimilated to it, all of which form an insulated group distinguished by the elongation of the muzzle beyond all others; the forehead very low, caused by the obliteration of the frontal sinuses, long and slender limbs, general lightness of make, and frequently by the want of the fifth toe, which is developed on the hind feet of the other varieties. To this group belong—

C. F. Grajus, *Lin. Sys. Nat.* i. 57.

Levrier, *Buffon*, v. The Grey-hound, *Shaw, Zool.* i. 283.

Icon. *Buffon*, v. *t* 27. *Ency. Méthod. t.* 98. *f.* 3. *Bewick. F. Cuv. Mam. Lithog.* xvi.

Sub.-var. a. The *Irish Greyhound*, from three to four feet in height; colour white, or cinnamon colour.

Icon. Lambert, Lin. Trans. Shaw, Zool. i. *t.* 77.

Obs. Nearly allied, if not identified with C. Liniarius.

Sub-var. b. The *Scotch Greyhound,* large, with the hairs slightly curled and rough ; called also the wiry-haired grey-ground.

Sub-var. c. The *Russian-Greyhound.* Body very thin, covered with long and thick hair; tail very long and twisted spirally.

Sub-var. d. The *Italian Greyhound,* a diminutive race, with very little hair ; white or isabella colour; trembling continually in the low temperature of our climate.

Sub-var. e. *Turkish Greyhound.* Skin naked ; suffering like the last from cold.

Buffon, in his hypothetical canine genealogy, deduces this group from the mâtin located in the warm climates of this part of the world. The larger greyhounds are used in coursing, but they are destitute of the powers of smell, which distinguish other races, and fit some of them more particularly for the chase.

SECTION II. Head moderately elongated; parietals not approaching each other from their insertion, but rather diverging, so as to enlarge the cerebral cavity and the frontal sinuses.

Var. a. *C. F. Extrarius* (the Spaniel.) Ears large and pendent; tail elevated; fur of different length, in different parts of the body, longer about the ears, under the neck, behind the thighs, and on the tail, than elsewhere; varying in colour, but most commonly white, with brown or black patches. Employed in the chase as a setter, for which it is qualified by its exquisite powers of smell.

C. F. Extrarius, *Lyn. Sys. Nat.* 1. 56.

The Spaniel. *English Authors.*

Icon. ——

The common spaniel, like the common greyhound, has several analogous breeds, all of which may form a group : of these are—

Sub-var. a. The *Alpine Spaniel*, very large and beautiful.

Sub-var. b. The *Newfoundland Spaniel*, large, with semi-palmate feet.

Sub-var. c. The *Calabrian Dog*, large, and participating in the distinctions of the Danish dogs and Spaniels from which it springs. Employed in hunting the wolf

Sub-var. d. The *Little Spaniel*. Head small and round; ears and tail covered with very long hair.

Icon. *Ency. Méthod. t.* 100, *f.* 3. *Buffon*, v. *t.* 38, *f.* 1.

Sub-var. e. The *King Charles Spaniel*, a black variety of the little spaniel.

C. F. Brevipilis, *Gmelin, Syst. Nat.* Le Gredin, *Buffon*, v. Icon. *Buffon*, v. *t.* 19, *f.* 1.

Sub-var. f. The *Pyrame*, like the preceding, but the black fur relieved with yellow over the eyes on the muzzle, throat, and limbs.

Le Pyrame, *Buffon*, v

Icon. *Buffon*, v. *t.* 39, *f.* 2. *Ency. Méthod, t.* 100. *f.* 2.

Sub-var. g. The *Maltese Dog*, very small; muzzle like that of the little Water Spaniel; fur all over very long and silky; generally white

Buffon conjectures this variety to be the produce of the alliance of the little spaniel and the little water spaniel; the latter also he conceives to be the offspring of the little spaniel and the great water spaniel.

Sub-var. h. The *Lion Dog* differs from the Maltese dog, only in having the hair short on the body and half of the tail; but long on the other parts, particularly the end of the tail, where it forms a tuft.

C. F. Leoninus, *Gmelin, Sys. Nat.*

Le Chien Lion, *Buffon*, v.

Icon. *Buffon*. v. *t.* 40, *f.* 2. *Ency. Méthod, t.* 100, *f.* 5.

Buffon attributes the same origin to this as to the preceding, with the genealogical addition of an ancestor with scattered hairs.

137

The group of spaniels seem originally to have been located in Spain, whence the name.

Var. b. *C. F. Aquaticus* (the Barbet, or Poodle.) Heaa large, and round ; cerebral cavity larger than in any other variety ; frontal sinuses very much developed ; ears large and pendent ; body thick ; tail nearly horizontal ; fur long and curly all over the body ; generally white, with black patches, or black with white patches.

C. F. Aquaticus, *Lin. Sys. Nat.* i. 57. Canis Aquaticus Aviarius, *Gesner.*

Great Water Spaniel, *Shaw, Zool.* 280. Water Dog, *Shaw, Zool.* Grand Barbet, *Buffon, H. N.* v. Caniche, or Chien Canard, *French.*

Icon. *Buffon, H. N.* v. t. 36.

Sub-var. a. The *Little Barbet* is bred, according to Buffon, from the great barbet and the little spaniel.

C. F. Minor, *Gmelin, S. N.* i.

Petit Barbet, *Buffon, H. N. v.* Little barbet, or water dog, *Shaw,* i. 280.

Icon. *Buff., H. N. v. t.* 38. *f.* 2. *Ency. Méthod. t.* 100. *f.* 1.

Sub-var. b. The *Griffon* is like the preceding, but the hair is not curled ; generally black, with yellow spots over the eyes and on the paws. It appears to have sprung from the barbet and the shepherd's dog.

Le Chien Griffon, *Desm. Mamm.* 193.

Icon.

Var. c. *C. F. Gallicus* (the Harrier ?) Muzzle as long, and thicker than that of the mâtin ; head thick and round ; ears large, long, and pendent ; limbs strong ; tail erect ; hair short, or varied with black spots, brown, or yellow, &c.

C. F. Gallicus, *Gmelin. S. N.* i.

Le Chien Courant, *Buffon, H. N.* v.

Icon. *Buffon, V. t.* 32.

Inhabits France.

Peculiar for its fine scent ; used in chasing.

138

Var. d. *C. F. Avicularius* (the Pointer.) Differs from the preceding only in having the muzzle a little shorter, and not so thick at the end; head thicker; ears shorter, not so large, partly erect, and partly pendent; legs rather longer; and body lighter; and the tail thicker and shorter.

C. F. Avicularis, *Lin. Sys. Nat.* i. 57.

Canis Pantherinus, *Aldr. Digit.* 555.

Le Braque, *Buffon, H. N.* v.

Icon. *Buffon, H. N.* v. *t.* 33. *Ency. Méthod. t.* 102. *f.* 2.

Sub-var. a. *C. F. Bengalensis* is like the former, but has the colours brighter : it has small black and yellow spots on a white ground.

Le Braque du Bengal. *Buff. H. N.* v.

Icon. *Buffon, H. N.* v. *t.* 34.

Var. e. *C. F. Vertagus* (the Turnspit.) Head similar to that of the two preceding; ears long and pendent; nose sometimes cleft; tail long; legs short, straight, and thick, generally white, with black and brown patches, or black, with yellow patches.

C. F. Vertagus, *Lin. Sys. Nat.* i. 57.

Le Basset à jambes droites, *Buffon, H. N.* v.

Icon. *Buffon, H. N.* v. *t.* 35. *Ency. Méthod. t.* 103. *f.* 3.

Sub-var. a. The *Crooked-legged Turnspit*, has the fore-legs bent outward.

The spaniels and turnspits produce a variety, with elongated body, short legs, and long and silky hair, the *Chien burgos* of Buffon.

Var. f. *C. F. Domesticus* (the Shepherd's Dog.) Head assimilated to that of the *Mâtin;* ears short and erect; tail directed horizontally behind, or curved upward, and sometimes pendent; fur long over all the body, with the exception of the muzzle and outer sides of the limbs; black is the most prevailing colour, oftentimes gray on the throat, chest, and belly, sometimes a yellow spot over the eyes.

139

C. F. Domesticus, *Lyn. Sys. Nat.*

Chien de Berger, *Buffon*, v. C. de Brie. French Shepherd's Dog. *Shaw.*

Icon. *Buffon*, v. *t.* 28, *Ency. Méthod. t.* 99, *f.* 1. *Shaw's* Zool. ɪ. *t.* 75.

Sub-var. a The drover's dog, with longer hair more curled, generally of a dingy colour.

Var. g. *C. F. Pomeranus*, (the Wolf Dog.) Ears erect and pointed ; head long ; muzzle also long and slender; tail high and curled before ; fur short on the head, feet, and ears, long and silky over the rest of the body, and particularly the tail ; white, gray, black, or yellow.

C. F. Pomeranus, *Gmelin, S. N.*

Wolf Dog, *Shaw.* Chien Loup. *Buffon, His. Nat.* v.

Icon. *Buffon*, v. *t.* 30. *Ency. Méthod. t.* 99. *f.* 2

Var. h. *C. F. Sibiricus*, (Siberian Dog.) Thick hair all over, even on the head and paws ; in other respects like the Wolf Dog.

C. F. Sibiricus, *Gmelin, Sys. Nat.* ɪ.

Icon. *Ency. Méthod. t.* 99. *f.* 3.

Var. i. *C. F. Borealis*, (the Esquimaux Dog.) Head similar to that of the wolf dog; tail turned in a circle; ears erect; silky hair, not abundant; woolly hair, on the contrary, excessively thick, very fine and undulated, capable of being plucked off in flocks; colour varied by great patches, irregularly distributed, of white, black, or gray; three black points on each cheek, whence proceed some long hairs.

C. F. Borealis, *Desmarest, Mam.* 194.

Chien des Esquimaux, *F. Cuv. Mam.*

Icon. *F. Cuv. Mam. Lithog. t.*

Inhabits Baffin's Bay (and perhaps Kamtchatka.)

Var. j. *C. F. Americanus*, (the Alco.) A small head ; back arched; tail short and pendent; fur long and yellow on the back, whitish on the tail, according to Fernandez.

140

This variety is placed by M. Desmarest in this subdivision, in consequence of the observation of the Baron Humboldt, that it is a variety of the Shepherd's Dog.

C. F. Americanus, *Gmelin, Sys. Nat.*

Michaucanens, *Fernand. Anim. N. Hisp.* 7. Teshichi, *Fernand.* i. *c.* 10. Peruvian Dog, or Alco, *Shaw's Zool.* 285.

Inhabits Peru and Mexico.

Obs. A specimen similar to Fernandez' description has lately been exhibited in London, as coming from Mexico.

SECTION III. Muzzle more or less truncated, cranium much elevated, frontal sinuses large, condyles of the lower jaw placed above the line of the upper cheek teeth.

Var. a. *C. F. Molossus* (the Bull-dog.) Muzzle thick, short, and flat; lips thick and pendent; head large, forehead flat, produced by the development of the frontal sinuses elevating the frontal bone above the nose, and reducing the capacity of the brain; ears pendent at their extremity; body thick, strong, and long; fur short; lips, extremity of the muzzle, and outside of the ears black, the rest of the body pale yellow; nostrils frequently cleft.

The *black variety of Thibet* has the skin excessively loose and plaited.

C. F. Molossus, *Lyn. Sys. Nat.*

Le Dogue, *Buffon,* v. The Bull-dog, *Shaw.*

Icon. *Buffon,* v. *t* 43. *Ency. Method. t.* 101. *f.* 3. *Bewick.*

Var. b. *C. F. Anglicus,* (the English Mastiff.) Head assimilated to that of the Bull Dog, but with the ears altogether pendent and never erected; upper lips falling over the lower jaw; end of the tail turned up, frequently having the fifth toe on the hind feet more or less developed. Bred between the Mâtin and the Bull-dog.

C. F. Anglicus, *Gmelin, Sys. Nat.*

Chien Dogue de fort Race, *Buffon,* v. The Mastiff, *Shaw.*

141

Icon. *Buffon*, v. *t.* 45. *Ency. Méthod. t.* 101, *f.* 4.

F. Cuv. Mem. Lithog. xviii.

Var. c. *C. F. Fricator*, (the Pug.) The Bull-dog in minia-ture, with the lip less extensive in thickness, and the tail more curled.

C. F. Fricator, *Lyn. Sys. Nat.*

Le Doguin, *Buffon*, v. *t.* The Pug-dog, *Shaw.*

Var. d. *C. F. Islandicus*, (the Iceland Dog.) Head round ; ears partly erect and partly pendent; fur soft and long, es-pecially behind the foreleg and on the tail. Described by Daubenton from a drawing.

C. F. Islandicus, *Gmelin, Sys. Nat.*

Le Chien d'Islande, *Buffon, His. Nat.* v.

Icon. *Buffon, His. Nat.* v., *t.* 31. *Ency. Méthod. t.* 99. *f.* 4.

Var. e. *C. F. Variegatus*, (the little Danish Dog.) Fore-head convex; muzzle thin, pointed; eyes very large; ears half drooping; legs thin; tail recurved; fur thin, spotted with white and black.

C. F. Variegatus, *Gmelin, Sys. Nat.*

Le Petit Danois, *Buffon,* v.

Icon. *Buffon,* v. *t.* 41, *f.* 1. *Ency. Méthod. t.* 100, *f.* 6.

Obs. When this variety is speckled with black on a white ground, it is called the Harlequin by the French.

Var. f. *C. F. Hybridus*, (the Shock Dog.) Head round; eyes large; ears small, partly erect and partly pendent; tail curved and bent forward ; muzzle like that of the pug ; fur generally patched black and white.

C. F. Hybridus, *Gmelin, S. N.*

Chien Roquet, *Buffon, H. N.*

Icon. *Ency. Méthod. F.* 101, *f.* i. *Buffon, H. N.* v. *F.* 41, *f.* 2.

Buffon attributes this to the little Danish Dog and the pug.

Var. g. *C. F. Britannicus.* (the Black and Tan Terrier ?) Forehead convex; eyes prominent; muzzle pointed, tail thin, arched horizontally; fur short; ears moderate, half erect, deep black, with a yellow spot over the eyes on the muzzle, on the throat and legs.

Britannicus, *Desmarest.*

Icon. ——

Obs. Desmarest thinks this variety is produced between the little Danish Dog and the Pyrame.

Sub-var. a.   The Scotch Terrier, white, with curly stiffish hair.

Var. h. *C. F. Fricator*, (the Artois Dog.)   Muzzle excessively short and flat.   Produced between the C. Hybridus and the Pug.

C. F. Fricator, *β. Gmelin, Sys. Nat.*

Chien d'Artois, *Buffon*, v. *t.* 253.

Chien Lillois, Islois, ou quatre vingts *of the French.*

Icon. ——

Inhabits Flanders.

Var. i. *C. F. Andalusia*, (the Alicant Dog.) With the short muzzle of the *Pug* and the long hair of the *Spaniel*, between which varieties this is produced.

C. F. Andalusiæ, *Desm. Mam.* 1. 196.

Chien d'Alicante, *Buffon*, v. 254. C. de Cayenne, French.

Icon. ——

Inhabits Alicant.

Var. j. *C. F. Ægyptius*, (the Egyptian Dog.)   Head very thick and round ; ears erect at the base, large and moveable, and carried horizontally ; skin nearly naked, black or dark flesh colour, with large patches of brown.

C. F. Ægyptius, *Gmelin, Sys. Nat.*

Chien Turc, *Buffon, Hist. Nat.* Barbary Dog.

Icon. *Buffon, Hist. Nat.* v. *t.* 42, *f.* 1. *Ency. Méthod. t.* 103, *f.* 1.

Inhabits.

Sub-var. a.   With a sort of mane behind the head formed of longish stiff hairs.

C. F. Ægyptius, var. a. *Desm. Mam.*

143

Chien Turc à crinière, *Buffon, Hist. Nat.* v.
Icon. *Buffon, Hist. Nat.* v. *t.* 42, *f.* 2,

371. 2. *C. Lupus,* (the Wolf). Head thick and oblong, terminated by a slender muzzle, tail with long hair and pendent; yellowish gray, with a black stripe across the forelegs of the adult; eyes oblique.

Canis Lupus, *Lin. Sys. Nat.* 58. Lupus, *Gesner, Quad.* 634. Wolf. *Penn. Quad.* 4. 248. Loup, *Buffon, Hist. Nat.* 7. Icon. *Schreb. tab.* 81 & 88. Buff. *pl.* 1. *Ency. Méthod. t.* 105, *f.* 3. *t.* 104, *f.* 3, 4. *t.* 105, *f.* 1 & 2. *Shaw's Zool.* ι. *t.* 75.

Inhabits the continent of Europe, and probably North America.

Var. a. *Albida.* Fur white.

Obs. Many varieties of this species are mentioned by the American zoologists, and as it seems, like most of the Arctic Mammalia, to vary much in colour, being whiter in high latitudes or in the winter season, not much dependance can be placed on its colour as a specific character.

372. 3. *C. Lycaon,* (Black Wolf). Tail straight; body black, without any white spots.

Canis Lycaon, *Gmelin, Sys. Nat.* ι. 73.

Loup Noir, *Buffon, Hist. Nat.* ιx. 362. Black Wolf, *Shaw's Zool.* ι. 297.

Icon. *Buffon, Hist. Nat.* ιx. *t.* 41. *Griffith's Anim. King.*

Inhabits mountainous parts of Europe

β *Americana,* black, with a white spot on the chest

Black Wolf, *Bartram.*

Inhabits Florida.

The Java Wolf is treated by Desmarest as distinct.

373. 4. *C. Jubatus,* (the Red Wolf.) Uniform brightish red colour, with a short black mane along the spine

Canis Jubatus, *Desmarest, Mam.* ι. 198.

Agouarà gouazou, *d'Azara, Quad. du Paraguay.* Loup
Rouge, *Cuvier, Reg. Anim.*

Icon. *Cuvier, Regnè Anim. tom.* 4. *pl.* 1. *Dict. des.Sciences,
Nat.* v. *pl.* 17. *Ency. Méthod. Sup. t.* 6, *f.* 1.

Obs. Cuvier identifies the *C. Jubatus* with the *C. Mexicanus* of *Gmelin;* but Desmarest makes a distinct species of
it, which he attributes also to the Xolsitzaintli of Hernandez, and the Cuetlachtle, or Lupus Indicus of Fernandez.

374. 5. *C. Antarcticus*, (Antarctic Wolf.) Fur reddish;
tail at base red, middle black, end white.

Canis Antarcticus, *Shaw's Gen. Zool.* i. 331. C. Culpeus
Molina.

Antartic Fox, *Pennant, Quad.* 840. Culpeu, *Molina,
Hist. Nat. Chili, p.* 259. Chili Fox, *Shaw,* i. 329.

Icon. *Pennant, Quad. t.* 29.

Inhabits Falkland Islands.

Obs. Desmarest and F. Cuvier, both are inclined to unite
the *C. Culpæus*, of Molina, and the *C. Antarcticus* of Shaw.

375. 6. *C. Cancrivorus*, (the Crab Wolf.) Fur ash-
coloured, varied with black above, yellowish white underneath; ears pointed, brown; sides of the neck and behind
the ears yellow; the tarse and extremity of the tail blackish;
muzzle pointed

C. Cancrivorus, *Desm. Mam.* 199. C. Thous. *Gmelin,
Sys. Nat.*

Chien des Bois de Cayenne, *Buff. Sup.* vii. Koupara,
Barrere. Surinam Dog, *Pennant, Quad.*

Icon. *Buff. Sup.* vii. *pl.* 38.

Inhabits Guiana.

376. 7. *C. Aureus,* (the Chacal, or Jackal.) Yellowish gray
above, whiter underneath; tail bushy, black at the extremity.

Canis Aureus, *Lin. Sys. Nat.* I. 59.   C. Barbarus, *Shaw Zool*, t. 311.

Lupus Aureus, *Kæmpf. Ann. Exot.* 413.   Schakal, *Penn. Quad.* 262, and Barbary Schakal, *Pen. Quad.* 260.   Jackal and Barbary Jackal, *Shaw*, I. 304 & 311.   Le Chackal, *Buff. Sup.* 6, 112.   Deab, or Dib, of *Barbary*.   Benawi, *Arabs*.   Nari, *Malabar*.   Jaqueparel, *Bengal*.

Icon. F. Cuvier, *Mam. Lithog.* II. *Ency. Méthod. t.* 107, *f.* 3.   *Kæmpf. Ann. Exot. t.* 407, *f.* 2.   *Buffon, Supp.* VI. *t.* 16.

Inhabits the warm parts of Africa and Europe, and Southern Asia.

377. 8. *C. Mesomelas*, (the Cape Jackal). Fulvous brown, with a large triangular patch, broad at the shoulders, and terminating in a point near the tail; dotted all over with white specks; flanks red, chest and belly white; tail descending to the ground.

Canis Mesomelas, *Gmelin, Sys. Nat.* I. 73

Cape Schakal, *Pen. Quad. p.* 265.   Cape Jackal, *Shaw, Zool.* I. 310.

Icon. *Schreb. tab.* 95.   *Griffith, Vest. Anim. t.   Shaw, Zool. t.* 79.   *Ency. Method. t.* 107, *f.* 4.

Inhabits the Cape of Good Hope.

Obs. The Baron we have seen, places this among the Foxes; but the individual whence the drawing here engraved from was taken, was distinguished, as we are assured, by round pupils, and we have therefore placed it, with M. Desmarest, among the Dogs and Wolves, whether as a separate species, or a mere variety of C. Aureus.

Baron Cuvier considers the Adive of Buffon (which is copied as the Jackal by Shaw) to be a factitious species, not differing from the *Chacal;* but M. F. Cuvier treats it as having a very great alliance to the *Corsac.*

146

378. 9. *C. Anthus* (Senegal Jackal.) Fur gray, sprinkled with yellowish spots ; yellowish above, whiter underneath ; tail yellow, with a longitudinal black line at its base, and some black hairs at the point.

Canis Anthus, *F. Cuvier, Mam. Lithog.*
Chacal du Senegal, *F. Cuvier, Mam. Lithog.*
Icon. *F. Cuvier, Mam. Lithog.* xvii.
Inhabits Senegal.

Foxes.   Pupils of the eyes long ; tail long and bushy.

379. 1. *Vulpes Vulgaris* (the Fox.) Yellow above, white underneath ; behind the ears black ; tail with long hairs, which are white at its extremity ; muzzle pointed ; ears erect, acuminated ; eyes diagonal.

Canis Vulpes, *Syst. Nat.* i. 59.
Vulpes, *Gesner. Quad.* 966. The Fox, *Ray.* Renard, *Buff.* vii. 75.
Icon. *Buffon.* vii. *t.* 6. *Ency. Méthod. t.* 106. *f.* 1. 2.

Var. a. *V. Alopex* has been treated as distinct, but is considered, as we have seen, by the Baron, to be a variety of the Common Fox.   It differs in having the fur thicker, and of a deeper red ; the additional blackness at the end of the tail, and the blackness of the paws.   Found in Alsace and Burgundy.

C. Alopex, *Gmelin, Sys. Nat.* i. 74.
Brant Fox, *Shaw. Zool.* i. 321.
C. Vulpes, *Desm. Mam.* 202.

Var. b. *Crucigera,* the colour deeper, with a black cruciform mark across the shoulders and down the spine.

Canis Crucigera, *Gesner, Quad.* 966.
Cross Fox *of European Naturalists.*
Inhabits the northern parts of the Old and New World.
According to Linnæus, the end of the tail is white ; and

147

by Desmarest, black ; the tip of the tail of the English species is usually white.

380. 2. *C. Cinereo-Argentius* (Tri-coloured Fox.) Upper part of the body gray-black ; head yellow-gray ; ears and sides of the neck bright-red ; throat and cheeks white ; under jaw black ; belly yellow ; tail of the same colour, mixed with black, which prevails exclusively at the end.

Canis Cinereo-Argenteus, *Gmelin, Sys. Nat.* i. 74.

Renard Gris, *Brisson, Quad.* Der Grisfuch, *Schreb.* Agouarachy, *d'Azara, Quad. du Paraguay.* Fulvous-necked Fox, *Shaw's Zool.* 324.

Icon. *Schreb. tab.* 92. A.

Inhabits the warm and temperate parts of America.

Schreber appears to doubt if it may not be a variety of the *Grey Fox* of Catesby.

381. 3. *C. Argentatus* (Silver Fox.) Black, with some of the hairs tipt with white; extremity of the tail white; forepart of the head and the flanks whitish ; sometimes a white spot under the throat; paws covered with short hairs.

C. Lycaon. var. *Gmelin. Sys. Nat.* i. Canis Argentatus *Desm. Mam.* i. 203.

Renard Noir ou Argenté, *Geoffroy, Collect. du Museé.* Renard Argenté, *F Cuv. Mam. Lithog.*

Icon. F. Cuvier, *Mam. Lithog.* v.

Inhabits America.

Obs. Gmelin confounded this with the Canis Lycaon or Black Wolf. A similar species is found in Asia, but M. F. Cuvier doubts the identity of the Asiatic and American Black Fox. The American Cross Fox, C. Decussatus of Geoff. Collect. du Mus., appears likely to be a mere variety of the Argentatus.

382. 4. *C. Lagopus* (Arctic Fox.) Fur very long, thick

and soft, uniformly brown in summer, white in winter; paws and soles of the feet protected by long hairs.

Canis Lagopus, *Lin. Sys. Nat.* I. 59. Vulpes Cœrulescens, *Lin. Faun. Suec.* 14. *t.* 13. Isates, *Act. Petrop.* 1760. *v.* 358. Renard Blue, *Buffon*, XIII. 272.

Icon. *Ency. Méthod. t.* 106. *f.* 3. *t.* 107. *f.* 2. *Bewick, Quad.*

Inhabits the Arctic regions.

383. *5. C. Corsac* (the Corsac.) Uniform yellow; gray above, lighter underneath; tail very long, touching the ground, and black at the extremity.

Canis Corsac, *Gmelin, Sys. Nat.* I. 74.

Corsac, *Guldenstaedt, Voyage.* Isatis, *Buffon, Sup.* III. 113, 114. L'Adive, *Buff.* II. ? Korsaki, *Pallas, Neu. Nord. Beytr.* I. 29. Corsac Fox, *Shaw, Zool.*

Icon. *Buff. Supp.* VI. 17. *Buff.* II. *t. cop. Ency. Méthod. t.* 107. *f.* 3. *Shaw, Zool.* I. *t.* 79.

Inhabits the Deserts of Tartary.

Obs. This is placed, by the Baron, among the Foxes; by Desmarest, it is transferred to the division with circular eye-pupils.

Var. a. *Karagan.* Gray; ears black.

Canis Karagan, *Gmelin, Syst. Nat.* I. 74

Karagan Fox, *Pennant.*

Inhabits Great Tartary.

Most probably a variety of the former.

384. *6. C. Decussatus* (Cross Fox.) Fur varied, with black-and-whitish above, with a black cross on the shoulders; muzzle, and lower parts of the body and legs black; tip of the tail white.

Canis Decussatus, *Geoff. Coll. Mus.—Sabine, Frank. Voy.*

American Cross Fox, *English Furriers.*

149   2 M

Inhabits North America. (Mus. Brit.)

M. F. Cuvier is inclined to consider it a variety of the C. Argentatus.

385. 7. *C. Virginianus* (Gray Fox.) Body entirely silvery-gray, with a cast of red about the ears.

C. Virginianus, *Gmelin, Syst. Nat.* i. 74.

Gray Fox, *Catesby, Carolina,* ii. 78. Virginian Fox, *Shaw. Zool.* i. 325.

Icon. *Catesby, Carolina.* ii. *t.* 78.

Inhabits warmer parts of North America.

386. 8. *C. Fulvus* (Fulvous Fox.) Fur reddish or fulvous; beneath the neck and belly white; chest gray; front part of the fore legs and feet black, with fulvous toes; top of the tail white.

Canis Fulvus, *Desm. Mam.* i. 303.

Renard de Virginie, *Palisot, Beauv. Bul. Soc. Phil.* Red Fox, *Sabine, Franklin, Voy.* 656.

Icon. *F. Cuv. Mam. Lithog.*

Inhabits North America.

Var. b. *Velong.*

Canis Velong. *Say, James, Exped. Rocky Mountains.*

The specimen of this animal, in the British Museum, presented by the Hudson's Bay Company, does not well agree with M. F. Cuvier's figure, but better with Desmarest and Say, and Mr. Beauvois' description.

387. 9. *C. Niloticus* (Egyptian Fox.) Body above reddish, beneath gray; behind the ears black; legs fulvous.

Canis Niloticus aut Ægyptiacus, *Geoff. Coll. Mus. Par Desm. Mamm.* 204.

Icon. ——

Inhabits Egypt.

The Red Fox of Bartram, the Ceylonese Dog of Pennant. and Shaw, and Vosmaer, and the Bengal, and the Sooty Fox of the same writers, are too doubtful for insertion.

The Dog, the Chacal, and the Fox, are intimately connected, and we could insert several others from drawings with pretensions to a distinct notice, as probable species or permanent varieties.

III. LYCAON, Brooks. *Head short; incisive teeth not forming a regular series, the central one in each jaw being placed more internally than the rest; the body, Hyæna-like, higher before than behind; joints of carpus very weak.*

388. 1. *C. Tricolor* (Burchel's Lycaon.) Ochreous-yellow, blotched and brindled with black, intermingled with white spots.

Hyena! Picta, *Temminck, Mem. de Bruxell.* Canis Pictus, *Desm. Mamm.* I. 538. Lycaon Tricolor, *Brook's Mus Anat.* Painted Hyena, *Griffith, Vert. Anim.* Loup. Peint. *Desm. l. c.* Icon. *Temminck, l. c. t. Griffith, Vert. Anim. Burchel, Trav.* Inhabits Cape of Good Hope.

The skeleton of this animal agrees exactly with that of the Dog and the Wolf, except in the want of the toe and in the placing of the cutting teeth; the head is also shorter, and, consequently, the teeth closer together; but the formation of the male organ of generation is said to be different from that of the Caninæ, and consequently the mode of copulation different; and it is understood that Mr. Brooks, who possesses a skeleton in his splendid museum, treats it therefore as a genus, under the name Lycaon.

IV. MEGALOTIS, Illiger. *Toes five; ears very large; teeth ——? tail tufted, head long, acute.*

389. 1. *Megalotis Lalandii* (Laland's Fennec.) Gray, the hairs of the dorsal line longer and blacker; tail very tufted, black, gray at its base; feet black.

Canis Megalotis, *Cuv. Ossemens Fossiles, Desm. Mamm.*
I. 538. Megalotes Lalandii, *Hamilton Smith, MSS.*
Icon. *Nobis.*
Inhabit Cape of Good Hope.

390. 2. *Megalotis Brucii* (Bruce's Fennec.) Dirty-white;
oelly whiter; ears thin, margined with white hairs.
C. Zerda, *Bodd.* Canis Cerdo, *Gmelin, Sys. Nat.* I. 75.
C. Megalotis. Fenneccus Brucii, *Desm. Mam.* 235. Mega-
lotis, *Illiger.*
Animal Anonyme, *Buffon, Sup. Hist. Nat.* VII. 128.
(1776.) Fennec. *Bruce, Voy.* I. 154. Zerda? *Sparman,
Voy.* IV. *Pennant, Quad.* I. 167.
Icon. *Buffon, Supp.* VII. *t.* 19. *Bruce, Voy. t.* 28. *Spar-
man, Voy.* II. *t.* 4? *Pennant, Quad. t.* 8. *Shaw, Zool. t.* 80.
Inhabits Abyssinia.

VIVERRA. Incisors $\frac{6}{6}$; canines $\frac{1-1}{1-1}$; cheek-teeth $\frac{6}{6}$. In the
upper jaw, three false molars, a little conical and com-
pressed; a large sharp-cutting carnivorous tooth nearly tri-
cuspidous, and two tuberculous teeth; in the lower jaw there
are four false molars, a large carnivorous bicuspidous tooth,
and one large tuberculous tooth behind; head long; muzzle pointed; nostrils pierced on the sides of the nose;
pupils of the eyes capable of contracting themselves almost
into a line; tongue aculeated; feet pentadactylous; claws
semi-retractile; anal pouch more or less deep.

I. VIVERRA. Cuv. *Anal pouch very deep, and divided
into two sacks, containing an unctuous, musk-scented secretion; cheek-teeth* $\frac{6}{6}$.

391. 1. *V. Civetta* (the Civet.) Gray, with brown or
black stripes and spots; tail, with four or five annuli,
shorter than the body; a mane along the dorsal line.

Viverra Civetta, *Gmelin, Sys. Nat.* I. 80.

Civetta, *Clusius.*

La Civette, *Buffon, Hist. Nat.* IX. The Civet, *Pennant, Perrault. Hist. des Anim. tab.* 23.

Icon. *Schreb. tab.* 111. *Buff.* IX. *pl.* 34. *Shaw, Zool.* I. *t.* 95. *F. Cuvier, Mam. Lithog.*

Inhabits Africa, especially Abyssinia.

392. 2. *V. Zibetha* (the Zibett.) Gray; legs transversely, spotted with brown; throat white, with two black bands on each side; no mane; tail long, with eight or ten annuli, black and white.

Viverra Zibetta, *Gmelin, Sys. Nat.* 1.

Zibet, *Buff.* IX. *t.* 31. Le Musc, *Lapeyronie, Mem. de l'Academie des Sciences,* 1731. Zibett? *Shaw, Zool.* I. 389.

Icon. *Schreb. tab.* 112. *Buff. Hist. Nat.* IX. *pl.* 31. *Ency. Method. t.* 88. *f.* 2. *Shaw. Zool.* I. *t.* 95. *F. Cuvier, Mam. Lithog.*

Inhabits both India and Africa, according to different writers.

393. 3. *Viverra Rasse* (the Rasse.) Yellowish-gray; ears close back, with eight parallel longitudinal blackish lines; neck obscurely banded; feet brown; hair of the body ridged, tail rather attenuated.

Viverra Rasse, *Horsf. Java,* VI.

Rasse, *Javanesse.*

Icon. *Horsf. Java,* VI. *t.* 2.

Inhabits Java.

Yields the *Dedes* of the Javanese, and the Zibet of the Malays.

II. Genetta, Desm. *Anal pouch reduced to a mere fold of the skin, containing very little excretion: tail straight; cheek-teeth* $\frac{6}{6}\frac{6}{6}$.

153

394. 1. *Genetta Vulgaris* (the Genet). Gray, with small round and elongated black spots; tail annulated with black

Viverra, Genetta, *Lin. Sys. Nat.* V. Tigrina, *Gmelin, Sys. Nat.* V. Malacensis, *Gmelin, Sys. Nat.*

Genette *Belon*, La Genette, *Buffon, Hist. Nat.* ix. La Civette de Malacca, *Sonnerat, Voy. des Indes,* ii. Le Genette du Cap, *Buff.* viii. Chat bizaam, *Vosmaer.* Genette de France, *Buff. Sup.* iii. *not the fig.*

Icon. *Shaw, Zool.* i. *t.* 96. *Schreb. Tab.* 113. *Buff.* viii. *t.* 58 and 59, *pl.* xxxvi. *Sonnerat's Voyage, pl.* 91. *F. Cuv. Mam. Lithog.* Cuv. *Menag. du Mus. Ency. Méthod. t.* 88, *f.* 1 and 3, *t.* 89, *f.* 1 and 3.

Obs. The above names are applied, by the authors cited, to species as distinct, all of which the Baron refers to the common Genet.

? Var. β. Pilosello, *Pennant, Quad.*

Genet Var. ? *Shaw. Zocl.* i. 401.

395. 2. *G. Fossa* (the Fossane.) Fur reddish-gray, marked with yellowish-brown spots, scattered on the flanks, four longitudinal lines on the back, rings on the tail reddish-brown, very obscure.

Viverra Fossa, *Gmelin, Sys. Nat.* 91.

La Fossane, *Buff. t.* xiii. Fossane, *Shaw, Zool., p.* 402. The Fossan, Weasel, *Pennant, Quad.* 75.

Icon. *Buff. Hist. Nat.* iii. *pl.* 20. *Schreb. tab.* 114. *Ency. Méthod. t.* 89, *f.* 2. *Shaw, Zool., i. t.* 96.

Inhabits Madagascar, and, as it is said, both Asia and Africa.

396. 3. *G. Geoffroii* (Geoffroy's Genette.) Fur clear, yellow, marked with brown spots, placed in longitudinal series; end of the nose white; and with a white cross-band over the eyes.

Viverra Fasciata, *Geoff. MSS. Desm. Mam.* 209, *not Lin.*

Icon. ——
Inhab. ——

397. 4. *G. Indica* (Indian Genet.) Fur yellowish white, with eight longitudinal narrow brown bands.

Viverra Indica, *Geoff. Mus. Hist. Nat. Desm. N. Dict. d'Hist. Nat.* VII. 170.

Icon. ——

Inhabits India, *Sonnerat.*

Var.? Junior? Smaller lines less apparent.

Petite Civette de Java, *Geoff. Mus. Par.*

398. 5. *G. Fasciata,* (Banded Genet.) Fur yellow-brown, marked with six brown broad bands.

Viverra Fasciata, *Gmelin, Sys. Nat.* 92. V. Striata, *Desm. Mam.* 210.

Le Chat Sauvage à bandes noires des Indes, *Sonnerat, Voy.* II. 193. Le Putois rayé des Indes, *Buffon, Hist. Nat. Sup.* VII. Fasciated Weasel, *Shaw, Zool.* I. 405.

Icon. *Sonnerat Voy.* II. *t.* 90. Buffon, *Sup.* VII. *t.* 57. *Shaw, Zool.* I. *t.* 97.

Inhabits Coromandel.

399. 6. *G. Bondar* (the Bondar.) Fur yellow, hairs tipped with black; dorsal band black, and two narrow parallel bands on each flank; feet and end of the tail black.

Viverra Bondar, *Blainville, MSS. Desm. Mam.* 210.

Icon. ——

Inhabits Bengal.

Obs. Described by Blainville from a drawing in the India House.

PROTELES (Isod. Geoff.) *Aspect of the hyæna. Toes five before and four behind; tail simple.*

155

400. 1. *Proteles Lalandii* (Lalande's Proteles). Fur gray; small mane; six or seven narrow transverse bands on the flanks, other small bands on the thighs and legs, and end of the tail black.

Viverra? Hyenoides. *G. Cuvier, MSS. Desm. Mam.* 538.

Proteles Lalandii, *Isod. Geoffroy, Mem. Mus.* ii. 5.

Icon. ——

Inhabits Cape of Good Hope.

Obs. This interesting animal appears to unite the Hyæna with the Viverra.

VI. MANGUSTA. *Anul pouch very large, with the vent situated at the bottom of it. Cheek-teeth $\frac{6}{6}\frac{6}{6}$; feet petadactylous.*

401. 1. *M. Ichneumon* (the Ichneumon.) Body dotted equally all over; dirty yellow and slate colour, each hair being annulated alternately with these tints; paws and muzzle black; tail long, and terminated by a diverging tuft.

Viverra Ichneumon, *Lin. Sys. Nat.* p. 84. Herpestes Pharonis, *Desm. Mamm.* 213.

Ichneumon, *Gesner, Quad.* 566. Nems and Pharaoh's rat. *Modern Egyptians*, Mangouste, *Buffon*, xiii. 150.

Icon. *Schreb. tab.* 116. *Buff. Sup. t.* 3. *pl.* 26.

F. Cuvier, Mam. Lithog.—Shaw, Zool. i. *t.* 92.

Inhabits Egypt.

402. 2. *M. Mungos* (the Indian Ichneumon,) similar to the last, but with the tail pointed, and with slightly indicated transverse bands on the body.

V. Ichneumon, B. *Lin. Sys. Nat.* 63. Mustela Glauca. *Lin. Sys. Nat. ed.* v. Viverra Mongoz, *Gmelin, Sys. Nat.* 184. Herpestes Mungo, *Desm. Mam.* 211.

Mungo, or Mungutia *of the Indians.*

Icon. *Buff. t.* 13. *pl.* 19. *A. Kæmpfer Amœnit. Exotic. tab.* 567.

Inhabits India.

403. 3. *M. Cafra* (the Caffrarian Ichneumon.) Brownish-gray, and uniformly speckled, paws of the same colour, tail pointed.

Viverra Cafra, *Gmelin, Sys. Nat.* 85.

Le Nems, *Buff. Sup. t.* 3.

Herpestes Griseus, *Desmarest, Ency. Méthod.*

Icon. *Buff. l. c. pl.* 27.

Inhabits Southern Africa and India according to M. Geoffroy.

404. 4. *M. Galera* (the Galera.) Deepish brown, speckled with yellow; tail of equal size, its whole length.

Mustela Galera, *Gmelin, Sys. Nat.* i. 95. Herpestes Galera. *Desm. Mam.* 212.

Vansire, *Buff. t.* 13. Mangouste Vansire, *Geoff. Mem. de l'Institut de l'Egypte.* Vohangshira, *of Madagascar.* Galera, *Shaw,* i. 428. Madagascar Weasel, *Pennant.*

Icon. *Buff.* xiii. pl. 21. *Quad.* ii. 51. Guinea Weasel, *Pennant,* ii. 53.

Inhabits Madagascar and the Isle of France.

405. 5. *M. Edwardsii* (Edwards' Ichneumon.) Back and tail olive, annulated with brown; muzzle reddish-brown; tail pointed.

Herpestes Edwardsii, *Geoff. Mem. De l'Egypte, His. Nat.* ii. 138.

Indian Ichneumon ? *Edwards' Birds.*

Icon. *Edwards' Birds, t.* 199.

Inhabit East Indies.

406. 6. *M. Javanicus* (Javanese Ichneumon.) Chestnut-brown, spotted with yellowish-white; head and legs chestnut; tail equal the whole length.

Ichneumon Javanicus, *Geoff. Desc. Egypt,* ii. 138.

Icon. *Horsfield, Zool. Researches. Horsfield, l. c.*

Inhab. Java.

407. 7. *M. Ruber*, (Red Ichneumon.) Fur very bright ferruginous red, especially on the head.

Ichneumon Ruber, *Geoff. Mem. Inst. Egypte, His. Nat.* II. 139.

Icon. ——

Inhab. ——

408. 8. *M. Major* (Large Ichneumon.) Chestnut brown, hair chestnut, finely ringed with yellow; tail brown, pointed.

Ichneumon Major, *Geoff. Mem. Inst. d'Egypte. His. Nat.* II. 139.

Icon. ——

Inhabits ——

VII. SURICATA, Desm. *Anal pouch very large, with the vent placed at its base; feet tetradactylous; cheek-teeth,* $\frac{6}{6}$.

409. 1. *Suricata Capensis* (the Surikate.) Hairs annulated with brown, white, yellow, and black, resulting into a dull brown; nose round; the eyes and ears black; underparts and tail yellowish; nails long, strong and black.

Viverra Tetradactyla, *Lin.* Viverra Zenick, *Gmelin, S. N.* Suricata capensis, *Desm. Mam.* 214. Ryzœna, *Illiger Prod.*

Zenick, *Sonnerat, Voyage to India.* Suricate, *Buff. t.* 13.

Icon. *Buff. l. c. pl.* 7. *Schreb. tab.* 117. *Sonnerat, l. c. pl.* 92. *Shaw, Zool.*—*Ency. Method. t.* 85. *f.* 4, *f.* 1.

Inhabits South Africa.

III. PARADOXURUS, F Cuvier. *Anal pouch none, plantigrade, claws half retractile, tail convolute, cheek-teeth* $\frac{6}{6}$.

410. 1. *P. Typus* (Common Paradoxurus.) Body blackish,

with some obscure vague longitudinal bands on the flanks; a white spot below the eye ; tail black.

Viverra Nigra, *Desm. Mam.* 208. Paradoxurus Typus, *F. Cuv. Mem. Soc. Phil. May,* 1822. Viverra Musanga, *Raffles Trans. Lin. Soc.* XIII., 253.

Genette de France, or Pougouna, *Buffon, Hist. Nat. Sup.* VII.? Indian Pine Marten, *Pennant?* Musang brilan, *Malay.* Lawack, *Javanese.*

Icon. *Buffon, Hist. Nat. Sup.* VII. *t.* 58. *Marsden, Sumatra,* F. ? 212. *Horsfield, Java.*

Inhabits Pondicherry.

411. 2. *P. Prehensilis* (Prehensile Paradoxurus.) Yellow-green, with dorsal line to the end of the tail, legs, two lines of elongated spots near the back, and many small orbicular spots on each flank, black.

Viverra Prehensilis, *Blain. MSS.* Paradoxurus prehensilis, *Desm. Mam. Sup.* 540.

Icon.

Inhab. Bengal.

412. 3. *P. Aureus* (Golden Paradoxurus.) Fur beautiful uniform golden yellow, hair very long.

Paradoxurus Aureus, *F. Cuv. Mem. Mus.* v.

Icon.

Inhab. ——? *Mus. N. Hist. Paris.*

IV. ICTERUS (*Valenciennes.*) *Anal pouch none; plantigrade, claws half retractile, tail convolute, cheek-teeth* $\frac{5.5}{5.5}$.

413. 1. *Icterus Albifrons*, (White-fronted Icterus.) Fur formed of a mixture of long white and black bristles, except

the head and limbs, where it is short; forehead and muzzle nearly white; tail and legs blackish; a black spot extending from the ear to the side of the nose enclosing the eye.

Viverra? Bentourong, *Raffles, Lin. Trans.* XIII. Paradoxurus Albifrons, *F. Cuv. Mem. Mus.* 540, Icterus Albifrons: *Valenciennes. Ann.*

Bentourong, *Raffles*, 1. c. *Anim. Hist. Nat.* I.

Icon. *Lin. Trans.* XIII. *Tab. Valenciennes, Ann. Sci. Nat.* II. *t.—F. Cuv. Mam. Lithog.—Horsf. Zool. Java, t. F. Cuv. Dents Mam.* Teeth.

Inhabits interior of India.

Obs. Varies greatly in colour.

V. PRIONODON, Horsf.   *Anal pouch none? head elongated; muzzle very pointed; cheek-teeth $\frac{5}{5}$; body and limbs slender.*

414. 1. *Prionodon Gracilis* (Slender Delundung.) Fur clear yellow, with four very large transverse brown bands; tail with two very narrow brown bands at its base, and seven broader annuli towards the end, bands and spots on the outer side of the shoulders and thighs.

Viverra? Lesang, *Hardwicke, Trans. Lin. Soc.* XIII. 253. Felis (Priodonta) gracilis, *Horsf. Zool. Java*, I. Viverra gracilis, *Desm. Mamm.* 539. Prionodon Gracilis, *Horsf. Zool. Java*, VII.

Lesang. *Hardwick.* Delundung. *Javanese.*

Icon. *Hardw. Lin. Trans.* XIII. *t.* 24. *Horsf. Zool. Java*, I. *t.* 2.

Inhab. India *and* Java. *Mus. Brit.* from General Hardwick.

III. Subdivision of Digitigrades,

Without any tubercular tooth behind the great carnivorous tooth in the lower jaw.

Hyæna. Incisors $\frac{6}{6}$ canine $\frac{1.1}{1.1}$, very strong; cheek-teeth $\frac{5.5}{4.4}$, three conical false molars; one very large, strong, carnivorous tooth, with three cutting edges on the outer side and a small tubercle within, and a little tuberculous tooth behind in the upper jaw; in the lower three false molars, the carnivorous tooth bicuspidous, without an inner tubercle, and no tuberculous tooth behind; jaws powerful, shorter than those of the Dog's but longer than in the Felinæ; tongue aculeated; ears large, tetradactylous; nails not retractile; a glandulous pouch at the anus; teats four.

415. 1. *Hyæna Vulgaris*, (the Striped Hyæna.) Dirty gray, or slate colour, with transverse darker stripes on the flanks and legs; a mane of stiff erect hair down the dorsal line.

Canis Hyæna, *Lin. S. N.* Hyæna vulgaris, *Desm. Mam*

Foadh, *Shaw's Travels in Barbary.* Abyssinian Hyæna, *Bruce's Travels. Hyæna of the Ancients.*

Icon *Buff. Sup. pl.* 46. *F. Cuvier, Mam. Lithog. No.* 10. *Pennant, Quad. Kœmph. Ameen. t.* 407, *f.* 4. *Bellon, aquat. t.* 34, *Ency. Méthod. t.* 108, *f.* 1.

Inhabits Barbary, Egypt, Abyssinia, Nubia, Syria and Persia.

Obs. Bruce's *Canis, Hyænomelas* does not appear to differ specifically from the common Hyæna, but is probably a larger variety.

416. 2. *Hyæna Crocuta* (the Spotted Hyæna.) Dingy

whitish-gray, yellow with round brown spots on the flanks and thighs; mane like the preceding.

Canis Crocuta, *Gmelin*, *S. N.* Hyæna Capensis, *Desm:* *Mam.* 216.

Hyæna, *Barrow's Cape of Good Hope.* Spotted Hyæna, *Pen. Quad.* Loup Tigre *of Kolbe.*

Icon. *Schreb. pl.* 96, B. *Pennant, Quad., pl.* 17. F. *Cuvier, Mam. Lithog. Ency. Méthod. Supp. t. v. f.* 4.

Inhabits South Africa.

Var. a. Fur thicker, of a decided gray-red colour; under-part of the throat and body whitish, with blackish indistinct spots. *Cuv. Ossemens Fossile,* iv. 385.

Obs. There is a specimen in the Museum at Paris described by the Baron Cuvier, *Ossemens Fossiles,* iv. 384; with long hair on the back and flanks, hanging down on each side of a deep brown colour, with transverse bands on the fore legs and hind feet. The Baron doubts, at present, whether to consider it a variety of H. Vulgaris, or a distinct species.

Felis. Incisors $\frac{6}{6}$; canine teeth $\frac{1.1}{1.1}$; cheek-teeth $\frac{4.4}{3.3}$, or $\frac{3.3}{3.3}$; two false conical and thick in the upper jaw; a large carnivorous tooth, with three lobes and a little tubercle, which is wanting in some of the species; the fourth cheek-tooth, in upper jaw, nearly flat, and situated transversely; the two anterior cheek-teeth in the lower false; head round; jaws short; tongue aculeated; ears generally short and triangular, in many species with a white spot on the back of them; pupils of the eyes in some circular, in others vertically oval; anterior extremities with five toes, posterior with four; nails retractile.

* *Large, yellow, spotless Cats.*

417. 1. *F. Leo,* (the Lion.) Yellow, with a tuft at the extremity of the tail. Neck of the male furnished with a long thick mane.

Felis Leo. *Lin. and of the Ancients.*

Le Lion, *Buff. Hist. Nat. t.* 9. Lion *of the English.*

Icon. *Schreb. tab.* 97, *A and B. Buff. l. c. pl.* 1 *and* 2, &c.

Var. a. Senegal Lion. Fur lighter and brighter. F. Cuier. *Mam. Lithog.*

Icon. *F. Cuvier, l. c.*

Var. b. South African Lion, with a black mane, and round, thick, bull dog-like head.

Var. c. Asiatic Lion, rather less than the African Lions; mane yellow, and much more scanty.

Inhabits Africa generally, and the southern parts of Asia, though more rarely.

418. 2. *F. Concolor* (the Puma.) Uniformly yellow, without mane or tuft at the end of the tail.

Felis Concolor, *Lin.* Tigris fulva, *Brisson, Regn. An.* 272. Le Cougouar, *Buff. Hist. Nat. t.* 9. Gouazouara, *d'Azara, Anim. du Paraguay.* Cougoua, Puma, or American Lion *of the English.*

Icon. *Schreb. tab.* 104. *Buff. l. c. pl.* 19, *Shaw, Pennant.*

Inhabits the warm and temperate parts of America.

** *Large Cats, with transverse stripes.*

419. 3. *F. Tigris* (the Tiger.) Yellow above; white underneath; striped on the body with irregular, narrow, black bands; hairs about the cheeks very long.

Felis Tigris, *Lin. and of the Ancients.*

Le Tigre, *Buff. Hist. Nat. t.* 9.

Tiger *of the English.*

Icon. *Schreb. t.* 98. *Buff. l. c. pl.* 9. *F. Cuvier, Mam. Lithog.* &c.

Inhabits Southern and Central Asia.

Var. a. White, with the stripes indicated by a more opaque white.

Icon. nobis.

163

*** *Large Cats of the Old World, the body covered with large, irregular patches.*

420. 4. *F. Nebulosa* (the Clouded Tiger.) Head small; body long, heavy, and cylindrical; legs thick, short, and muscular; tail very thick, long, and annulated; body covered with large, irregular patches, forming enclosures, deeper than the ground-colour, but lighter than the edge.

Felis Nebulosa, *Griffith, Animal Kingdom.* Felis Microcelis, *Horsfield, Zool. Journal, No.* 4.

Tortoiseshell Tiger, *Griffith, l. c.* Remau-daham, *Sumatress.*

Icon. *Griffith, l. c. and Horsfield, l. c.*

Inhabits Sumatra, China?

Obs. It is not improbable that the individuals here specifically identified, may belong to separate varieties located in Sumatra and China.

**** *Large Cats, with roundish black spots, or clusters.*

421. 5. *Felis Jaguar* (the Jaguar.) Yellow above, and white about the belly; body marked, with open circles of black, containing a central black dot; the circles disposed in five or six parallel horizontal lines.

Felis Onça, Gm. Tigris Americana, *Bolivar;* Yagouarété *d'Azara,* Quad. of Paraguay; Onza, *Marograve;* Panthère femelle, *Buff. t.* 9.

The Panther of Exhibitors and Furriers.

Icon. *D'Azara, Voyage au Paraguay, f. F. Cuvier, Lithog. Geoffroy, Ann. Mus. t.* 4. *tab.* 94.

Inhabits America.

Var. a. Differing from the above in being larger and stouter, the Jaquarété Popé of d'Azara.

Var. b. Brow-nblack, with the spots blacker.

422. 6. *F. Pardus* (the Panther.)   Pale yellow above, with six or seven lines of rose-formed dots, which form clusters of five or six spots on each flank ; tail longer than that of the Jaguar, with the latter part black above, white underneath, having three or four white annuli on the black part *.

Felis Pardus, *Lin.*

Panthére, *Cuv. Menag. du Mus.*

Icon. *Cuv. Menag. du Mus. f. Buff. Hist. Nat. t. 9. tab.* 11.

Inhabits Northern Africa.

423. 7. *F. Leopardus* (the Leopard.)   Fur bright yellow on the upper part, white underneath, with at least ten ranges of small black clusters of spots on each flank ; lower part of the tail, for about one-third of its length, black above, white underneath, with five or six white annuli on the black part, rather smaller than the Panther.

Leopard, *Cuv. Ann. du Mus. t.* 14. 148.

Icon. *Buff.* ix. *pl.* 14.   *Schreb. pl.* 101.   *Shaw. Zool. Vol.* 1. *f.* 2. *pl.* 85.

Inhabits Central Africa, or *Sunda* only, according to the *Baron, Oss. Foss.* iv. *pl.* 426.

424. 8. *F. Pardus Antiquorum* (Panther of Antiquity.) Spots assimilated to those of the Common Panther, but the ground colour entirely buff yellow ; spotted to end of tail.

* As Buffon does not distinguish the Jaguar, and has insufficiently characterised the Leopard, it is difficult to distinguish the synonymy with certainty ; but after a minute comparison of the figures, and of the descriptions of Daubenton, I think that his *Panthére mále,* ix. *pl.* 11. is our Panther ; that his *Panthére femelle (ib. pl.* xii. *c. Schreb. pl.* xcix. and *Shaw, Gen. Zool.* i. *part* ii. *pl.* 84.) is a Jaguar ; and that his *Leopard (ib. pl.* xiv. *c. Schreb. pl.* ci. and *Shaw, pl.* 85.) is in fact our Leopard ; but the character of the tails are ill expressed in these figures. *Cuv. Oss. Fos. t.* iv. *pl.* 425

Felis Pardus Antiquorum, *Hamilton Smith, MSS.*
Icon. Nobis.
Habitat ?
Obs. The circumstances and peculiarities of these three species have been already noticed, v. 2. 466. at some length. The names Panther and Leopard are sometimes applied by different writers to the two first interchangeably.

425. 9. *F. Jubata* (the Maned Hunting Leopard.) Light yellow, covered with small black round full spots ; legs long ; a slight mane upon the neck ; nails semiretractile.
Felis Jubata, *Lin.* Felis Guttata, *Herman ?*
The Jaguar of *Buff. Sup.* 3. Hunting Leopard, *Pen. Quad.* 1. 284.
Icon. *Schreb. t.* 105. *Buff. Sup.* 3. *pl.* 38. under the name of Jaguar, or Leopard, *Pennant's Quad. t.* 56.
Inhabits Africa.

426. 10. *F. Venatica* (Maneless Hunting Leopard) ? Brighter yellow than the last ; head smaller, without any appearance of a mane ; nails semiretractile ; legs longer than the last, and the whole appearance more canine.
Chetah of India ?
Icon. *Hamilton Smith's MSS.*
Inhabits India.

427. 11. *F. Uncia* (the Once.) Tail long ; body whitish, covered with irregular black simple spots ?
Felis Uncia, *Gm. Syst. Nat.*
Once, *Buff.*
Icon. *Buff. Nobis.*
Habitat. Persia.

428. 12. *F Chalybeata* (the Liver-coloured Cat.) Wholly

grayish liver-colour, with numerous dark brown simple spots; two feet nine inches long, tail darker than the body ; annulated, one foot three inches long.

Felis Chalybeata, *Hamilton Smith, MSS. from Bullock's Museum.* Not Chalybeata of Herman.

Icon. Nobis.

Inhabits Chili.

Middle-sized Cats, with tail rather long, and generally with Stripes and Spots.

429. 13. *F. Serval* (the Serval.) Bright fulvous, more or less inclining to gray, and some hues to yellow round the lips ; the throat, under part of the body, and inside of the thighs whitish. Spots full, black.

Felis Serval and Capensis, *Lin.* Felis Serval, *Cuv. Oss. Foss.* iv. 433. *Mam. Lithog. No.* 1.

Chat pard. *Perrault's Mem. de l'Acad.* iii. *part* 3. *t.* 13. *and Ib. part* iii. *t.* 3. under the name Panthére.

Icon. *Mam. Lithog. No.* 1.

Habitat. Southern Africa.

430. 14. *F. Pardalis* (Ocelot of Linnæus.) Fulvous on the nose, forehead, shoulders, fore-arm, back, rump, and paws ; temples ochrey ; ground colour of the animal white ; on the shoulders and flank four or five long open fulvous spots ; rest of the back, rump, and hams, small open spots ; tail annulated ; tip black ; no streak on the forehead.

Felis Pardalis, *Linn.*

Ocelot, *Buff. Hist. Nat. t.* 13. Ocelot ♂. or *No.* 4, *Hamilton Smith, MSS.*

Icon. *Buff.* 13. *t.* 35 and 36. *Shaw, Gen. Zool. t.* 1. *f.* 88. Nobis.

Inhabits Mexico.

431. 15. *F. Chibi-gouazou* (the Chibigouazou). Rufous

on the nose, face, neck, and shoulders; general ground-colour, reddish, with open spots, and patches, bordered with black, with specks within them.

Ocelot α, or *No.* 1, *Hamilton Smith, MSS.* F. Pardalis, *Lin.* Felis mites, *Desm. Mam.* 221.

Chibi-gouazou, *d'Azara Quad. du Paraguay*, II. 152. Mexican Tiger, *Pen.* 1. 288?

Icon. Nobis.

Habitat. South America.

432. 16. *F. Ocelot* β. (Ocelot β. of *Hamilton Smith.*) Like the last, but the spots more numerous, and smaller; large spot on each cheek; no specks within the open patches.

Felis Ocelot β, or *No.* 2, *Hamilton Smith, MSS.*

Icon. Nobis.

Inhabits South America.

133. 17. *F. Ocelot* γ. (Ocelot γ. of Hamilton Smith.) Ground-colour, ashy mixed, with ochrey, parallel streak from the eye to the ear, with spots within it; tail annulated; tip white.

Felis Ocelot γ, or *No.* 3, *Hamilton Smith, MSS.*

Icon. Nobis.

Inhabits Mexico.

434. 18. *F. Catenata* (the Linked Ocelot.) Ground-colour, reddish yellow; temples ochrey; on the temples, cheeks, throat, belly, and inside of legs white; body marked with long chain-like markings; belly and throat black streaks; tail with imperfect annuli.

Felis Catenata, *Hamilton Smith, MSS.*

Icon. Nobis.

Inhabits America.

435. 19. *F. Macrourus* (Neuwied Cat.) Ground-colour,

ochrey gray, streaked with long patches; tail semiannulated, tip black ; two streaks from the eye to the jaw; spots on the forehead and cheeks ; stands higher than the last-mentioned.

Felis Macrourus, *Prince Maximilian of Neuwied, MSS. and Hamilton Smith, MSS.*

Icon. Nobis.

Inhabits Brazil.

436. 20. *F. Chati* (the Chati of F. Cuvier.) Ground-colour brownish-gray ; white on the cheeks and under the belly ; spots on head and ears as in *F. Chibi-gouazou;* three series of black spots on the neck ; spots on the shoulders, formed into an oblique band.

Felis Wiedii, *Sching's* translation of the *Regné Animal?*

Chati. *F. Cuvier, Mamm. Lithog. No.* 18.

Icon. *F. Cuvier, Mamm. Lithog. No.* 18.

Inhabits America.

Obs. Desmarest identifies this with F. Chibi-gouazou.

437. 21. *F. Colocolo* (the Colocolo of Hamilton Smith.) Head flat and broad ; body slender ; legs strong; ground-colour whitish-gray ; body covered with lengthened streaks of black and tawny ; legs from toes to knees dark gray or slate colour.

Felis Colocolo, *Hamilton Smith, MSS.* F. Colocolo, *Molina?*

Icon. Nobis.

Inhabits America.

438. 22. *F. Margay* (the Margay.) Upper part of body yellowish-gray ; under part white; four black lines pass from the vertex to the shoulders, and then change into series of long streaks ; tail irregularly annulated.

Felis Tigrina, *Lin.*

Margay, *Buff Quad.* t. 12.
Icon. *Buffon, t.* xii. 37, *Schreb. tab.* 106.
Inhabits Brazil.

439. 23. *F. Javanensis* (the Kuwuk of Java.) Ground-colour grayish brown; throat, neck, and belly whitish; ears small and distant from the eyes; body slender; four regular series of elongated full spots along the sides, smaller spots towards the belly.
F. Javanensis, *Horsfield, Zool. of Java, No.* 1.
Icon. *Horsfield, l. c.*
Inhabits Java.

440. 24. *F. Capensis* (Cape Cat of Forster.] Yellow, with black spots, of different sizes, and bands on the shoulders, back, and fore legs; forehead elevated abruptly from the muzzle.
Felis Capensis, *Forster, Phil. Trans. v.* 71. Cape Cat, *Pen. Quad.* 1. 291.
Icon. *Pen. Quad.* 1. *pl.* 1. *Muller, Cemelia Physica, t.* 39. Nobis.
Inhabits South Africa.

441. 25. *F. Yagouaroundi* (the Yagouaroundi.) Black brown, spotted with pale white.
Yagouaroundi, *d'Azara, Quad. du Paraguay.*
Icon. ——
Inhabits South America.

442. 26. *F. Pageros* (the Pageros or Pampa Cat.) Bright gray brown above, with reddish transverse bands on the throat and belly, and dark rings on the paws.
Chat Pampa *and* Pageros, *d'Azara, Quad. du Paraguay.*
Icon. *Hamilton Smith, MSS.*
Inhabits the continent south of Buenos Ayres.

443. 27. *F. Eyra* (the Eyra.) Uniformly brightish red; tail more tufted than that of the Domestic Cat; lower jaw and mustachios whitish; and a white spot on each side the nose.

Icon. *Hamilton Smith, MSS.* Eyra, *d'Azara, Paraguay.*
Inhabits Paraguay.

444. 28. *F. Nigra* (the Black Cat of America.) Uniformly black.

Chat Nègre *of Azara's; Paraguay.*
Icon. ——
Inhabits Paraguay.

445. 29. *F. Undata* (the Waved Cat). Pale gray, with numerous transverse brown or black bands; under part of the body reddish white; lower jaw pure white; inner part of foream, and humerus, and hind part of the tarsus black; convexity of the ears reddish.

Felis Undata. *Desm. Nov. Dict. de l'Hist. Nat.* vi. *Young Cuvier, Ossemens Fossiles,* iv. 435. *adult.*
Chat. Sauvage Indien *Vosmaer, Monog.*
Icon. *Vosmœr, Monog. t.* 13.
Inhabits South Africa. Java?
Obs. The Felis Obscura of M. F. Cuvier, Dict. des Sciences, Nat. viii. is probably a variety of the above.

446. 30. *F. Sumatrana* (the Rimau Bulu of the Malays), assimilated to the last, but the spots much more irregular, both in disposition and shape.

F. Sumatrana, *Horsfield, Zool. Java, No.* 2.
Icon. *Horsfield, l. c.*
Inhabits Sumatra.

447. 31. *F. Diardi* (Diard's Java Cat.) Larger than the Kuwuk; ground-colour, yellowish-gray, covered with nu-

merous full small spots; throat and back with longitudinal bands.

Felis Diardi, *Cuvier, Ossemens Fossiles*, 4.

Icon. *Hamilton Smith, MSS.*

Inhabits Java.

448. 32. *F. Nigripes* (the Black-footed Cat.) Tawny, entirely covered with black spots, elongated on the back and neck, under parts of the feet black, tail spotted, not annulated.

Felis Nigripes, *Burchell's African Travels, Vol.* ii.

Icon. ——

Inhabits South Africa.

Obs. This may possibly be no other than the booted Lynx of Bruce.

449. 33. *F. Catus Ferus* (the Common Wild Cat). Yellowish gray-brown, with a black dorsal line diverging into four on the neck; sides, flanks, and thighs, covered with obscure transverse bands; tip of the tail black.

F. Catus, *Lin.*

Chat Savage, *Buff. Hist. Nat. t.* 6. Wild Cat. *Pen.*

Icon. Schreb. *t.* 107.

Habitat. of the wild variety, the forests of Europe and Asia; of the tame, all the civilized parts of the world.

Var. a. Domestic Cat, differing infinitely in external marks; forehead and cheeks generally preserving a vestige of the stripes proper to the wild variety; intestines rather larger than in the wild variety.

Sub-var. b. Chartreuse Cat (F. Catus cœruleus, *Lin. t.* 6. *pl.* 4.) Fur long and fine, generally of an uniform colour.

Sub-var. c. Spanish Cat (F. C. Hispanicus and F. C. maculatus, *Bod. t.* 6. *pl.* 3.) Fur short; feet and lips flesh-colour; the female only, as it is said, spotted with white, bright red, and deep black patches.

Sub-var. d. Angora Cat (F. C. Angerousis, *Lin. Buff. t.* 6. *pl.* 5.) Fur long, soft, and silky ; generally white, but sometimes patched ; the eyes frequently of different colours.

The New Spain Cat is probably a variety of the common species.

Lynxes, or middle-sized Cats, with shortish tails, and generally pencilled ears.

450. 34. *Felis Caracal* (the Caracal.)   Uniformly vinous red above, white underneath ; tail reaching to the heels ; ears strongly pencilled, black on the outside, white within.

Lynx of the Ancients.   Felis Caracal, *Lin.*

Caracal à longue queue, *Buff. Sup.* III.

Icon. *Schreb. tab.* 110.   *Buff. t.* 9. *pl.* 24. *Sup.* III. *t.* XIV.

Inhabits the northern parts of Africa, south-eastern parts of Asia, and Bengal.

There appear to be several hereditary, as well as casual varieties of this animal, as the Caracal of Algiers of Bruce, without pencil to the ears, and with longitudinal stripes. The Caracal of Nubia, of the same traveller, with a rounder head, and the ears black on the outside mixed with white ; and the Caracal of Bengal, of Edwards, with a longer tail than the ordinary variety.

451. 35. *F. Chaus* (the Booted Lynx.)   Uniformly yellowish gray ; hind part of the legs black ; tail reaching to the heels, annulated with black to its extremity ; ears brown without, white within, terminated with a pencil of black hairs.

Felischaus, *Guldenst. Nov. Annum. Petrop.* xx. *p.* 483.

Felis Lybicus, *Olivier, Voyage en Egypt.*

Booted Lynx, *Bruce, Travels,* vol.

Icon. *Guldenst. as above, pl.* 14 *and* 15.   *Bruce as above, pl.* 30.   *Olivier, pl.* 41.

Inhabits Abyssinia, Nubia, and the adjoining parts of Asia.

452. 36. *F. Lynx* (the Lynx.) Reddish-yellow, with small dark brown spots ; long pencilled ears, and short tail, black at the extremity.

Felis Lynx, *Lin.*

Loup cervier *of the French furriers.*

Icon. *Schreb. tab.*109. *Buff. t.* 9. *pl.* 21. *Perrault, Act. des Sc.* III. *p.* 1. *t.* 17.

Inhabits Asia and Africa, and the eastern parts of Europe.

Obs. There appear to be at least three varieties in this species, known in Sweden by the names Cat Lynx, Wolf Lynx, and Fox Lynx.

453. 37. *F. Canadensis* (the Canada Lynx.) Covered with fine long hair all over the body, the sides of the face, with longer hair, like the tiger ; of a very pale ash colour, with a tint of yellow on the upper parts, whiter underneath.

Felis Canadensis, *Geoff.*

Le Lynx du Canada, *Buff. t.* 3.

Icon. *Ib. pl.* 44.

Inhabits Canada.

454. 38. *F. Rufa* (the Red Lynx.) Reddish-yellow, spotted with brown ; tail short, white at the extremity.

Felis Rufa, *Guldenstaedt.*

Chat cervier *of the French furriers.*

Icon. *Screb. tab.* 109. *B.*

Inhabits the United States, but not so far north as the preceding.

M. Rafinesque, in the American Monthly Magazine, has designated as distinct species the following Lynxes: —

Lynx Fasciatus, discovered by Lewis and Clarke, and described in their Travels, differing apparently from the Canada Lynx only in being reddish-brown, with blackish bands and spots. Lynx Montanus, Mountain Cat of the Americans, probably the Lynx du Mississippi of *Buff. t.* 8.

Among the more uncertain insufficiently described species may be included :

F. Manul of Pallas, *Travels, t.* 3. It seems strongly to resemble the Red Lynx, though it is not spotted. The Mountain Cat of Pennant, which Cuvier refers to the common Lynx. The Tiger Cat of Collinson, *Buff. Sup. t.* 3, which seems to be the Margay. F. Varia, of Schreber, which the Baron considers to be a Leopard. F. Chalybeata. *Schreber from Herman.* M. F. Cuvier refers this to the Serval. F. Guttata, of *Herman,* figured by *Schreber,* is stated by *M. F. Cuvier* to be a young Panther.

Barrow speaks of two Cats of the Cape, with numerous black spots on a yellow ground ; one inhabiting the mountains, and the other the plains ; and also of the Cape Leopard, with a mane like a lion, &c.

TRIBE III. AMPHIBIA. Feet short, enveloped in the skin, shaped like fins, and fitted for swimming, those behind horizontal ; cutting teeth variable, mostly $\frac{4}{4} \frac{6}{4} \frac{4}{2}$.

Gen. 1. PHOCA.

Teeth varying greatly in the different species. Incisors, $\frac{6}{4}$, or $\frac{6}{2}$, or $\frac{4}{4}$; also, varying in form ; canine $\frac{1 \cdot 1}{1 \cdot 1}$, strong conical, slightly curved ; cheek-teeth, $\frac{5 \cdot 5}{5 \cdot 5}$ or $\frac{5 \cdot 6}{5 \cdot 5}$, or $\frac{6 \cdot 6}{6 \cdot 6}$, generally similar to the anterior-cheek teeth, or false molars of the carnivora, trenchant, triangular, but more conical and more obtuse ; nose sometimes elongated into a proboscis ; nostrils capable of being completely closed at the will of the animal; eyes large ; external ears wanting, or merely rudimentary ; pentadactylous, the anterior extremities consisting only of hands, and the posterior only of feet; fingers enveloped in the skin ; tail short and thick ; teats four, abdominal ; mustachios very strong and numerous.

175

Inhabit all the seas, and perhaps Lake Baikal; live on the sea-shore, and visit the water occasionally, eat fish, mollusca, and algæ.

455. 1. *P. Vitulina* (Common Seal.) Fur yellowish-gray, more or less variegated and spotted with brown, according to its age; hair abundant; thick nails, black and strong; whiskers waved.

Phoca Vitulina, *Lin. Sys. Nat.*

Sea Dog. Sea Wolf. Sea Calf, *Pen. Brit. Zool.* Common Seal, *Pennant. Quad. p.* 270. Phoque Commune, *Buffon,* XIII.

Icon. *Pennant, Brit. Zool. Rondelet, Pisc. Marin.* 453. *Buffon,* XIII. *t.* 45. *Supp.* VI. *t.* 46. *F. Cuv. Dent. Mam. t.* 38. *Mam. Lithog.* XVI. *Shaw's Zool.* 1. *t.* 70.

Inhabits the North Sea.

This species is the type of M. F. Cuvier's genus, *Callocephale.*

Var. a. Bothionica. Nose larger; claws longer; fur deeper.

Phoca Vitulina Bothionica, *Lin. Faun. Sue.*

Var. b. Sebrica, silvery.

Phoca Vitulina Sibirica. *Gmelin. Syst. Nat.*

Inhabits Lakes Oronn and Baikal.

Said by Peron most probably to be an Otter.

Var. c. Caspica, variegated with black, yellow, ash-coloured, and white.

Phoca Vitulina Caspica. *Gmelin, Syst. Nat.*

Var. d. Maculata, adult, spotted; when young, black above, white beneath.

Phoca Maculata, *Bodd.*

456. 2. *P. Leporinus,* (Hare Seal.) Fur yellowish, with a white half collar, forming a cross on the neck; claws of the fore-feet very strong.

Phoca leporina, *Lepechin, Act. Acad. Petrop.*

Phoque commune, *F. Cuv. Mam. Lithog.*

Icon. *Lepechin, l. c. t.* 8, 9. *F. Cuv. Mam. Lithog.* IX.

Inhabits the North Sea and the White Sea.

Obs. According to Desmarest, the cutting teeth of this species are $\frac{4}{4}$.

457. 3. *P. Discolor* (Two-coloured Seal.) Fur blackish, marked with torquious yellowish gray lines.

Calocephale discolor, *F. Cuv. Mem. Mus.* XI.

Icon. ——

Inhabits the North Seas.

458. 4. *P. Lagura* (White-Tailed Seal.) Fur gray, clearer on the sides and belly ; back and flanks sometimes variegated with small irregular black spots ; whiskers black ; tail long, thin, white; claws of the fore-feet long, strong, and compressed.

Phoca lagura, *Cuv. Ossemens Fossiles,* v. 206.

Icon. ——

Inhab.   (Mus. Paris.)

459. 5. *P. Grœnlandica* (Greenland Seal.) Fur of the adult males whitish, with the forehead and muzzle, and a large subcontiguous lunated blotch on the sides, crossing at the shoulders ; claws strong, black ; females and young covered with unequal distant angular spots.

Phoca Grœnlandica, *Muller, Pro.* 8. P. Semilunaris, *Bodd.* P. oceanica, *Lepechin, Act. Petrop.*

Svartside, *Egede,* 46. Attarsoak, *Crantz.* 163—169. Harp Seal, *Shaw's Zool.*

Icon. *Shaw's Zool. t.* 71. *Egede, l. c. t. Lepechin, Act. Petrop.* v. *t.* 78.

Inhabits the North Sea.

Lepechin describes *P. oceanica* as having only $\frac{4}{4}$ cutting-teeth ; but we have followed *M. F. Cuvier* in placing it with *P. Grœnlandica.*

177

460. 6. *P. Fætida* (Fœtid Seal.)   Fur pale brown, variegated with white above, and dirty white beneath; hair rough; claws strong.

Phoca fœtida, *Muller, Bod.* 8. *p.* Hispida, *Schreb. Staught,* Halychoerus Griseus, *Nilson, His.* 1824, 810.

Phoque nutsoak, *Buffon, Supp.* vi. 7. Neitsek, *Crantz,* 164.

Icon. *Buffon, Supp.* vi. *Ency. Méthod. t.* 111. *f.* 2. *Crantz. t.* 152.

Inhabits the North Sea.

This species appears to be from the genus *Halychœrus* of Nilson.

461. 7. *C. Barbata* (Bearded or Great Seal.)   Fur blackish; thumb of the hand shorter than the fingers.

Phoca Major, *Parson's Phil. Trans.* xlvii. 121. P. barbata, *O. Fab. Grœnland.* 18.

Urksuk, *Crantz.* 168. Grand Phoque, *Buffon, Hist. N. Supp.* vi.

Icon. *Buffon, Supp.* vi. *t.* 45. *Ency. Méthod, t.* 3. *f.* 1. *Phil. Trans.* xlvii. *t.* 5.

Inhabits the North Seas.

Obs. Parsons gives a very slight notice of this species, which Buffon identifies with that of Crantz. It is twice the size of the Common Seal; Fabricius identifies his Barbata with the Urksuk of Crantz; this species is ten feet long, and is distinguished by the shortness of the thumb.

462. 8. *P. Leptonyx* (Small-clawed Seal.)   Claws small, especially those of the hind feet; fur above gray, variegated with yellow; beneath yellowish white; whiskers rigid.

Phoca Leptonyx, *Blainv. MSS. Desm. Mam.* 247.   Monochus, *Flemming, Phil. Zool.* iv. 187.

F. Cuv. Mem. Mus. iv. 193.   Le Phoque à ventre blanc.

178

Buffon, Supp. vi. Foca a ventre bianco. *Ranzani Men. di Stor. Nat.* i. 102. Cowled or Mediterranean Seal. *Pennant, Quad.* vi. 273. Red Seal. *Pennant, Quad.* vi. 273. Greek Seal.

Icon. *Buffon, Supp.* v. *f.* 44. *Shaw's Zool.* i. *t.* 70, 71. *Herman. Oct. Nat. Seut. Bertol. p. t.* 12, 13.

Inhab. Adriatic Sea.

Johnstein has published a work on the comparative Anatomy of this Seal. It is the type of M. F. Cuvier's genus *Pelage.*

MIROUNGA* Gray. *Cutting teeth $\frac{4}{2}$ or $\frac{6}{2}$, canine teeth $\frac{1-1}{1-1}$, cheek teeth short, broad, roots simple ; crown striated, nearly flat, external ears none ; fur crowned on the nose, elongated into a trunk.*

463. 1. *P. Cristatus,* (Crested Sea-Lion.) Top of the forehead furnished with a moveable hood, susceptible of erection, and of covering the eyes and muzzle.

Phoca cristata, *Gm. Sys. Nat.* i. P. Leonina, *Fab. Faun. Grœnl.* P. Mitrata, *Camper.*

Klap Migosen. *Egede, Grœnland.* 62 ? Klap. Mutz. *Egede, Grœnland.* 62? Neitsersoak, *Grœnlander,* Fabricius.

Icon. *Egede, Grœnland.* 62?

Inhab. North Seas, Greenland.

According to Peron several Seals have been confounded with this species. It forms the genus *Stematope* of M. F. Cuvier.

---

* The species included in this subgenus are treated by M. Desmarest, *Ency. Méthod,* as constituting merely a group in the genus Phoca, distinguished by the cutaneous appendage to the head, or a sort of trunk to the snout; but as the cheek teeth of all the species appear to be cetaceous, Mr. Gray proposes separating them into a subgenus, which we have adopted.

464. 2. *M Proboscidea* (Peron, Sea Elephant, or Proboscis Seal.) Hair very thinly scattered, gray; claws of the fore feet small; the occipital and sagittal ridges very prominent; the mastoid apophysis slightly developed; cutting teeth $\frac{4}{2}$.

Phoca Proboscidea, *Peron and Leseur, Voyage aux Terres Australes,* vi. 34.

Macrorhine. *F. Cuv. Mem. Mus.* vi. 200.   Miouroung, *Native of New Holland.*

Icon. Peron and Leseur's Atlas, *t.* 32. *F. Cuv. Mem. Mus.* vi. *t.* 3. *f.* 1. *ab. c.*

Head. *Desm. Dent. Mam. t.* 39. *A.* Teeth.

Ency. Méthod. Sup. t. 6. *f.* 4.

Inhabits the Seas of New Holland.

Obs. The type of M. F. Cuvier's genus, *Macrorhyna.*

465. 3. *M. Patagonica* (Patagonian Sea-Elephant.) The skull convex, the cerebral cavity more extended, and the nasal region shorter than in the M. Peronii; cutting-teeth $\frac{4}{2}$.

Phoque des Patagons. *F. Cuv. Mem. Mus.* iv. 203.

Icon. *F. Cuv. l. c. t.* 14. *f.* 2. Skull.

Described by M. F. Cuvier from a skull only of a young specimen.

466. 4. *M. Ansonii* (Anson's Sea-Elephant.) Hair short, clear yellow ; feet and tail black ; claws of the fore-feet strong. The occipital and sagittal crests slightly developed ; the mastoid apophyses not prominent ; the cutting-teeth $\frac{6}{2}$.

Phoca Leonina, *Gmelin, Syst. Nat.* i. 63. *Shaw. Zool.* i. 268.   Phoca Ansonii, *Desmarest,* 239.

Sea-Lion, *Anson's Voyage,* 122; *Dampier's Voy.* i. 118 ?
Loup Marine, *Pernetty, Voy. aux Isles Malouines ;* Lion

Marine, Phoque a museau ride *Buffon*. Bottle-nosed Seal, *Pennant*, *Quad*. vi. 286.

Inhab. Pacific Ocean.

Obs. The skull of this species is in the College of Surgeons.

467. 5. *M. Byronii* (Byron's Sea-Elephant.) The occipital and sagittal ridges, and the mastoidal apophyses very prominent ; the cutting-teeth $\frac{6}{2}$; the upper ones next the canine, larger than the rest.

Phoca Byronii. *Blainville, Desm. Mam.* 240.

Sea-Lion. *Byron's Voyage.*

Icon. ———

Inhab. Island of Tinian.

Obs. The skull of this species is also in the College of Surgeons.

The Sea-Lion of Cox's Island of St. Paul, (*Phoca Coxii Desm. Nov. Dict. d'Hist. Nat. and Fleurieu Voyage du Capitaine Marchand,* t. 3. 17) ; the Wolf-Seal, (*Phoca Lupina of Melina's Chili,* 260) ; the long-necked Seal of Parsons, *Phil. Trans. Vol.* 47. *pl.* 6 ; the speckled Seal of the *Encyclopædia ;* the spotted Seal of the same, and the black Seal of the same ;   the Tiger-Seal of Krachenninikon, and the Grumm-selur or King of Seals of Olassen ; the Phoca Testudinea or Tortoise-headed Seal of Parson's *Phil. Trans. Vol.* 48 ; the Ribbon-Seal of Pennant, (*Phoca fasciata,* Shaw),—may all be considered as doubtful, and we have therefore merely made a marginal reference to them.

The Phoca Vitulina, or Common Seal, P. Grœnlandica of Fabricius, (P. Oceanica of Lepechin) ; the P. Leporina of Lepechin ; the P. Leptonyx of Blainville ; the P. Monachus of Herman ; the P. Cristata of Gmelin ; the P. Ansonii ; and the P. Lagura (with the exception, as to this last species of the head and hind-feet),—are the only species of seals, properly speaking, whose osteological characters have been examined by the Baron Cuvier.

OTARIA Peron. *Cutting-teeth* $\frac{6}{4}$. *canine-teeth* $\frac{1-1}{1-1}$ *large ; cheek-teeth* $\frac{6-6}{5-5}$; *root simple ; crown with a principal conical point, and one little conical lobe before and behind it; external ears distinct.*

468. 1. *O. Ursina* (Common Sea-Bear.) Fur brown ; males maneless. The hind feet furnished with long flaps of skin.

Phoca Ursina, *Lin. Gmelin, Syst. Nat.* Otaria Ursina, *Desm. Mam.* 249. Ursus Marinus, *Steller, Nov. Comm. Petrop.* v. 1751. 331.

Ours Marin, *Buff. Suppl.* vi. Chat Marin, *Krachenninikon, Hist. du Kamtschatka.*

Icon. *Nov. Com. Petrop.* v. *t.* 15. *Schreb.* 182. *Buffon, t.* 47.

Inhab. Kamtschatka.

The type of M. F. Cuvier's genus, *Arctocephale.*

468. 2. *O. Peronii* (Peron Sea-Bear.) Fur soft, blackish ; the hind feet having only three claws apparent in the middle, ending in a five-lobed membrane ; bristles of the whiskers round and smooth.

Phoca pusilla, *Lyn. Sys. Nat.* P. Peronii, *Bodd. Elen.* Otaria Peronii, and O. pusilla, *Desm. Nov. Dict. Hist. Nat.* xxv. 598 and 602.

Petit Phoque, *Buffon*, XIII. L'Ours Marine du Cap de Bonne Esperance, *F Cuv. Dent. Mam.*

Icon. *F. Cuv. Dent. Mam. t.* 39. *Ency. Méthod. t.* III. *f.* 2. *Buffon*, XIII. *t.* 53.

Inhab. Cape of Good Hope.

469. 3. *O. Coronata* (Crowned Sea-Bear.) Fur black, variegated with yellow spots ; with a yellow band on the head, and a spot on the muzzle ; the hind-feet with five claws.

Otaria Coronata, *Blainville, MSS. Desm. Mam.* 251.

Icon. ——
Inhab. —— ? Bullock's Museum.
Obs. Like *Phoca fasciata*, Shaw, in colour, but has ears.

470. 4. *O. Cinerea* (Ash coloured Sea Bear.) Fur ash-coloured, hard, ridged, gray, without any downy fur; ears conical, short.

Otaria Cinerea, *Peron and Leseur, Voy. aux Ter. Aust.* vi. 75.

Inhab. Coast of New Holland.

Obs. To the obscure description of Peron, we have added the account of the fur and ears, from the skin of the head in the Linnean Society, presented by Capt. P. King, and described by Mr. Gray.

471. 5. *O. Flavescens* (Yellowish Sea-Bear.) Fur uniform pale yellow; fore-feet clawless; the hind-feet with claws; the middle toes longer than the rest; ears long, pointed foliaceous.

Phoca Flavescens, *Shaw, Zool.* 260. Otaria Flavescens, *Desm. Mam.* 250.

Eared Sealed, *Penn. Quad.* 278.

Icon. *Shaw, Zool.* i. *t.* 73.

Inhab. Strait of Magellan.——Leverian Museum.

472. 6. *O. Falklandica*, (Falkland Island Seal.) Fur ash-gray, shaded with white; fore feet clawless; hind feet with four claws?

Phoca Falklandica, *Shaw, Zool.* 256. Otaria Falklandica, *Desm. Mam.* 250.

Falkland Island Seal, *Penn. Quad.* 275.

Icon.

Inhab. Falkland Islands.

Obs. There is a specimen in the Museum from New South Shetland, which, as observed by Mr. Gray, agrees with the description of this Seal ; but it has five claws on the hind feet ; its ears are long, filiform, and it has downy fur under its hair.

473. 7. *O. Jubata* (Sea-Lion.) Fur yellow; neck of the male with a large mane ; hind feet, with the ends deeply lobed.

Phoca Jubata, *Gmelin, Sys. Nat.* i. 63. Otaria Jubata, *Desm. Mam.* i. 248. Leo marinus, *Steller, Nov. Com. Act. Petrop.* v. 418.

Leonine Seal, *Pennant. Quad.* v. 288. Sea-Lion, *Forster, Cook's Second Voyage,* iv. 54. Lion Marine, *Buffon, Supp.* vi. *t.* 418.

Icon. *Perrette's Voyage, aux iles Malouines,* vi. *t.* 10. *Forster, Voyage, t.   Buffon, Supp.* vi. *t.* 48. *Ency. Mé-thod. t.* 109. *f.* 3. *Shaw, Zool. t.* 174.

Inhab. Pacific Ocean, Coast of Patagonia. *Forster.* Bering's Strait. *Steller.*

This species is the type of *M. F. Cuvier's* genus *Platyrhinque.*

The Otaria Albicollis of Peron and Leseur, ii. 118 ; and the P. Porcina of Molina, an eared species, if distinct, have not yet been examined with sufficient accuracy to fix their osteological characters.

Genus II. TRICHECUS.

Cutting-teeth $\frac{4}{4}$; small deciduous canine teeth $\frac{1.1}{0.0}$, very large, longer than the head, oval, laterally compressed, arched, blunt ; cheek-teeth $\frac{5.5}{3.3}$, small, cylindrical, crown oblique, truncated ; body long, conical ; head round ; muzzle large ; external ears none, tail short ; fore-feet paddle-shaped, armed with five short claws; hind feet horizontal ; toes, five, united in the skin.

474. 1. *T. Rosmarus* (Sea-Elephant.) Lips very thick ; bristles ridged, strong ; fur thin, short, reddish.

T. Trichechus Rosmaris. *Lin Sys. Nat.* i. 49. Manati

184

Trichechus, *Bodd. Elen. Anim.* 173. Rosmarus, *Scopili.*
Odobenus, *Lin. Sys. Nat. ed.* 10. 38. *Brisson.* Rosma-
rus, *Jonston, Pisc. t.* 44. Equus Manuus, *Rai, Quad.* 191.
Mors. Morfo. Rosmarus, *Gesner, Aquat.* 211.

Wallruss, *Marten, Spits. b.* 78. 'Le Morse, *Buffon, Hist.
Nat.* xiii. Sea Elephant. Arctic Wallrus, *Shaw's Zool.*

Icon. *Hout. Nat.* v. *t.* 11. *f.* 1. *Schreb.* v. *t.* 79. *Jon-
ston, Pisc. t.* 4. ix. *Lev. Mus. t.* 23. *f.* 3. *Bonanu. Mus.
f.* 27. *Gesner, Aquat.* 211. *Ellis, Hudson, t.* 6. *f.* 3.
Marten, Spits. t. 1. *f.* 13. *Buffon, Hist. Nat.* xiii. *t.* 54.
Ency. Method. t. 112. *f.* 1. *Cook's Third Voyage,* iv. *t.* 8.
Shaw's Zool. t. 78.

Inhab. Northern Ocean.

The two species proposed by Dr. Shaw, appear to be
only varieties.

Family IV.—MARSUPIATA.

Teeth varying very considerably in the different subdivisions; all the species born prematurely, at least in a very
early stage of development, and become attached to the
mammæ of the mother, in a manner not known. The
mammæ and young of most of the species enclosed
within an external pouch, or second womb; the pouch
supported by two marsupial bones; thumb of the hind
extremities mostly distinct and opposable to the fingers,
in a few species wanting.

Inhabiting South America, New Holland, and the Indian
Islands ?

Sub-Division I., *with Canine and insectivorous Cheek-teeth.*

Gen. I.—Didelphis.

Incisors $\frac{10}{8}$; the two inte mediate in the upper jaw
longer than the rest, and separated from them; in the
lower jaw they are very small and equal. Canine $\frac{1-1}{1-1}$,
strong, compressed, and a little projected outward. Cheek-

teeth $\frac{11}{77}$, or $\frac{64}{44}$; the three first in the upper jaw, false teeth, being very small and triangular, the remainder insectivorous, or furnished with sharp points ; the four first in the lower jaw, also false and small, and the three others furnished with points. Head long and conical, and muzzle pointed; ears large, rounded, and almost naked; tongue aculeated. Locomotion plantigrade ; pentadactylous, nails long and bent ; thumbs of the hind feet long, opposable to the fingers, and destitute of nail ; tail long, half hairy, and scaly, hair hard, close ; stomach simple in all.

Inhabiting South America.

* Female with a Pouch.

475. 1. *D. Virginiana* (Virginian Opossum.) Fur silky, mixed black and white bars, partly black and partly white ; head nearly white ; size of a Rabbit.

Didelphis Virginiana, *Pennant, Quad.* 11. *pa.* 18.

Sariguesa Oreilles Bicoleres, *Cuv. Rég. Anim.* Sarigue des Illinois, *Buff. Sup.* VI. S. a long poil, *Buff. Sup.* VI. Micouré premier Manicou, *Bonn. Ency.* Opossum, *Americans, d'Azara, Quad. du Paraguay.* Virginian Opossum, *Shaw, Zool.* vol. I.

Icon. *Buff. Sup.* VI. *f.* 35. *t.* 34. *Ency. Méthod. t.* 264, *Sup. t.* 7. *f.* 1. *Bonn. Ency. t.* 24. *f.* 6. *Shaw, Zool.* I. *t.* 107. *Phil. Trans. Abridg.* VI. *t.* 13. *Mus. Lever. t.* 6.

Inhabits most of the hot and temperate parts of America.

476. 2. *D. Azaræ* (Azara's Opossum.) Fur of two sorts, one cottony underneath, white at the base, and black at the tip; the other long, stiff, and white ; black round the eyes ; all the legs black ; first half of the tail hairy, the rest scaly. Total length about thirty inches, of which the tail is fourteen.

Didelphis Azaræ, *Temminck, Monographie de Mammalogie, pa.* 30.

Micouré premier, *d'Azara, Quad. du Paraguay.*
Icon. ――――
Inhabits South America.

477. 3. *D. Cancrivora* (the Cayenne, or Crab-eating Opossum.) Yellowish, mixed with white; a brown line upon the forehead.

D. Marsupialis, *Lin. Sys. Nat.* i. 71. D. Cancrivora, *Gm. Sys. Nat.* Philander maximus orientalis, *Seba.*

Crabier, (the male) *Buff. Sup.* vii. D. Carcinophaga, (the female) *Bodd.* Molucca Oppossum, *Shaw.*

Icon. *Buff. Sup.* vii. *pl.* 54. *Seba. Thes.* v. 1. 38.
Inhabits Cayenne and Surinam.

478. 4. *D. Quica* (Quica Opossum.) Tail longer than the body and head, nearly half hairy, and white; above blackish-yellow, with a silvery hue; upper part of the head blackish, with three white bands; chin white; belly reddish; size of a Pole Cat.

Didelphis Quica, *Temminck, Monog. pa.* 36. Quica of the *Brazilians.*

Icon. ――――
Inhabits Brazil.

479. 5. *D. Opossum* (the Four-eyed Opossum.) Yellowish above, almost white underneath; a pale white spot above each eye; tail hairy, part only of its length.

D. Opossum, *Lin. Sys. Nat.* i. 105.

Sarigue, or Opossum, *Buff.* x. 279. Philander, *Seba,* iv. Carigueia *of the Brazilians;* Le Quatreœil on moyen Sarigue de Cayenne, *Cuv. Rég. Animal.*

Icon. *Buff.* x. *t.* 45. *t.* 46. *Seba* i. *t.* 36. *Shaw, Gen. Zool.* i. *t.* 108.

Inhabits Cayenne commonly, and probably other parts of America.

480. 6. *D. Philander* (Philander Opossum.) Head very short ; muzzle blunt; head marked with a red central longitudinal band; eyes placed in a gray spot. Tail naked, longer than the head ; and body spotted with brown.

Didelphis Philander, *Schreb. Temminck, Monog.*

Icon. Schreb. vii. t. 147.

Inhab. ——

* *Females without a Pouch.*

481. 7. *D. Cayopollin* (Mexican Opossum.) Yellowish gray above, whitish underneath, round the eyes, and upon the nose ; tail brown, spotted with blackish, and much longer than the body.

D. Cayopollin, *Gm.* i. 106. Mus Africanus Cayopollin dictus, *Seba, Thes. t.* 1. Philander Africanus, *Brisson, Reg. Anim.*

Philandre de Surinam, *Seba.* Cayopollin, *Buff.* x. 350. Mexican Opossum, *Pennant.*

Icon. *Seba, Thes.* i. *pl.* 31. *f.* 3. *Buff.* 10. *f.* 55. *Ency. Méthod. t.* 24. *f.* 5.

Inhabits Mexico.

482. 8. *D. Murina* (Murine Opossum.) Yellow gray above, very pale yellow underneath ; eyes surrounded with brown ; tail naked, as long as the body.

D. Murina, *Gm.* 107.

La Marmose, *Buff.* x. 335.

Murine Opossum, *Shaw,* i. 484.

Icon. *Buff.* x. *t.* 52 *and* 53.

Inhabits Cayenne and Surinam.

483. 9. *D. Cinerea* (Gray Opossum.) Muzzle blunt ; head small ; ears contracted at the base ; tail much longer than the head and body, very hairy at the base, end white ;

fur short, cottony above, ash-gray in the males, yellowish-gray in the female ; beneath white.

Didelphis Cinerea, *Temminck, Monog.*

Icon. ——

Inhab. Brazils.   Prince Maximilian de Neuwied.

484. 10. *D. Dorsigera* (Back-bearing Opossum.)   Head long ; tail hairy at the base, of a uniform brown colour.

Didelphis Dorsigera, *Temminck, Monog.* 1. *Gmelin, Sys. Nat.* i. 107.

Icon. ——

Inhabits. ——

485. 11. *D. Nudicaudata* (Naked-Tailed Opossum.) Gray brown above, whitish underneath ; a pale yellow spot over each eye ; tail longer than the body, and entirely naked ; no marsupial pouch, but a longitudinal fold of the skin along each side of the belly.

D. Nudicaudata, *Geoff. Collect. du Mus. d'Hist. Nat.* Didelphis Myosuros, *Temminck, Monog.* 38 ?

Inhabits Cayenne.

Obs.   This species appears to differ from D. Opossum, principally in the fold of skin, instead of the marsupial pouch, and the total nudity of the tail.

M. Temminck (*Monog.*) describes his D. Myosuros as having a pouch in the female.   He doubts the accuracy of M. Geoffroy's M. Nudicaudata being destitute of the pouch, and refers both conditionally to one species.

486. 12. *D. Crassicaudata* (Thick-Tailed Opossum.) Deeper yellow above, brighter over the eyes ; tail nearly the length of the body, very thick at its insertion, and covered with a fold of skin, instead of a marsupial pouch, for about a third of its length.

Didelphis Crassicaudata, *Desm. Nat. Dict. d'Hist. Nat.*

ix. 425. Third Micouré, with a thick tail, *d'Azara, Quad.
du Paraguay.*

In the female specimen described by d'Azara, there were
four mammæ on one side, and but two on the other, which
fact, in addition to other observations made by M. Desmarest, seems to have induced that naturalist to suspect that
the teats in these animals are developed at the time only
of suckling, and then only in number corresponding with
the young.

487. 13. *D. Lanigera* (Woolly Opossum.) Fur woolly,
snuff colour above, whitish underneath : tail nearly triangular at its insertion, longer than the body, and naked on
the upper side only, for about a third of its length ; a fold
of skin instead of a marsupial pouch.

Didelphis Lanigera, *Desm. Mam.* D. Cayopollin, *Desm.
Dict. Hist. Nat.*

The second, or woolly Micouré of *d'Azara, Quad.* of
Paraguay.

Inhabits Paraguay.

Obs. Only two individuals of this species appear to have
been seen and described.

488. 14. *D. Brachyura* (Touan, or Short-Tailed Opossum.)
Blackish-brown on the back, brightish red on the flanks,
and white on the belly ; tail short, covered with hair to the
end ; a fold of skin instead of a marsupial pouch.

D. Brachyura, *Pallas, Act. Petrop.* 1780. D. Tricolor,
Geoff. Col. Mus. Desm.

Touan, *Buff. Sup.* vii.

Icon. *Pallas, Act. Petrop. t. 5. Buff. Sup.* vii. *pl.* 4.

Inhabits Cayenne and Paraguay.

489. 15. *D. Sebæ* (Seba's Opossum.) Fur deep red,
brown above, and on the flanks, beneath whitish ; tail half
as long as the body.

D. Brachyura, *Gmelin*, 108, not. *Pallas*. D. Sebæ, *Gray*, *MSS.*

Short-Tailed Opossum, *Pennant*, vi. 26. Mus Sylvestris Americana, *Seba, Mus.* i. 50.

Icon. *Seba, Mus.* i. *t.* 31. *f.* 6.

Inhabits South America.

Obs. M. Temminck preserves the specific name Brachyura for the species, and Tricolor for the last.

490. 16. *D. Pusilla* (the Dwarf Opossum.) Mouse-colour ; tail longer than the body, naked, of a white colour.

Didelphis Pusilla, *Desm. Nov. Dict. d'Hist. Nat.* 430.

The sixth, or Dwarf Micouré of *d'Azara, Quad.* of *Paraguay.*

Icon. ——

Inhabits Paraguay.

Obs. D'Azara inspected only a male, so that the existence or non-existence of the marsupial pouch in this species is not ascertained.

CHEIRONECTES. *Incisors* $\frac{10}{8}$, *canines* $\frac{1\cdot1}{1\cdot1}$, *cheek-teeth pointed and trenchant ; muzzle pointed ; ears naked ; round feet, with five toes ; the posterior plantigrade and palmated ; thumb without a nail ; nails sharp and bent ; tail long, cylindrical, naked, scaly, and prehensile ; abdominal pouch in the female.*

Obs. In the *Régne Animal* this is printed as a subgenus. It is perhaps immaterial whether it be treated as a subgenus of Didelphis or a genus of Marsupiata.

491. 17. *D. Palmata* (the Yapock, or Palmated Opossum.) Brown above, with three transverse gray patches or bands, white underneath.

Didelphis Palmata, *Geoff.* Didelphis Yapock, *F. Cuv.* Chironectes Memina, *Illiger.* Lutra Memina, *Zimmerman, Bodd.* Chironectes Yapock, *Desm. Mam.*

Le Yapock, *Cuv. Rég. Anim.* Petit Loutre de la Guyane, *Buff. Hist. Nat. Sup.* vii. 159.

Icon. *Buff. Sup.* vi. *t.* 22. nobis.

Inhabits the banks of the Yapock, a river of Guyane.

Obs. This species, which has been very generally treated as an Otter, has been separated by Illiger into a distinct genus, under the name of Cheironectes, or hand-swimmers.

Genus II.—Dasyurus.

Incisors $\frac{8}{8}$, small and regular, canines $\frac{1\cdot1}{1\cdot1}$, large; cheek-teeth $\frac{6\cdot6}{6\cdot6}$, the two first compressed and trenchant, the remainder with points on their crowns; head conical; five toes on the fore-feet, armed with crooked nails; four on those behind, unguiculated; thumb without a nail, short, distant from the fingers, being little else than a mere tubercle; tail long, covered all over with hairs; a marsupial pouch in the females.

492. 1. *Das. Cynocephalus* (the Dog-faced Dasyurus.) Yellowish-brown, crupper marked with transverse black bands; tail compressed. Size of a wolf.

Didelphis Cynocephala, *Harris, Transactions of the Lin. Soc. vol.* ix. Dasyurus Cynocephalus, *Geoff. Ann. Mus.* xv. Thylacinus Harrisii, *Tem. Monog.* 63.

Icon. *Lin. Trans.* ix. *t.* 19.

Inhabits Van Diemen's Land.

Mr. Brooks, it is understood, proposed to make this species a type of a new genus, to be named *Paracyon.* M. Temminck has since done so, and applied to it the name Thylacynus.

493. 2. *Das. Ursinus* (the Ursine Dasyurus.) Black, with occasionally a very few white spots; tail not very long, but naked on the under side, and subprehensile.

192

Didelphis Ursina, *Harris, Lin. Trans. vol.* ix. Dasyurus Ursinus, *Geoff. Ann. Mus. t.* 15.

The Devil of the Colonist.

Icon. *Harris,* as above, *pl.* 19.

Inhabits Van Diemen's Land.

494. 3. *Das. Macrourus* (Spotted Dasyurus.) Marron-colour, spotted irregularly with white ; the hairs of the tail not so long as in the other Dasyuri, but spotted like the body. Size of a Cat.

Viverra Maculata, *Shaw, Gen. Zool. vol.* i. Dasyurus Macrourus, *Geoff. Ann. Mus. t.* 3.

Spotted Marten, *Phillips, Voyage to New Holland,* 276. Dasyure tachete. *Peron and Lesueur, Voy. aux Ter. Aust.*

Icon. *Peron and Leseur, Atlas, pl.* 33, *Ency. Méthod. Sup. t.* 762.

Inhabits the vicinity of Port Jackson, in New Holland.

495. 4. *Dasyurus Maugei* (the Dasyure of Maugé). Olive colour, spotted with white, except the tail, which is spotless.

Dasyurus Maugei, *Geoff. Ann. du Mus. t.* 3.

Dasyure gutté, *Desmarest, Nov. Dict. d'Hist. Nat. t.* 24.

Icon. —— ?

Inhabits New Holland.

495 5. *Das. Viverrinus* (the Viverrine Dasyurus.) Black, spotted with white; without spots on the tail.

Didelphis Viverrina, *Shaw's Gen. Zool.* v. 1. Dasyurus Viverrinus, *Geof. Ann. Mus.* iii.

Spotted Opossum, *Philips' Voy.* Tapoa tafa, or Tupha, *White's Journal of a Voyage to New Holland.* Dasyure ta-cheté, *Cuv.,* and Das. de White, *Cuv.*

Icon. *White, Jour. Shaw, Gen. Zool. pl.* iii.

Inhabits the vicinity of Port Jackson, in New Holland.

This species is the type of the genus Dasyurus, as established by Illiger.

497. 6. *Dasyurus Tafa* (the Tapoa tafa.) Uniformly grayish brown, without spots.

Dasyurus tafa, *Geoff. Ann. du Mus.* III. Tapoa Tafa, *Var. White, Journal,* Viverrine Opossum, *Shaw.*

Icon. *White, Jour. t.* 281. *Shaw, Zool. t.* 1. *p.* 491. *pl.* 111, upper figure.

Inhabits the vicinity of Port Jackson, in New Holland.

498. 7. *Das. Penicillatus* (Brush Dasyurus.) Ashy-gray ; the tail with long stiff black hairs toward the end.

Didelphis penicillatus, *Shaw, Gen. Zool.* v. 1. 502. Dasyurus penicillatus, *Geoff. Ann. du Mus.* Phascogale penicillata, *Tem. Monog.* 58.

Icon. *Shaw, Zool. t.* 1. *pl.* 3. *Schreb.* 152.

Inhabits New Holland.

499. 8. *Das. minimus* (the Dwarf Dasyurus.) Uniformly ashy-red ; each hair red at the point, dark cinerous at the base ; thumb of the posterior extremities larger, and the teeth more regular than in the other species.

Dasyurus minimus, *Geoff. Ann. du Mus.* III. Phascogale minimus, *Tem. Monag.* 59.

Dasyure nain, *Cuv. Reg. Anim.*

Icon. ——

Inhabits the southern part of Van Diemen's Land.

Obs. These two species form the *Phascogale* of M. Temminck.

Genus III. PARAMELES.

Cutting-teeth $\frac{10}{6}$ or $\frac{10}{8}$; the last on each side of the upper jaw very long ; of the lower, divided in half by a

194

groove; canine $\frac{1.1}{1.1}$, long; cheek-teeth $\frac{7.7}{7.7}$ or $\frac{8.8}{8.8}$; crowns acutely tubercular; head very long; toes of fore-feet five, distinct, the three middle largest, and the thumb nearly rudimentary; hind-feet longer than the fore; toes four; two internal, very small, united and enveloped in the skin, so that the claws only are to be seen; the third very long, with a strong claw, and the outer very small; tail long, pointed; base thick and naked beneath; not prehensile.

500. 1. *P. Nasuta* (the Long-nosed Pouched Badger.) Head very long; muzzle thin; nose prolonged beyond the jaws; cutting teeth $\frac{o}{8}$; fur above gray-brown; beneath white.

Parameles nasuta, *Geoff. Ann. Mus.* iv. 62. Thybais nasuta, *Illiger, Prod.*

Icon. *Geoff. Ann. Mus.* iv. *t.* 44, with skull and toes.

Inhab. New Holland.

501. 2. *P. Bougainvillia* (Bougainville's Pouched-Badger.) Head long; acute ears; ovate, long; body above red; beneath gray.

Parameles Bougainvilliæ, *Quoit and Gaimard.*

Icon. *Quoit and Gaimard, Freycinet's Voyage.*

Inhab. ——

502. 3. *P. Obesula* (the Fat Pouch-Badger, or Porculine Opossum of Shaw.) Head rather short; forehead convex; cutting teeth $\frac{1o}{8}$; fur above reddish-yellow; beneath white.

Didelphis Obesula, *Shaw, Gen. Zool.* i. 490. Isoodon, *Geoff. Dict. Hist. Nat. ed.* vi. Thylaris Obesula, *Illiger, Prod.* 76. Parameles Obesula, *Geoff. Ann. Mus.* iv. 64.

Icon. *Shaw, Nat. Misc. n.* 96. *t.* 298. *Geoff.* l. *c. t.* 45, with skull. *Ency. Méthod. Supp. t.* 9. *f.* 5.

Inhab. New Holland.

SUB-DIVISION II. *Cutting-teeth* $\frac{6}{2}$; *the lower very long;
canines in the lower-jaw, very small or wanting.*

Genus IV. PHALANGISTA.

Cutting-teeth $\frac{6}{2}$; canine $\frac{1-1}{0-0}$ or $\frac{0-0}{0-0}$; false grinders $\frac{2-2}{3-3}$ or $\frac{3-3}{2-2}$;
grinders $\frac{5-5}{5-5}$ or $\frac{6-6}{5-5}$; head elongate; forehead convex; feet
five-toed, not united to the body by the skin of the sides;
tail naked or covered with hair.

   * *Tail naked or scaly, prehensile.*

503. 1. *P. Maculata* (Spotted Phalanger.) Fur whitish,
spotted with brown or black, size of a Cat.

 Didelphis Orientalis, *Gmelin, Sys. Nat.* Cuscus Amboi-
nensis, *Lacepede.* Phalangista Maculata, *Geoffroy, Col.
Mus. Par. Desmar. Mam.* 266. Balantia Orientalis, *Illiger,
Prod.* 78.

 Phalanger male, *Buff. Hist. Nat.* XII. Surinam-Rat.
Cosceoes of the *Natives.*

 Icon. *Buffon, H. N.* XIII. *t.* 11. *Ency. Method. t.* 24. *f.* 1.
Inhab. Molluca, Java.

504. 2. *P. Rufa* (Red Phalanger.) Fur reddish or
whitish, with a darker dorsal line.

 Didelphis Orientalis, Var. *Gmelin, Sys. Nat.* Phalangista
Rufa, and Alba, *Geoff. Col. Mus. Par.* P. Rufa, *Desm.
Mam.* 266.

 Phalanger, femelle, *Buffon, Hist. Nat.* XIII.
Icon. *Buffon, H. N.* XIII. *t.* 10. *Ency. Method. t.* 24. *f.* 2.
Inhab. Molluca, Java.

505. 3. *P. Papuensis,* body above gray ; beneath yellowish-white ; upper part of the head gray ; throat and chest
white ; upper part of the extremities brown ; ears very
small, hairy.

Phalangista Papuensis, *Desm. Mam.* 541. P. Quoicy, *Gaimard, Bull Sci.* vii. 64.

Icon. ——

Inhab. New Guinea.

Obs This seems to be considered by M. Temminck as a variety of the last.

506. 4. *P. Ursina* (Bear-like Phalanger.) Fur thick; black, with a yellowish cast, caused by the end of the hairs being tipped with that colour; tail very furry; size of a Civette.

Phalangista Ursina, *Temminck, Monag. de Mamm.*

Icon. *Temminck, l. c. t.* 8. *and Osteology, t.* 4.

507. 5. *P. Chrysorrhos* (Yellow-tufted Phalanger.) Fur short, thick, cottony; head pale ash-gray; ears white above; blackish-gray, with a black line on each flank; rump and end of the tail bright golden yellow; beneath white; hinder part reddish.

P. Chrysorrhos, *Temminck, Monag. de Mamm. p.* 12.

Icon. ——

Inhab. Island of Celebres.

** *Tail hairy, prehensile.*

508. 6. *P. Vulpina* (Fox-like Phalanger.) Fur gray-brown above, passing into yellow-gray on the head and shoulders; gray beneath; tail tufted; base like the back, end black.

Didelphis Lemurina, *Shaw, Zool.* i. 487. D. Vulpina, *Shaw, l. c.* 363. D. Peregrinus, *Bod. Elench. Anim.* Phalangist Vulpina, *Desm. Nouv. Dict. Hist. Nat.* xxv. 475.

Wha Topoa Voo, *White, Jour. Voy. to New South Wales*, 278. Lemurine Opossum, *Shaw*, i. 487. New South Wales Opossum, *Bewick, Quad.* 376. New Holland Bear, *Pennant, Quad.* vi. 13. Vulpine Opossum, *Phillips, Voy.* 150. Le Bruno, *Vicq. d'Azyr. Sys. Anat. des Anim.* vi. 251.

Icon. *Shaw, Zool.* i. *t.* 110. *Phillips' Voy. t.* 16. *Bewicki, Quad.* i. 376. *F. Cuv. Dent. Mamm.*

Inhab. Port Jackson.

509. 7. *P. Cookii* (Cook's Phalanger.) Fur brown or reddish-gray above ; white beneath; tail brown, and white at the end ; size of the Pole cat.

Phalangista Cookii, *Desm. Mamm.* 268. Petaurista Cookii, *F. Cuv. Dent. Mamm.*

Opossum. *Hawskworth, Voy.* vii. 586. *Cook's last Voyage,* i. 108. White-tailed Opossum, *Shaw, Zool.* i. 504. Phalanger de Cook, *Cuv. Rég. Anim.* i. 779.

Icon. *Cook's Voy. t.* 4. *Ency. Méthod. Supp. t.* 8. *f.* 3.

Inhab. Van Diemen's Land.

This species differs from the former in its teeth according to M. F. Cuvier.

510. 8. *P. Nana* (Dwarf Phalanger.) Fur reddish-gray above ; white beneath ; tail brown.

Phalangista Nana, *Geoff. MSS. Desm. Mamm.* i. 268.

Icon. ——

Inhab. East Coast of Van Diemen's Land.

GENUS V.—PETAURISTA.

Cutting-teeth, $\frac{6}{2}$; lower horizontal ; canine $\frac{1-1}{0-0}$ or $\frac{1-1}{1-1}$; cheek-teeth $\frac{6-6}{6-6}$ or $\frac{7-7}{7-7}$; head rather long ; eyes small ; ears long; feet short; five toes ; the hinder with a large nailless thumb ; the two first toes short, united by a common skin ; claws compressed, arched ; skin of the sides extended and uniting the extremities, so as to form a parachute. Tail long ; hairs not prehensile.

* Tail round.

511. 1. *P. Taguanoides* (Petaurine Opossum.) Fur very soft; gray-brown, or shining-brown above ; throat and chest white ; tail brown, yellowish-brown at the base.

Didelphis Petaurus, *Shaw, Gm. Zool.* Petaurus Tagua-noides, *Desm. Nouv. Dict. Hist. Nat.* xxv. 400. Phalangista Petaurus, *Illiger, Prod.* 78. Petaurista Tauguanoides, *Desm. Mamm.* i. 269. Phalanger Tauguanoides, *Geoff. Col. Mus. Var.*

The Southern Petaurus, *Shaw, Zool. Misc.* Hepoona Koo, *White, Journal of a Voy. N. S. W.* 208. Grand Phalanger Volant, *Cuv. Rég. Anim.*

Icon. *Shaw, Zool.* i. *t.* 112. *White, Journal, t. Shaw, Nat. Misc. t.* 60.

Inhabits Port Jackson.

512. 2. *P. Macroura* (Large-tailed Petaurista). Fur gray-brown above, whitish beneath; tail thick; longer than the body; base brown, end black.

Didelphis Macroura, *Shaw, Gen. Zool.* i. 113. Petaurus Macrourus, *Desm. Dict. Hist. Nat.* xxv. 402.

Phalanger Volant a grande queue, *Cuv. Rég. Anim.* 1. 180.

Icon. *Shaw, New Holland Zoology, t.* 12. *Shaw, Gen. Zool. t.* 113. *Ency. Méthod. Supp. t.* 8. *f.* 4.

Inhab. New Holland.

513. 3. *P. Flaviventer* (Yellow-bellied Petaurista). Fur chesnut-brown above, yellowish-white beneath; tail chesnut-brown, round; a little longer than the body

Petaurus Flaviventer, *Desm. Dict. Hist. Nat.* xxv. 403. Petaurista Flaviventer, *Geoff. MSS. Desm. Mam.* 269.

Icon. ——

Inhab. New Holland.

514. 4. *P. Sciurea* (Squirrel Petaurista). Fur above ash-gray, edges of the parachute and dorsal line deep-brown; beneath white; head, yellow-gray; tail, reddish gray at its base, end black, crown revolute.

Didelphis Sciurea, *Shaw, Zool. of New Holland*, 29 ; *Gen. Zool.* 1 ; Petaurus Sciureus, *Desm. Dict. Hist. Nat.* xxv. 403. Phalangista Sciurea, *Illiger's Prod.* 78. Petaurista Sciurea, *Desm. Mamm.* i. 270.

Norfolk Island Squirrel, *Pennant, Quad.* Squirrel Opossum, *Shaw, Zool.*

Icon. *Shaw. New. Hol. Zool. t.* ii ; *Gen. Zool.* i. *t.* 113.

Inhab. New Holland.

515. 5. *P. Peronii* (Peron's Petaurista). Body brown above, white beneath ; parachute above brown, varied with gray, legs white ; end of tail white.

Didelphis Sciurea Var. *Shaw, Gen. Zool.* Petaurus Peronii, *Desm. Dict. Hist. Nat.* xxv. 404. Petaurista Peronii, *Desm. Mamm.* i. 270.

Icon. ——

Inhab. New Holland.

*** Tail feathery.

516. 6. *P. Pygmea* (Pygmea Petaurista). Fur uniform, mouse-gray ; with a reddish cast on the back, and a white one beneath.

Didelphis Pygmea, *Shaw, Zool. Nat. Hist.* 5. *Gen. Zool.* i. Petaurista Pygmæus, *Desm. Dict. Hist. Nat.* v. 405. Phalangista Pygmea, *Geoff. Col. Mus. Var. Desm. Mam.* 270. Acrobata Pygmea, *Gray, King's Voy. to New Holland.*

Icon. *Shaw, Gen. Zool. t.* 114. *Ency. Method. Suppl. t.* 8. *f.* 5.

Inhab. New Holland.

SUB-DIVISION III.—*Cutting-teeth* $\frac{6}{2}$; *the lower very long, shelving ; canine* $\frac{0.0}{1.1}$.

GENUS VI.—POTOROUS.

Cutting-teeth $\frac{6}{2}$; canine $\frac{1.1}{0.0}$ small ; grinders $\frac{5.5}{5.5}$. Head long, pointed. Ears large, upper-lid cut. Fore-legs short.

Toes five, sharply clawed; the third very large, with a large claw; the fourth moderate. Tail long, rather thick.

517. 1. *P. Murinus* (Kanguroo Rat). Fur brownish above, gray beneath.

Macropus Minor, *Shaw, Gen. Zool.* Hypsiprymus Murinus, *Illiger, Prod.* 79. Potorous Murinus, *Desm. Mamm.* 271. Kangurus Gaimardi, *Desm. Mam.* ii. 540.

Kanguroo Rat, *Philips' Voy. Bot. Bay*, 277. Potooroo or Kanguroo-Rat, *White's Journal of a Voyage to New South Wales*, 286. Lesser Kanguroo, *Penant, Quad.* ii. 32.

Icon. *Philips, l. c. t.* 47. *White, l. c. f.* 60. *Shaw's Gen. Zool. t.* 116. *Ency. Méthod. Suppl. t.* 962.

Inhab. New Holland.

M. F. Cuvier observes that there are three or four species of this genus. Messrs. Quoy and Guimard have named two, *P. Lesuerii*, from the head only, and *P. Peronii*, from the skeleton in the Paris Museum.

Genus VII.—Kangurus.

Cutting-teeth $\frac{6}{2}$, canine $\frac{0}{0}\frac{0}{0}$; cheek-teeth $\frac{5}{5}\frac{5}{5}$. Head elongated. Ears large, pointed; eyes large. Fore-legs very short; toes five, strongly clawed. Hind-legs very long, strong; toes four; the two inner small, united; the central very large, strongly clawed; the outer moderate; the metatarsus very long, thin; sole applied the whole length to the earth. Tail long, very strong; not prehensile, used in jumping.

518. 1. *K. Labiatus* (Large Kanguroo). Fur ash-gray above, white beneath, with an ash gray line across the chin; the legs and upper part of the tail blackish.

Didelphis Gigantea, *Gmelin, Sys. Nat. f.* 109. Macropus Gigantea, *Shaw, Zool.* 505. Halmaturus Gigantea, *Illiger. Prod.* Kangurus Labiatus, *Geoff. Mus. Desm. Mamm.* 273.

Kanguroo, *Cook's First Voyage.* vi. 277. Great Kanguroo, *Shaw, l. c.*

Icon. *Cook's First Voyage,* vi. *f.* 20. *Philips' Voyage,* t. 10. *White, Jour. of a Voy.* t. 54. *Shaw. Zool. Ency. Method.* t. 21. *f.* 4.

Inhab. Botany Bay. Discovered by Captain Cook in 1770.

519. 2. *Kangurus Fuliginosus* (Sooty Kanguroo). Fur sooty, brown above ; clear gray beneath the legs, and tail blackish.

Kangurus Fuligiosus, *Peron and Leseuer, Desm. Mamm.* 273. Macropus Giganteus, *F. Cuv. Desm. Mamm.*

Kanguroo Giant, *F. Cuv. Mamm. Lithog.*

Icon. *F. Cuv. l. c.*

Inhab. South Coast of New Holland, and near Port Jackson.

520. 3. *Kangurus Rufus* (Red Kanguroo). Fur woolly, clear read bove, white beneath.

Kangurus Rufus, *Desm. Mam.* i. 541. K. Lanosus, *Guimard, Sor. Hist. Nat. Par. Bul. Sci.*

Icon. ——

Inhab. the Blue Mountains in the interior of New Holland.

521. 4. *K. Griseus* (Gray Kanguroo). Fur reddish gray above, paler beneath ; legs and end of the tail becoming brown ; lower part of the tail reddish gray.

Kangurus Rufogriseus, *Peron and Leseuer.* K. Griseus, *Desm. Mamm.* 273. Macropus Rufogriseus, *F. Cuv. Desm. Mamm.*

Inhab. New Holland.

522. 5. K. Ruficollis (Red-necked Kanguroo). Fur hare-gray above, pure white beneath. Neck and upper part of the shoulders red, variegated with gray; beneath the tail red.

Kangurus Ruficollis, *Peron and Lesseuer, Desm. Mam.*, I. 274. Macropus Ruficollis, *F. Cuv.*

Icon. ——

Inhab. King's Island.

523. 6. K. Eugenii (Eugene's Kanguroo.) Fur gray-brown above; front of the fore-legs variegated with red; below whitish; lower part of the tail reddish-white.

Kangurus Eugenii, *Peron and Leseuer, Voy.* VI. 117. *Desm. Mam.* 274.

Icon. ——

Inhab. Isle of Eugenia, New Holland.

Considered by most of the French naturalists as the young state of the former species.

524. 7. K. Fasciatus (Banded Kanguroo.) Fur gray, with brown band across the back and loins.

Kangurus Fasciatus, *Peron and Leseuer, Voy. aux Tems Aust.* I. 114. Halmaturus Fasciatus, *F Cuv. Desm.*

Kanguroo, *Dampier Voyage to New Holland,* IV. 111. Kanguroo Elegant, *Cuv. Col. Mus. Par.*

Icon. *Peron. l. c. Atlas, t.* 27. *Dict. Hist. Nat.* XVII. *t.* 22.

Inhab. East Coast of New Holland.

M. F. Cuvier considers it the type of a genus distinct from *Macropus*, for which he has adopted the name of *Halmaturus,* used by Illiger for the Kanguroos.

525. 8. Kangurus Billardierii (Labellardiere's Kanguroo.) Ears short, oval, rounded; fur uniform; gray-brown above; reddish beneath; upper-lip reddish.

Kangurus Billardierii, *Desmarest, Mamm.* 542.

Icon. ——
Inhab. Van Diemen's Land.

526. 9. *Kangurus Brunii* (Le Brun's Kanguroo.) Fur brown above; yellow beneath.

Didelphis Brunii, *Gmelin, Sys. Nat.* i. 109. D. Asiatica, *Pallas, Art. nov. Petrop.* Kanguros Bicolor, *Mus. Paris.* Kangurus Brunii, *Desm. Mam.* 278. Halmaturis Brunii, *Illiger, Prod.* 80.

Felander, *Valentyn Amboyne,* vi. 275. *Le Bruyn, Voy. aux Indes,* 374. Javan Opossum, *Pennant, Quad.* vi. 22. *Shaw,* i. 402.

Icon. *Le Bruyn, Voyage aux Ind.* 374. t. 213.
Inhab. Aroe Islands.

527. 10. *Kangurus Pencillatus* (Tufted-Tailed or Mountain Kanguroo.) Fur above gray, variegated with darker tint; beneath rufous-brown; feet dark; tail as long as the body, and tufted at the end; head dark gray, with a dark longitudinal dorsal line, and a pale spot on the cheek and under the throat.

Icon. nobis, from a drawing by Lewin, made in New Holland.
Inhab. New Holland. In the collection of the Linnean Society.

The Macropus Lanigerus (Woolly Kanguroo) noticed at Vol. III. p. 49, is not inserted in the table, in the existing absence of more decided particulars.

Genus VIII.—Phascolarctos.

Cutting-teeth $\frac{6}{2}$, false-grinders $\frac{2\cdot2}{2\cdot2}$, grinders $\frac{4\cdot4}{4\cdot4}$, with two tubercles; ears large and pointed; feet with five toes, the fore-feet parted into two groups; the thumb and under-finger on one side ; and the three others on the opposite ; the hind-feet with a large distinct clawless thumb, and the two inner fingers small, united to the claws. The Baron Cuvier describes the animal as thumbless.

Mr. Gray observes, that the skull of specimen which he examined " had a short canine-tooth in the upper jaw, and grinders $\frac{4.4}{5.5}$, all with two fangs ; the first on each side small, rather compressed, and the rest depressed with acute tubercles, so that they exactly agreed with the Potorus in the number, but the length of the jaws were more equal, and the skull was compressed and depressed, so as to be subquadrangular. The temporal fossæ larger." Mr. Gray refers the animal to the group of Phalangers.

528. 1. *K. Koala*, (The Koala.)   Of an uniform chocolate colour ; fur long, thick, and harsh ; size of a moderate dog.

Phascolarctos Fuscus, *Blainville? Desm. Mamm.* 276 ? Lepurus Cinereus, *Goldfuss Schreb. South.*

Koala, *Cuv. Reg. Anim.* ii. 184.   Coala or Koala, *Blainville?* Koala, or New Holland Sloth.

Icon. *Cuv. Reg. Anim.* iv. *t.* 1. *Ency. Method. Supp. t.* 9. *f.* 4. Inhab. New Holland.

Sub-Division IV.—E. *Cutting-teeth* $\frac{2}{2}$, *canine-teeth* $\frac{0.0}{0.0}$.

PHASCOLOMYS.

Cutting-teeth $\frac{2}{2}$, very strong and thick, short; the upper ones converging at their tips, canine $\frac{0.0}{0.0}$, grinders $\frac{5.5}{5.5}$, separated from the cutting teeth by an empty space; crown oval, flat, separated into two by a groove; body thick; head large, flat; ears short; feet with five toes; claws of the forefeet strong; the thumb of the hind-feet small, indistinct, clawless; tail very short or reddish-brown.

530. 1. *P. Wombat*, (The Wombat.) Fur uniform, grayish ; about as big as a Badger.

Phascolomys Wombat, *Peron and Leseuer, Voy. aux Terres Aust.* Wombatus, *Fossor Geoff. Ann. Mus.* vi. 364. Phascolomys fusca, *Desm. Dict. Hist. Nat.* xxv.

Wombat, *Colonists.*

Icon. *Peron and Leseuer Atlas, t.* 28. *Ency. Method. Supp. t.* 9. *f.* 1. *Leach, Zool. Misc. Cuv. Reg. Anim.* iv. *t. skull.*

Inhab. King's Island, and near Port Jackson, New Holland.

Obs. The Wombat described by Bass and Flinders is said to have six cutting and two canine-teeth in each jaw, from which Illiger formed his genus *Amblotis.* Is it different from the above ?

Since the Synopsis of Felinæ was printed, M. Temminck's monograph of the genus came to hand. We are unable therefore to avail ourselves of the result of that eminent zoologist's observations, further, than by inserting here a marginal notice of the several species which, on repeated inspection and comparison, he admits into the genus. These are, 1. *F. Leo ;* 2. *F. Tigris ;* 3. *F. Jubata ;* 4. *F. Leopardus,* (apparently our Panther.) 5. *F. Pardus,* (apparently our Leopard.) 6. *F. Macrocelis,* (the Nebulosa of Hamilton Smith.) 7. *F. Serval and Capensis ;* 8. *F. Cervaria,* (probably the Siberian Lynx figured in this work.) 9. *F. Borealis,* (probably the Canada Lynx.) 10. *F. Lynx ;* 11. *F. Pardina,* (the Loup Cervier of Perrault.) 12. *F. Caracal ;* 13. *F. Aurata,* (an inedited species, bay-red above, sprinkled with little spots on the sides ; tail half as long as the body, with a brown band down the upper side ; tip black.) 14. *F. Chaus ;* 15. *F. Caligata,* (the Booted Lynx.) 16. *F. Catus ;* 17. *F. Maniculata,* (apparently the F. Catenata of Hamilton Smith.) 18. *F. Minuta,* (the Kuwuk of Java.) 19. *F. Puma ;* 20. *F. Onca ;* 21. *F. Jaquaramdi ;* 22. *F. Celidogaster,* (apparently the F. Chalybeata of Hamilton Smith.) 23. *F. Rufa ;* 24. *F. Pardalis ;* 25. *F. Macroura ;* 26. *F. Mitis ;* 27. *F. Tigrina.*

ORDER IV.—RODENTIA.

Two large incisors in each jaw, separated from the cheek-teeth by a void space, and which wear by use, and grow again on the inner side. No canine teeth; cheek teeth in some genera with flat or ridged crowns, in others, with blunt tubercles. Lower jaw articulated by a longitudinal condyle; orbits not separated from the temporal fossæ; zygomatic arches small; toes variable in number; nails unguiculated; stomach simple; intestines long; cæcum large.

Eats in general vegetable matter, but the species with tuberculated teeth are nearly omnivorous.

Habits various, generally timid.

Inhabits the Continents, and larger islands, but not those of the South Sea.

SECTION I.—*With Clavicles.*

GENUS I.

CASTOR.—Incisive teeth, $\frac{2}{2}$; canines, $\frac{0}{0}\frac{0}{0}$; cheek-teeth, $\frac{4}{4}\frac{4}{4}=20$. Incisive teeth very strong, with the anterior surface flat, and the posterior, angular; the cheek-teeth with a sort of fold or ridge of enamel on the internal edge, and three similar folds on the external edge of the upper teeth, which apparent folds are inversed in the lower teeth; eyes small; ears short and round; five toes on all the feet, the anterior short and close, the posterior longer and palmated; tail large, flat, naked, and scaly; a pouch, into which an unctuous matter is secreted near the genitals of the male.

531. 1. *Castor Fiber* (the Beaver.) Rather larger than the Badger, uniformly reddish brown, with a shorter downy gray fur.

207

Castor fiber, *Lin* Καστωρ, *Arist. Hist. Anim.* Fiber, *Pliny, l.* 8, *c.* 30.

Beaver, *Ray, Synop.* 209. Le Castor ou le Biévre, *Buff. t.* 8. 282.

Icon. *Screb. tab.* 175. *Buff. l. c. pl.* 36. *F. Cuvier, Mam. lithog. Pennant, Br. Zool.* I. *pl.* 9, *and Hist. Quad. pl.* 71.

Inhabits North America, and the vicinity of some of the larger rivers of Europe, as the Rhine, the Rhone, and the Danube.

Obs. The Beaver is considerably subject to vary in colour, thus M. Geoffroy notes that the Beaver of France is generally of an olivaceous yellow, Brisson *(Regne Anim.* 135) describes the white Beaver. Black, spotted, and yellow varieties, have also been noticed.

Genus II.—Mus, Lin.

Incisive teeth, $\frac{2}{2}$; canine teeth, $\frac{0-0}{0-0}$; cheek-teeth varying in the different subgenera; anterior toes, four or five; posterior, five; anterior limbs furnished with clavicles.

Sub-Genus I.—Fiber, *Cuv.—Incisive teeth,* $\frac{2}{2}$; *canines,* $\frac{3-3}{3-3}$; *cheek-teeth,* $\frac{3-3}{3-3}$; *the lower incisors sharp pointed, and convex in front, cheek-teeth with flat tops, furnished with scaly transverse zigzag laminæ; anterior feet with four toes, and the rudiment of a thumb; posterior, with five, with the edges furnished with stiff hairs, used in swimming like the membrane of palmated feet; tail long, compressed laterally; an oderiferous unguent secreted in both sexes.*

532. 1. *F. Zibethicus* (the Ondatra, or Musk Arvicola.) About the size of a Rabbit; reddish gray, ashy underneath.

Castor Zibethicus, *Lin.* Mus Zibethicus, *Gm.*

L'Ondatra, *Buff. t.* 10. Rat musqué, *Sarrazin Mem. de l'Acad.* 1725, 323, Rat Musqué de Canada, *Briss. Régn. Anim.* 136. Musk Rat, *Lawson's Carolina,* 120. Musquash,

Josselyn's New England. Massascus, *Smith's Virginia,*27. Musk Beaver, *Pen. Quad.* ii. 118.

Icon. *Sarrazin, l. c. tab.* ii. *Buff. l. c. pl.* 1.

Inhabits Canada, and other parts of North America.

Sub-Genus II.—Arvicola, Lacep. (Campagnols ordinaires, Cuv.) *Teeth like those of last sub-genus; but the hind feet have not the stiff hair or swimming apparatus; tail round and hairy.*

533. 1. *A. Amphibius* (Water-Rat.) Blackish gray, slightly tinted with yellow, lighter underneath; tail black; rather larger than a common Rat.

Mus Amphibius, *Lin. Syst. Nat.* 82, *and Faun Suec, No.* 32. Mus aquaticus, *Briss. Régn. Anim.* 175. Mus aquatilis, *Ray, Synops.* 217.

Rat d'Eau, *Buff. t.* 7. Wasser-maus, *Kramer, Austr.* 316. Water-Rat, *Pen. Br. Zool.* i. *No.* 27.

Icon. *Belon.* 30, *tab.* 31, *Buff. l. c. tab.* 43, *Screb.* 186.

Inhabits the whole of Europe, Northern Asia, and North America.

Var. a. Niger, Lin. Inhabits Siberia.

Var. b. Maculata, Pallas, yellowish, with a large white spot on the shoulder.

Var. c. Paludosa, Lin. black, feet white.

534. 2. *A. Arvalis* (the Field-Mouse.) About as big as the common Mouse; reddish ash colour; ears small and round.

Mus agrestis, *Ray, Syn.* 218. Mus terrestris, *Lin. Syst.* 82. Mus campestris minor, *Briss. Regn. Anim.* 176. Mus arvalis, *Pallas, Nov. sp. fasc.* i. 78.

Campagnol, *Buff. Hist. Nat. t.* 7. Short-tailed Field Mouse, *Pen. Br. Zool.* i. *No.* 31, and Meadow Rat, *Quad.* ii. 205.

Icon. *Buff. l. c. tab.* 47, *Screb.* 191.
Inhabits Europe and Northern Asia

535. 3. *A. Œconomus* (Œconomic Rat.) Rather larger than the last, and the females larger than the males; brown above, yellowish underneath ; ears short ; tail about one-fourth the length of the body.

Mus œconomus, *Pallas, Nov. Spec. Glires, n.* 125.

La Fegoule, *Vicq-d'Azyr, Syst. des Anim. t.* 2. 389. Œconomic Rat, *Pen. Quad.* 194.

Icon. *Pallas, l. c. pl.* 14, *A. Screb. tab.* 190.

Inhabits Siberia and Eastern Asia, in the deep and humid valleys. M. Bosc found a specimen in the forest of Montmorency, which he refers to this species.

536. 4. *A. Saxatilis* (the Rock-Rat.) About four inches long ; brown, mixed with gray above, deep gray on the sides, and whitish underneath ; tail as long as the body.

Mus saxatilis, *Pallas, Glires, p.* 80 *and* 256.

Le Saxin, *Vicq-d'Azyr, Syst. des Anim. t.* 2, 452. Rock-Rat, *Pen. Quad.* 192.

Icon. *Pallas, l. c. pl.* 23, *B. Screb.* 185.

Inhabits Siberia.

537. 5. *A. Alliarius* (Garlic Mouse.) About four inches long ; tail, one and a half ; ashy-gray above, white underneath ; ears large, nearly denuded.

Mus alliarius, *Pallas Glires,* 251.

L'Alliaire, *Vicq-d'Azyr, Syst. des Anim. t.* 2, 393. Garlic Rat, *Pen. Quad.* ii. 197.

Icon. *Pallas, l. c. pl.* 14. *C. Schreb.* 187.

Inhabits Siberia.

538. 6. *A. Rutilus* (Red Mouse.) Rather less than the last ; reddish above, pale white underneath ; ears moderate.

Mus rutilus, *Pallas, Glires*, 248.

Le rona, *Vicq-d'Azyr, Syst. des Anim. t.* 2, 402. Campagnol doré ou roux, *Desm. Nov. Dict. d'Hist. Nat.* Red Rat, *Pen. Quad.* ii. 198.

Icon. *Pallas, l. c. pl.* 14, *B. Screb.* 188.

Inhabits Siberia, and extensively in Northern Asia.

539. 7. *A. Gregalis* (Baikal Mouse.) About three inches long ; pale gray on the back, with long black hairs intermixed, sides paler, belly white ; tail black, one fourth the length of the body ; ears large.

Mus gregalis, *Pallas Glires,* 238. Le Gregari, *Vicq-d'Azyr, Syst. des Anim. t.* 2, 400. Baikal Rat, *Penn. Quad.* ii. 204.

Icon. *Pallas, l. c. pl.* 17, *Screb. t.* 189.

Inhabits Eastern Siberia.

540. 8. *A. Socialis* (Social Rat.) About three inches long ; tail an inch ; pale gray above, white underneath ; ears short, but broad.

Mus socialis, *Pallas, Glires.* 218. Mus gregarius, *Lin. Syst. Nat. Ed.* 11, 84. Mus terrestris, *var. Erxleban, Syst. Nat.* 397.

Campagnon, *Vicq-d'Azyr, Syst. des Anim. t.* 2, 397. Social Rat, *Pen.* 203.

Icon. *Pallas, l. c. t.* 13, *B.*

Inhabits the vicinity of the Caspian Sea.

541. 9. *A. Pumilio* (Lineated Mouse.) Bright brown above, marked with four longitudinal black bands.

Mus pumilio, *Gmel. Syst.* 130, *Sparman, Voy. t.* 2. 376.

Lineated Mouse, *Pen. Quad.* 2, 191. *Rat nain du Desmarest, Nov. Dict. d'Hist. Nat.*

Icon. *Sparman, l. c. pl.* 9, *and Act. Stock.* 1784, *t.* 6, *Pennant, l. c. pl.* 82.

Inhabits South Africa, eastward of Cape of Good Hope.

542. 10. *A. Albicaudatus* (White-tailed Mouse.) About five inches long; brown, with the paws and upper side of the tail white; tail half the length of the body.

Lemnus albicaudatus, *Geoff. Catt. de la Collection du Mus.*

Icon. ——

Inhabits

543. 11. *A. Niloticus* (Egyptian Arvicola.) Brown, intermixed with yellow on the upper parts, yellowish gray underneath; tail brown, nearly as long as body; ears large, denuded, brownish.

Lemmus niloticus, *Geoff. Descript. de l'Egypte.*

Icon. *Geoff. l. c.*

Inhabits Egypt.

544. 12. *A. Fulvus* (Yellow Arvicola.) About four inches long; reddish yellow; belly and paws more yellow tail less than half the length of the body.

Lemmus Fulvus, *Geoff. Catal. de la Coll. du Mus.*

Icon.

Inhabits France.

Desmarest inserts a mark of doubt on the following species.

545. 13. *A. Argentoratensis* (the Schermaus or Strasbourg Arvicola.) Six inches long; tail about two; dusky gray; eyes small; external ears scarcely visible; edge of the mouth fringed with white.

Schermaus, *Herman.* Schermaus, *Buff. Sup. t. 7.* F. Cuvier, *Dict. des Sciences, Natural, t. 6.* Arvicola argentoratensis, *Desm. Ency. Méthod. Mam. sp. 436.*

Icon. *Buff. l. c. pl. 70.*

Inhabits the vicinity of Strasbourg.

546. 14. *A. Xanthognatus* (Yellow-cheeked Arvicola.)
Yellow, varied with black on the upper parts; ashy-gray
underneath; cheeks yellow. Length five inches.

Lemmus Xanthognatus, *Leach, Nat. Miscel.* i. Arvicola
Xanthognatus, *Desm. Ency. Méthod. sp.* 441. A. Xan-
thognata, *Harlan.*

Icon. *Leach, l. c. t.* 26.

Inhabits the shores of Hudson Bay.

Obs. Perhaps a variety of the common species.

547. 15. *A. Hortensis* (Garden Campagnol.) Body above
ferrugineous brown; sides lead-coloured; underneath yellow; hairs coarse, standing more or less obliquely from
the body, giving the animal a shaggy appearance; ears
broad, oval; head globular; snout contracted, conical;
tail more than half as long as the body. Length of body
and head five inches and a half, tail two inches and a half.

Arvicola hortensis, *Harlan, Faun. Amer.* 138.

Sigmodon, *Say et Ord. Jour. Acad. Nat. Sci. Phil.* iv.

Icon. *Jour. Acad. N. S. Phil. t.*

Inhabits Florida.

Obs. This species is the type of Mr. Say and Ord's genus
Sigmodon, which only differs from Arvicola in some slight
variation in the form of the plates of the teeth.

548. 16. *A. Palustris* (Marsh Campagnol.) Body above
dark grayish-brown; beneath pale lead-coloured; snout
rather elongated, reddish-brown at its extremity; ears moderately long, slightly edged with hair; tail short, slightly
hairy.

Arvicola Palustris, *Harlan, Faun. Amer.* 136. Arvicola
Riparius, *Ord. Jour. Acad. Sci. Phil.* iv.

Icon. ——

Inhabits the shores of the Delaware, living on the seed
of the wild rice.

549. 17. *A. Pennsylvanica* (Pennsylvanian Campagnol.) Fur above brownish-fawn; beneath grayish-white; eyes very small; ears short and round.

Arvicola Pennsylvanica, *Ord. Guthrie's Geogr.* Myonotes Pratensis, *Raffinesque.* Campagnol of Pennsylvania, *Warden's Descrip. Unit. States,* v. 625.

Icon. *Wilson. Ornith.* vi. *t.* 50, *f.* 3.

Inhabits Pennsylvania.

Obs. Perhaps a variety of *A. Xanthognatus,* or the common species.

550. 18. *A. Astrachanensis* (Astracan Arvicola.) Yellow above; ashy underneath. Length four inches; tail one.

Mus Astrachanensis, *Erxleb. Syst.* 493. Arvicola Astrachanensis, *Desm.* 485. Maus-gottung, *S. G. Gmel. Rei.* ii. 173.

Icon. *Gm. l. c. tab.* 11.

Inhabits the vicinity of Astracan.

551. 19. *A. Floridanus* (Florida Campagnol.) Lead colour, mixed with black on the dorsal line; yellowish on the flanks; ears large and membranous; fur very soft and fine; tail little more than half the length of the body. About eight inches long.

Mus Floridanus, *Ord. Nouv. Bull. de la Société Philomatique.* 1818. Arvicola Floridanus, *Harlan, Faun. Amer.* 142. Neotoma Floridana, *Say et Ord. Jour. Acad. N. S. Phil.* iv.

Icon. *Jour. Acad. N. S. Phil. t.*

Inhabits Florida.

Obs. This is the type of the Genus Neotoma lately established by Messrs. Say and Ord, which differs from Arvicola in the teeth being furnished with roots.

The Guangue of Molina, 281 (Mus Cyanus, Gmel. 132, the Sky-coloured Rat of Pennant, Quad. 183), blue above and white underneath, and the Mus Microuros of Erxleben, Syst. Mam. 403, are referred conditionally by M. Desmarest to the Arvicolæ, as are also, by the same writer, the Lemmus Vittatus, the Lemmus Talpoides, and the Lemmus Novaboracenis of Rafinesque, Annals of Nature, Nov. 1820, Nos. 9, 10, and 11.

LEMMUS. *Incisives* $\frac{2}{2}$; *canines,* $\frac{0.0}{0.0}$; *cheek-teeth* $\frac{3.3}{3.3}$. *This sub-genus differs from the preceding only in the character of the fore-feet, which, in some species, have five, and in others four toes; but the nails are fitted for digging in all, whence Illiger named them Georychus, or Diggers.*

552. 1. *Lemmus Norvegicus* (the Lemming.) Reddish-yellow, black, and tawny, irregularly spotted or clouded; five toes on the fore-feet; thumb-nail large and strong. Length of body five or six inches, of the tail about half an inch.

Mus Norvegicus, *Ray. Syn.* 227. Cuniculus Norvegicus, *Briss. Quad.* 100. Mus Lemmus, *Lin. Syst. Nat.* 80. Glis Lemmus, *Erxl.* 371. Lemmus Norvegicus, *Desm.* 287.

Le Lemming, *Buff. Hist. Nat. t.* 13. Lemmar vel Lemmus, *Olaus Magnus de Gent. Septent.* 358.

Icon. *Pallas, Glires, tab.* 12, A. *Schreber,* 195. *Ency. Méthod. t.* 67, *f.* 6.

Var. a. One-fourth less than the other; dark above, white underneath, with a lighter band passing from the nose to each ear; tail short. *Pallas, Glires, tab.* 12, B. Inhabits Lapland.

553. 2. *L. Aspalax* (The Zokor Lemming.) Body reddish-gray; tail short; fore-feet pentadactylous; the three intermediate nails very long and arched; eyes very small.

Mus Aspalax, *Pallas, Glires,* 165.

Le Zokor, *Vicq. d'Azyr. Syst. des Anim. t.* 11. 585. Daurian Rat, *Pen. Quad.* 216.

Icon. *Pallas, l. c. t.* 10. *Schreb. t.* 205.

Inhabits the Altaic mountains and the vicinity of Lake Baikal.

554. 3. *L. Lagurus* (Hare-tailed Lemmus.) Ashy-gray, with a black dorsal line and no collar ; fore-feet with five toes ; nails not very strong, that of the thumb short and round ; tail very short ; ears moderate.

Mus Lagurus, *Pallas, Glires*, 210. Glis Lagurus, *Erxleb. Syst. Mam.* 375. Lemmus Lagurus, *Desm. Ency. Mam. sp.* 455.

Hare-tailed Rat, *Pen. Quad.* 202. Le Lagure, *Vicq. d'Azyr. Syst. des Anim.* 11. 363.

Icon. *Pallas, l. c. tab*. 13, A. *Schreb. tab.* 193.

Inhabits about the river Irtish, in Siberia, and the Deserts of Tartary.

555. 4. *L. Talpinus* (Talpine Lemming.) Dusky or gray-brown above, whitish underneath ; fore-feet with five toes, armed with moderate digging claws ; tail very short ; eyes small.

Mus Talpinus, *Pallas, Glires*, 176. Spalax Minor, *Erxleb· Syst. Mam.* 377. Lemmus Talpinus, *Desm. Ency. Mam.* 288.

Le Sukerkan, *Vicq. d'Azyr. Syst. des Anim.* 490.

Icon. *Pallas, l. c. tab.* 11, A.

Inhabits temperate parts of Russia and Western Siberia.

Var. A. Of an uniform black colour.

556. 5. *L. Hudsonius* (the Hudson's Bay Lemming.) Cinereous, tinged with tawny on the back, with a dusky stripe down the middle ; belly pale ash ; four toes, and the rudi-

ment of a thumb on the fore-feet; the two middle nails very large, and apparently double or divided; tail very short.

Mus Hudsonius, *Pallas, Glires,* 208.   Lemmus Hudsonius, *Desmarest, Mam. sp.* 453.

Hudson's Bay Rat, *Pen. Quud.* 201.

Icon. *Pallas, l. c. tab.* 26, *fig. A, B,* and *C.   Schreb.* 194.

Inhabits Labrador, Canada.   Mus. Brit.

557. 6. *L. Torquatus* (Ringed Lemming.) Ferruginous, with a black dorsal line, and a white collar, imperfect underneath; ears very short; five toes before; nails moderate, that of the thumb short and rounded.

Mus Torquatus, *Pallas, Glires,* 206. Lemmus Torquatus, *Desm. Ency. Mam. sp.* 454.

Ringed Rat, *Pen. Quad.* 201.   Le Collier, *Vicq. d'Azyr. Syst. des Anim. t.* ii. 368.

Icon. *Pallas, l. c. pl.* 11, *B.   Schreb.* 194.

Inhabits the vicinity of the Oby, in Siberia.

558. 7. *L. Terrestris* (Land Lemming.) Blackish-gray, slightly variegated with yellow; paler beneath; tail black.

Mus Amphibius Terrestris, *Lin. Syst. Nat.* 82. Lemmus Terrestris, *F. Cuv.*

Icon. *F. Cuv. Mam. Lithog.*

Inhabits Europe.

ECHIMYS *.   Geoff. *Incisive teeth* $\frac{2}{2}$; *canines* $\frac{0\,0}{0\,0}$; *cheek-teeth* $\frac{4\,4}{4\,4} = 20$. *Head long; eyes large; ears shortish; no cheek-pouches; four toes, and the vestige of a thumb on the fore-feet, five on those behind; tail long, and generally scaly; back covered with shortish spines, more or less abundant.*

* This word (Spiny Rats) is equally applicable to a few species of the Rats proper, not included in this subgenus.

559. 1. *E. Chrysurus* (Gilt-tailed Echimys.) Brown-red, head deep brown, with a narrow white band down the middle, white underneath; tail longer than the body, black, with the posterior half yellowish or white, woolly.

Myoxus Chrysurus, *Bodd. Elench.* 122. Histrix Chrysurus, *Schreb.* Echimys Cristatus, *Desmarest, Ency. Méthod. Mammalogie, p.* 291. Loncheres Chrysuros, *Lichtenstein, Tr. Acad. Berl.* 1818.

Le Lerot à queue dorée, *Buff. Sup.* VII. 283. Gilt-tailed Dormouse, *Pen. Quad.* 162.

Icon. *Buff. l. c. tab.* 72. *Ency. Méthod. t.* 78, *f.* 4.

Inhabits Surinam.

560. 2. *E. Rufus* (Red Echimys.) Fur dark brown, mixed with red above, and white beneath. Length about eight inches; tail not quite half that length. Males larger than the females.

Loncheres Rufa, *Lichtenstein, Tr. Acad. Berl.* Echimys Spinosus, *Desmarest, Ency. Méthod. Mammalogie,* 291.

Spiny Rat, or first Rat of *d'Azara, Quad. du Paraguay,* 73. Echimys Roux, *Cuv. Règn. Anim.* 175. Angouya-y-bigoni of Paraguay.

Icon. *D'Azara's Voyage, pl.* 13.

Inhabits South America.

M. Geoffroy distinguishes the following as species.

561. 3. *E. Dactylinus* (Long-toed Echimys.) Fur of the back deep brown, mixed with gray and yellow, red on the flanks; two middle toes of anterior feet much longer than the rest ; tail longer than the body.

Echimys Dactylinus, *Geoff. Nouv. Dict. d'Hist. Nat. t.* 10, 57. Loncheres Myosuros? *Lichtenstein, Trans. Acad. Berl.* 1818, 192.

Icon. *Lichtenstein, l. c. t.* 1, *f.* 2.

Inhabits South America.

562. 4. *E. Hispidus* (the Rough-haired Echimys.) Brown-red; lighter underneath; head reddish; tail as long as the body; scaly; hairs of the back very rough. About seven or eight inches long.

Echimys Hispidus. *Geoff. Nouv. Dict. d'Hist. Nat.* x. 58. Loncheres Paleacea, *Lichtenstein, Tr. Acad. Ber.* 1818. 191.

Icon. *Lichtenstein, l. c. t.* 1. *f.* 1.

Inhabits South America.

563. 5. *E. Didelphoïdes.* Brown on the back; lighter on the flanks; yellowish underneath; tail as long as the body; the tip, and for about one-seventh of its length, scaly; the rest hairy.

Echimys Didelphoides, *Geoff. Nouv. Dict. d'Hist. Nat.* 10. 58.

Inhabits America.

564. 6. *E. Cayennensis* (Cayenne Echimys.) Red, passing into brown, toward the middle of the back; belly white; hind-feet with long tarsi, and with the three middle toes of equal length.

Echimys Cayennensis, *Geoff. Coll. du Mus.*

Inhabits South America.

565. 7. *E. Setosus* (Bristly Echimys.) Fur red, soft, and but little intermixed with spines; under-part white; end of the feet white; tail rather longer than the body; posterior tarsi long.

Echimys Setosus, *Nouv. Dict. d'Hist. Nat. t.* 10. *pa.* 59, from Geoffroy.

Inhabits South America.

Myoxus, (Dormice.) *Incisors,* $\frac{2}{2}$; *canines,* $\frac{0 0}{0 0}$; *cheek-teeth,* $\frac{44}{44} = 20$, *divided by transverse bands; eyes large and prominent; ears large, round; long mustachios; no cheek-pouches; fore-feet with four toes, and the rudiment of a thumb; posterior with five; tail long, more or less villose; fur soft. No cæcum, or large intestines.*

566. 1. *M. Glis* (the Fat Dormouse.) Gray-brown-ashy above, whitish underneath, with brown round the eyes; tail very villose its whole length; about six inches long.

Glis, *Brisson, Règne Anim.* 113. Sciurus Glis, *Lin.* 12 *ed.* Mus Glis, *Pallas, Glires,* 88. Sciurus Epilepticus, *Klein, Quad.* 54. Myoxus Glis, *Gmelin.*

Loir, *Buff.* VIII. Fat Dormouse, *Pen. Quad.* II. 159.

Icon. *Buff. l. c. pl.* 24. *Schreber tab.* 22. *Ency. Méth. t.* 78. *f.* 1.

Inhabits Southern parts of Europe.

Mus Brit.

Note. According to the Baron Cuvier, the Myoxus Dryas of *Gmelin, Schreb. t.* 225. B. is only a variety of this species.

567. 2. *M. Nitela* (Garden Dormouse.) Fur gray-brown above, white underneath; black round the eyes to the shoulders; tail tufted, black. with the tuft white.

Mus Avellanarum Major, *Ray.* Mus Quercinus, *Lin.* Mus Nitidula, *Pallas, Glires, t.* 88. Myoxus Nitela, *Gm.* Sciurus Quercinus, *Erxleb.*

Greater Dormouse, or Sleeper, *Ray, Quad.* 219. Garden Dormouse, *Pen. Quad.* II. 159. Lerot, *Buff.* VIII. *tab.* 25.

Icon. *Buff. l. c. tab.* 24. *Ency. Méth. t.* 78. *f.* 3.

Inhabits the temperate parts of Europe.

Mus. Brit.

568. 3. *M. Avellanarius* (Common Dormouse.) Brownish yellow above, white underneath; hairs of the tail disposed like a feather; tail as long as the body, and flatted horizontally.

220

Mus Avellanarius Minor, *Ray.* Mus Avellanarius, *Lin.* Sciurus Avellanarius, *Erxleb.* Myoxus Muscardinus, *Gm.*

Muscardin, *Buff.* Croque-noix, *Brisson, Règ. Anim.* 162. Dormouse, or Sleeper, *Ray, Quad.* 220.

Icon. *Buff. l. c. tab.* 26. *Schreb.* 227. *Ency. Méth:* *t.* 70. *f.* 5.

Inhabits Europe, including England.

Mus Brit.

Var. b. *Lalandii.* Twice the size of the Common Dor- mouse.

Inhabits Cape of Good Hope. Perhaps the same as *M. Africanus.*

569. 4. *M. Murinus* (Murine Dormouse.) Gray, rather paler beneath, and some of the hairs white, especially under the belly; tail as long as the body, flattened horizontally, and covered with two-rowed hairs.

Myoxus Murinus, *Desm. Supp.* 544. Myoxus Compei, *Cuv. Dict. Sci. Nat.* xxvii.

Icon. *F. Cuv. Mam. Lithog. n.* 17. *t.* 4.

Inhabits Cape of Good Hope.

570. 5. *M. Africanus* (African Dormouse.) Fur above pale ferrugineous, beneath whitish, with a white line above each eye; head flat; nose blunt; upper lip cut; tail mode- rate, black in the middle, gray on the sides; eyes large, black; whiskers long; ears very short.

Myoxus Africanus, *Shaw, Zool.* ii. 172.

Icon. —— ?

Inhabits Africa.

The Dégu of Molina's Chili, 269, (Sciurus Dégus of Gmelin, and Chilian Squirrel of Shaw,) appears to be an *Arvicola.* M. Desmarest suggests that the *Musculus Fru-*

givorus and *Musculus Dichrurus* of Rafinesque belong, probably, to this sub-genus.

HYDROMYS. Geoff. *Incisive teeth $\frac{2}{2}$; canine $\frac{0.0}{0.0}$; cheek-teeth $\frac{2.2}{2.2}=12$; tops of the cheek-teeth flat, furnished with enamelled ridges, in the shape of the figure 8, with two excavations corresponding with the spaces in that figure; ears small and round; pentadactylous, but the thumb of the fore-feet extremely small; hind-feet palmated; tail as long as the body, cylindrical, but pointed at the end, and covered with thick hair.*

571. 1. *H. Leucogaster*, (the White-bellied Hydromys.) Brown above, white underneath; rather more than a foot in length.

Hydromys Leucogaster, *Geoff. Ann. Mus.* VI.

Icon. *Geoff. l. c. tab.* 36. *fig. B, C, D. Ency. Méthod. Supp. t.* 10. *f.* 3.

Inhabits Van Dieman's Land.

572. 2. *H. Chrysogaster* (the Yellow-bellied Hydromys.) Red-brown above, orange-yellow underneath.

Hydromys Chrysogaster, *Geoff. Ann. Mus.* VI. 86.

Icon. *Geoff. l. c. pl.* 36.

Inhabits Van Dieman's Land.

Obs. This and the former are the only true *Hydromys;* they are peculiar to *Australasia.*

Whether the above two constitute more than varieties of one species is doubtful.

573. 3. *H. Coypus* (the Coypus or Racoonda.) Reddish-brown on the back; red on the flanks, and light-brown on the belly; fur soft and downy, except on the tail. About two feet long; tail eighteen inches.

222

Mus Coypus, *Gmelin.* Hydromys Coypus, *Geoff. Ann. Mus.* vi. 90. Myopotamus Canariensis, *t.* 167, *from Commerson MSS.*

Carpon, *Molina Chili.* Quocuya, *d'Azara, Paraguay,* ii. 5.

Icon. *Geoffroy, l. c. f.* 35.

Inhabits parts of South America. Mus. Brooks.

Obs. Specimens are found which vary considerably in colour.

This is the type of M. F. Cuvier's genus *Myopotamus,* which has been adopted by Desmarest, and all modern authors. The genus was first proposed by Commerson. It is peculiar to South America, and very nearly allied to the Beaver.

The fur of the animal is known to the furriers, by the name of *Racoonda,* and is used in the place of Beaver-fur to make hats.

Mus. *Incisors* $\frac{2}{2}$; *canines,* $\frac{0\ 0}{0\ 0}$; *cheek-teeth,* $\frac{3\ 3}{3\ 3} = 16$; *cheek-teeth furnished with tubercles; ears oblong, or round, nearly naked, without cheek-pouches; anterior feet with four toes, and a wart, covered with an obtuse nail, in the place of a thumb; posterior feet pentadactylous; tail long, naked, and scaly; fur, with a few long scattered hairs, extending beyond the rest, which, in some species, become spines, like those on the Echymys.*

* *Spineless Rats of the Old Continent.*

574. 1. *M. Giganteus* (the Malabar Rat.) Dark brown on the back, gray on the belly; feet black; body above a foot long.

Mus Giganteus, *Hardwick, Lin. Trans. t.* vii. Mus Malabaricus, *Pen. Quad.* vi. *n.* 377.

Icon. *Hardwick, l. c. tab.* 8.

223

Inhabits the Coasts of Malabar and Coromandel, and in the Mysore and Bengal.

The *Mus Indicus* of M. Geoffroy, *Catal. de la Collect. du Mus.* appears greatly assimilated to this species.

575. 2. *Mus Javanus* (Javanese Mouse.) Fur above red-brown; end of the legs white; tail shorter than the body; feet not webbed.

Mus Javanus, *Herman, Obs. Zool.* 63. Mus Sumatrensis, *Raffl. Lin. Trans.* XIII.

Icon. ——

Inhabits Java.

576. 3. *M. Caraco* (Caraco Rat.) Fur mixed gray and reddish, deeper on the back than on the sides; paws and belly whitish; tail rather more than half the length of the body; feet semi-palmate; length about seven inches.

Mus Caraco, *Pallas, Glires,* 335. *Pennant, Quad.*

Icon. *Pallas, l. c. tab.* 23. *Ency. Méthod.* t. 67. *f.* 8. *Schreb. t.* 177.

Inhabits eastern Siberia.

577. 4. *M. Decumanus* (Norway Rat.) Gray-brown above, dirty white underneath; tail nearly as long as the body; feet of a dirty flesh-colour, not webbed. Body nine inches long.

Mus Sylvestris and Mus Norvegicus, *Brisson, Règn. Anim.* 170. *c.* 173. Mus Decumanus, *Pallas, Glires,* 91. Mus Griseus, *Pen. Syn. Quad.* 300.

Brown Rat, *Pen. Quad.* 178. Surmulot, *Buff.* VIII. Le Pone, *Buff.* XV?

Icon. Schreb. *tab.* 178. *Buff. l. c. tab.* 27. *Ency. Méthod. t.* 67. *f.* 9.

Habitat. Originally Persia or India, but the species now spread to all parts of the civilized world.

578. 5. *M. Rattus* (the Black Rat.) Black above, deep
ashy underneath; tail rather longer than the body ; about
seven or eight inches long.

Mus Rattus, *Lin. Sys. Nat.* 1.   Mus Domesticus Major,
Ray, Syn. Quad. 217.   *Lin. Syn. Nat. ed.* 2.

Black Rat, or Common Rat.

Icon. *Schreb.* 179.   *Buff. t.* 7. *tab.* 36.   *Ency. Méthod. t.*
67. *f.* 4.

Habitat. Originally Persia or India, but now spread to
all parts of the civilized world; destroyed by the Norway
Rat, and, consequently, now becoming rare in England.

579. 6. *M. Alexandrinus* (Alexandrian Rat.)   Reddish-
gray above, ashy beneath; tail one-fourth part longer than
the body; feet not webbed.

Mus Alexandrinus, *Geoff. Egypt.*

Icon. *Geoff. l. c. pl.* 5. *f.* 1.

Inhabits the vicinity of Alexandria in Egypt.

580. 7. *Mus Indicus* (Indian Rat.)   Fur reddish-gray.
above, and gray beneath ; legs reddish-gray ; tail a little
shorter than the body ; feet not webbed.

Mus Indicus, Geoff. Cat. Mus. Par. Desm. vi. 299.

Icon. ——

Inhabits Pondicherry.

581. 8. *M. Sylvaticus* (Field Mouse.)   Reddish-gray
above, white underneath; tail shorter than the body, which
is nearly five inches long.

Mus Agrestis Major, *Gesner.*   Mus Domesticus Medius,
Ray. Mus Campestris Major, *Briss. Règ Anim.* 171. Mus
Sylvaticus, *Lin.*

Mulot, *Buff.* vii.

Icon. *Schreb. tab.* 180. *Buff. l. c. pl.* 41. *Ency. Méthod.*
t. 68. *f. B.*
Inhabits all Europe. Brit. Mus.

582. 9. *M. Campestris* (Field Mouse.) Ears short,
rounded; fur yellow-gray above, white beneath.
Mus Campestris, *Desm. Mam. Supp.* 453.
Petit Mulot, or Mulot des Champs, *Buff. Hist. Nat.* vii.
Icon. *Cuv. Mam. Lithog.*
Inhabits France.

583. 10. *M. Musculus* (the Mouse.) Dusky-gray above,
ashy underneath; tail about as long as the body, which is
nearly four inches long.
Mus, *Aristotle, Hist. of Animals.* 1. *c.* 2. Mus Domesticus Vulgaris, *Ray, Synop.* 218. Mus Musculus, *Lin. Syst.*
83. Mus Sorex, *Brisson, Règne Anim.* 169.
Icon. *Schreb.* 181. *Buff. Hist. Nat.* vii. *tab.* 39. and
Sup. viii. *tab.* 20.
Inhabits all Europe, the Colonies of Europeans, and
most parts of the world.
It varies, white, black, and black and gray and white,
mixed.

584. 11. *M. Messorius* (Harvest Mouse.) Mouse-gray,
mixed with yellowish above; belly and feet white; tail a
little shorter than the body, which is but little more than
two inches long.
Mus Messorius, *Shaw, Zool.* vi. 62.
Harvest Mouse, *Pen. Quad.* ii. 384.
Icon. *Shaw, Zool.* ii. *p.* 1, *frontispiece.*
Inhabits England, observed in Hampshire   Mus. Brit.
Var. β. Black-gray.

Mus Pendulinus, *Herman, Obs. Zool.* 61.
Inhabits Alsace and Germany.

585. 12. *M. Minutus* (Minute Mouse.) Fur ferruginous above, whitish underneath; muzzle slightly elongated; tail rather shorter than the body; length under three inches.

Mus Minutus, *Pall. Glires*, 345. Mus Parvulus, *Herman, Obs. Zool.* 64?

Rat Fauve, *Desm. Nouv. Dict. d'Hist. Nat. t.* 29. 60.

Icon. *Pallas, l. c. pl.* 24, *B.*

Inhabits Russia generally.

586. 13. *M. Agrarius* (Sitnic Mouse.) Fur reddish-gray above, with a narrow black dorsal line; tail about half the length of the body; about three inches long.

Mus Agrarius, *Pallas, Glires*, 341.

Rat Sitnic, *Vicq. d'Azyr. Syst. des Anim.* ii. 455. Rat à bande noire, *Ency. Méthod.*

Icon. *Pallas, l. c. pl.* 24, *A. Schreb. tab.* 182. *Ency. Méthod. t.* 67. *f.* 10.

Inhabits Northern Germany, Russia, and parts of Siberia.

587. 14. *M.? Subtilis* (Subtle Mouse.) Yellow or ashy above, with a black dorsal line; ears folded; tail rather longer than the body, about three inches long.

Mus Subtilis, *Pallas, Itin.* vi. *A.* 70. *N.* 11. *A. B.*

Icon. *Pallas, Glires, pl.* 22. *f.* 1, and *f.* 2. *Schreb. tab.* 284. *f.* 1. *c.* 2. *Ency. Méthod. t.* 68. *f.* 2 and 5.

Inhabits Tartary and Siberia.

Var. *α. Vagus.* Ground colour of the fur gray; tail black.

Mus Vagus, *Pallas, Glir.* 327. Sikistan, *Pallas*, 1.

Icon. *Pallas, t.* 22. *f.* 2.

Var. β. *Betulinus.* Ground colour of the fur yellow-gray, tail brown above, gray beneath.

Mus Betulinus, *Pallas, Glir.* 332.

Icon. *Pallas, l. c. t.* 22. *f.* 1.

These animals live in trees, which they climb with ease, by the assistance of their large hands. They have very great analogy with the Dormice, and, like them, want the gall-bladder, but differ from them in having a cæcum. Mr. Gray has formed them into a distinct genus, under the name of *Sicista.*

588. 15. *M.? Striatus* (the Striated Mouse.) Red-gray above, marked with several longitudinal lines of little white spots; tail as long as the body, about as big as a Mouse.

Mus Orientalis, *Seba, Thes.* ii. 22. Mus Striatus, *Lin. Mus. Adolph. Frider.* i. 10. Striated Mouse, *Shaw, Zool.* vi.

Icon. *Seba, l. c. f.* 2. *Shaw, Zool.* vi. *t.* 133. *Ency. Méthod. t.* 68. *f.* 6.

Inhabits the East Indies, according to Seba.

Obs. This has been considered by some as the young of Sciurus Getulus, but apparently without foundation.

589. 16. *M.? Barbarus* (Barbary Mouse.) Fur above brown, marked with ten longitudinal whitish lines; three toes only on the anterior feet.

Mus Barbarus, *Lin. Syst. Nat. ed.* 12.

Icon. ——

Inhabits Africa.

590. 17. *M. Soricinus* (Soricine Mouse.) Fur yellowish-gray above, whitish underneath ; muzzle elongated; tail as long as the body; ears orbicular, hairy; length about three inches.

Mus Soricinus, *Herman, Obs. Zool.* 57.

Icon. *Schreb. tab.* 18. *B.* *Shaw, Zool.* ii. *t.* 133.· *Ency. Méthod. t.* 68. *f.* 4.

Inhabits the vicinity of Strasbourg.

591. 18. *M. Frugivorus* (Frugivorous Mouse.) Fur reddish brown, with scattered long brown hairs above; beneath white; ears naked, rounded; tail as long as the body, brown ringed, ciliated.

Musculus Frugivorus, *Raff. Précis de Découvert.* 5.

Inhabits Sicily, living on trees.

592. 19. *M. Dichrurus* (Two-coloured Tailed Rat.) Fur gray, varied with brown above and on the sides; head with a brown band; belly whitish; tail as long as the body, ringed, ciliated, above brown, beneath white, rather squared.

Musculus Dichrurus, *Raff. Précis de Découv.* 5.

Inhabits Sicily, living in fields.

593. 20. *Mus Setifer* (Bristle-bearing Rat.) Fur bristly, blackish-brown beneath, especially the hinder part; gray back; with nearly erect rigid bristles; ears large, rounded, nearly naked; tail long.

Mus Setifer, *Horsf. Zool. Java.* Tckus Urrok, *Javanese.*

Icon. *Horsf. l. c. t.*

Inhabits Java.

Dr. Hamilton has described and figured a somewhat similar species, under the name of *Mus Icria.*

594. 21. *M. Islandicus* (Iceland Rat.) Fur of the back black; red-gray spotted with yellow on the sides; tail nearly naked, with scales, a little longer than the body.

2 R

Mus Islandicus, *Thienemann, Natur Bemerk.* 1.
Icon. *Thienemann, l. c. t.* 22.
Inhabits North of Europe.

595. 22. *M. Donovani* (Donovan's Rat.) Fur blackish-
gray, varied with brown ; back with three pale dorsal
bands ; tail moderate, rather hairy.
Rattus Donovani, *Donovan, Nat. Repos.*
Icon. *Donovan, l. c. t.* 35.
Inhabits Cape of Good Hope.

* * *American Spineless Rats.*

The dentition of the following American species of
d'Azara has not been ascertained with certainty ; and it is
only on the authority of external characters, therefore, that
they are placed in this subdivision.

596. 23. *M. Angouya* (Angouya Rat.) Yellow-brown
above, whitish beneath; tail rather longer than the body ;
ears rounded, moderate.
Angouya Rat, or third Rat of *d'Azara, Quad. of Para-
guay.* Mus Braziliensis, *Geoff. Collect. du Mus.?* Mus
Caugouya, *Desm.* 305.
Icon. ——
Inhabits Paraguay.

597. 24. *M. Rufus* (Red Rat of d'Azara.) Yellowish-
red, darker on the head and back ; belly yellowish ; tail
more than half as long as the body; about six inches in
length.
Red Rat, or fifth Rat of *d'Azara, Quad. of Paraguay.*
Icon. ——
Inhabits Paraguay.

230

598. 25. *M. Cephalotes* (Great-headed Rat.) Head very large, muzzle short; brown above, lighter on the sides, whitish underneath; tail as long as the body.

Mus Cephalotes, *Desmarest, Nouv. Dict. d'Hist. Nat.* 305.

Great-headed Rat, or second Rat of *d'Azara. Quad. of Paraguay.*

Icon. ——

Inhabits Paraguay.

599. 26. *M. Auritus* (Long-eared Rat.) Head thick; ears long; mouse-colour, lighter underneath; tail shorter than the body; about five inches long.

Mus Auritus, *Desmarest, Nouv. Dict. d'Hist. Nat.* 305.

Long-eared, or fourth Rat of *d'Azara, Quad. of Paraguay.*

Icon. ——

Inhabits South of Buenos Ayres.

600. 27. *M. Nigripes* (Black-footed Rat.) Head thick; ears short and round; yellow-brown above, whitish underneath; paws deep black; tail shorter than the body; about four inches long.

Mus Nigripes, *Desmarest, Nouv. Dict. d'Hist. Nat.* 305.

Black footed Rat, or sixth Rat of *d'Azara, Quad. of Paraguay.*

Icon. ——

Inhabits Paraguay.

601. 28. *M. Laucha* (Laucha Rat.) Head moderate; muzzle pointed; lead colour above, whitish underneath; tail rather shorter than the body; about three inches long.

Mus Laucha, *Desmarest, Nouv. Dict. d'Hist. Nat.* 305.

Laucha Rat, or seventh Rat of *d'Azara, Quad. of Paraguay.*

Icon. ——

Inhabits Buenos Ayres.

The three following are referred by their describers to this division of the Glires, but their dentition is not ascertained.

602. 29. *M. Leucopus* (White-footed Mouse.) Brownish-yellow above, white underneath, head yellow; ears large; tail as long as the body, pale brown above, gray underneath; the paws white; about five inches long.

Raffinesque, Journey to the Westward of the United States, and American Monthly Mag. t. 3, 444.

603. 30. *M. Nigricans* (Wood Rat of Raffinesque.) Black above; gray on the belly; tail black; longer than the body; about eight inches long.

Mus Nigricans, Black, or Wood Rat, *Raffinesque, Journey to the Westward of the United States*, and *American Monthly Mag. Oct.* 1818.

₊ *Spiny Rats.*

604. 31. *M. Perchal* (Perchal Rat.) Reddish-brown above, with spiny hairs intermixed; grayish underneath; tail not so long as the body; about seventeen or eighteen inches long.

Mus Perchal, *Gmel.* Echimys Perchal, *Geoff.*

Rat Perchal, *Buff. Hist. Nat. Supp. t.* i. 276.

Icon. *Buff. l. c. pl.* 69.

Inhabits the town and vicinity of Pondicherry.

605. 32. *M.?* *Cahirinus.* (Egyptian Rat.) Ashy gray, deeper on the upper than on the under parts, composed of rough spiny hairs; tail as long as the body; about four inches long.

Mus Cahirinus, *Geoffroy, Collect. du Mus.*   Echimys, d'Egypte, *Ejusdem, Egypt. partie d'Hist. Nat.*
Icon. *Geoffroy's Egypt. pl. 5. f. 2.*
Inhabits Egypt.

CRICETUS. *Dentition like that of Mus; cheek pouches; body low on the legs; head thick; ears oval and round; toes like those of the Mus, or with five toes on the fore-feet.*

606. 1. *C. Vulgaris* (the Common Hamster.) Grayish fawn-colour above ; black underneath, with three large yellowish spots on each side, one white spot on the throat, and another under the chest.
Mus Cricetus, *Pallas, Glires,* 83. Glis Cricetus, *Erxleben.* Glis Marmota Argentoratensis, *Brisson, Quad.* 166.
Hamster, *Buff. t.* 13. Hamster Rat, *Pennant, Quad.* ii. 206.
Icon. *Schreber, tab.* 198, A. *Buff. l. c. pl.* 14.
Inhabits the central and northern parts of Europe and Asia.
Var. a. Black, with a little white round the mouth, on the nose, edge of the ears, feet, and end of the tail.

607. 2. *C. Migratorius* (Yaik Hamster.)   Ashy-gray above ; white underneath ; muzzle, round the nostrils, and feet, white ; ears indented.   About four inches long ; tail less than an inch.
Mus Accedula, *Pallas, Glires, pa.* 74. Mus Migratorius, *ejusd. Voyage.* Cricetus Migratorius, *Desm.* 318.
Yaik Rat, *Pen.* ii. 210. Le Hagri, *Vicq. d'Azyr. Syst. des Anim.* ii. 395.
Icon. *Pallas, Glires, pl.* 18, *A. Schreber, tab.* 197. *Ency. Méthod.* 70, *f.* 2.
Inhabits the vicinity of the Yaik, in Siberia.

608. 3. *C. Arenarius* (Sand Hamster.) Whitish ash-colour above ; pure white beneath ; feet and tail white ; ears round. About four inches long ; tail about an inch.

Mus Arenarius, *Pallas, Glires*, 86.

Sand Rat, *Pen. Quad.* II. 211. Le Sablé, *Vicq d'Azyr. Syst. des Anim.* II. 407.

Icon. *Pallas, l. c. tab.* 16, *A. Ency. Méthod. t.* 70, *f.* 4. Inhabits the sandy vicinity of the Irtisch, in Siberia.

609. 4. *C. Phæus* (Astracan Hamster.) Brownish ash-colour on the back and upper part of the tail, the under side of which is white, together with the under part of the body and internal sides of the limbs ; ears oval and large.

Mus Phæus, *Pallas, Glires*, 86. Mus Alpinus, *Hablitz, Gm. Voy.* 172.

Zaryzin Rat, *Pen. Quad.* II. 211. Astracan Mouse, *Shaw, Zool.* II. *p.* 2, 103. Le Phé, *Vicq. d'Azyr. Syst. des Anim.* 405.

Icon. *Schreb.* 200. *Pallas, Glires, tab.* 15. *A. Ency. Méthod. t.* 70, *f.* 2.

Inhabits the Deserts of Astracan, the temperate parts of Persia, &c.

610. 5. *C. Songarus* (Songar Hamster.) Ashy on the back, with a black dorsal line ; sides varied with white and brown ; belly white ; tail very short ; about three inches in length of body.

Mus Songarus, *Pallas, Glires*, 86. Glis Æconomicus ? *Erxleben.*

Songar Rat, *Pen. Quad.* II. 212.

Icon. *Pallas, l. c. tab.* 16, *B. Schreber, tab.* 201. *Ency. Méthod. t.* 71, *f.* 1. *t.* 70, *f.* 5.

Inhabits Siberia.

611. 6. *C. Furunculu s*(Baraba Hamster.) Ashy above with a black dorsal line ; belly and paws white. Body about four inches long; tail one inch.

Mus Furunculus, *Pallas, Glires*, 86, and Mus Barabensis, *ejusd. Voyage*, ii. 704. Furunculus Myoides, *Messer-schmid, Mus. Petrop*. 343.

Baraba Rat, *Pen. Quad*. ii. 213. L'Orozo, *Vicq. d'Azyr. Syst. des Anim*. ii. 412.

Icon. *Pallas, Glires, tab*. 15, *B. Schreber*, 202. *Ency. Méthod. t*. 71, *f*. 2.

Inhabits the sandy plains of Baraba.

612. 7. *C. Bursarius* (Canada Hamster.) Gray; anterior feet pentadactylous, armed with long digging nails ; ears short ; body cylindrical; about as big as the Norway Rat.

Mus Bursarius, *Lin. Trans*. v. 227. Mus Saccatus, *Mitchell, New York Medical Repository, Jan*. 1821. Geomys Cinereus, *Raffinesque, Amer. Monthly Mag*. 1817, 45.

Canada Rat, *Shaw's Zoology*, ii. *pl*. 1, 100. Sand Rat, *Geoff*.

Icon. *Lin. Trans. l. c. Shaw*, i. *t*.

Inhabits Canada.

Obs. Raffinesque appropriates this to a distinct genus, *Geomys*, distinguished by the five toes on the fore-feet, and the subterranean habits. Its teeth are still uncertain, and its location consequently conditional. Say formed it into a genus, under the name of *Pseudotoma*.

613. 8. *C. Laniger* (the Chinchilla Hamster.) Gray and white, waved ; ears large and round ; tail short, furnished with longish stiff hairs ; fur of the body extremely soft and downy.

Mus Laniger, *Molina, Chili*, 283. Cricetus Laniger, *Geoff. Collect. du Mus*.

Chinchilla, *Acosta, Nat. History of India,* 199.

Icon. ——

Inhabits Chili, according to Molina, Peru, according to Acosta; and probably the whole chain of the Andes.

Obs. The teeth of this species are still unknown to Naturalists; and M. Geoffroy's location of it with the Hamsters is conditional.

614. 9. *C. Anomalus* (Anomalous Hamster.) Reddish-brown above; white underneath; some flat spines on the back; tail nearly as long as the body, nearly naked, scaly, and black.

Mus Anomalus, *Thompson, Trans. Lin. Soc.* Cricetus Anomalus, *Desmarest, Nouv. Dict. d'Hist. Nat. t.* 14, 180.

Icon. ——

Inhabits the Isle of Trinity in the Gulf of Mexico.

Obs. Desmarest has proposed to consider this species as a new genus, under the name of *Heteromys.*

DIPUS. *Incisors $\frac{2}{2}$, those below sharp-pointed; canines $\frac{0.0}{0.0}$; check-teeth $\frac{3.3}{3.3}$ or $\frac{4.4}{4.4}$, simple, with tuberculous crowns; eyes large; ears long, pointed; anterior feet short, with four toes, and tubercle with a nail in the place of a thumb; hind feet five or six times longer than those before, terminated by three or five toes, with one metatarsus for the three middle toes.*

615. 1. *D. Sagitta* (the Jerboa.) Bright yellow above; white underneath; tail longer than the body, with a tuft at its extremity. About six inches long.

Mus Ægyptius, *Hasselquist.* Mus Jaculus, *Lin.* Mus Sagitta, *Pallas, Glires,* 306. Dipus Gerboa, *Gm.*

Daman, *Shaw, Travels in Barbary.* Gerboa, *Bruce's Travels, var.* Gerbo ou Gerboise, *Buff. Hist. Nat. Sup. t.* VI.

Icon. *Buff. l. c. pl.* 39, *and* 40. *Pallas, l. c. pl.* 21.
Inhabits Barbary, Egypt, and Western Asia.

616. 2. *D. Jaculus* (Siberian Jerboa.) Pale yellow
above ; white underneath ; muzzle white, and a white stripe
across the buttocks ; five toes on the fore-feet, of which the
lateral are very small. About seven inches long ; tail
nearly a foot.

Mus Jaculus, *Pallas, Glires,* 275. Dipus Alagtaga, *Oliv.
Bull. Soc. Phil.*

Siberian Jerboa, *Pen. Quad.* II. 166.

Icon. *Pallas, l. c. tab.* 20.

Inhabits Deserts of Tartary, and probably a considerable
part of South Western Asia.

617. 3. *D. Brachyurus* (Striped Jerboa.) Pale yellow,
varied with brown above, and white underneath ; a white
stripe across the buttocks ; muzzle white at the extremity,
brown above ; tail and limbs rather thick ; ears short ; hind
feet with five toes, the three internal of equal length.
Nearly five inches long ; tail rather longer.

Mus Jaculus, var. B. *Pallas, Glires,* 297. Dipus Brachyurus, *Blainville. Desmarest, Nouv. Dict. d'Hist. Nat.*

Icon. ⸺

Inhabits Eastern Tartary and Siberia.

618. 4. *D. Minutus* (Little Jerboa.) Pale yellowish-
gray, varied with brown above, and white underneath ; the
extremities, and a transverse stripe on the buttocks, white ;
muzzle like the back ; hinder feet pentadactylous, with the
three intermediate nails equal in length. Length under
five inches ; tail rather longer.

Dipus Jaculus, var. minor, *Pallas, Glires,* 296. Dipus
Minimus, *Blainville. Desmarest. Nouv. Dict. d'Hist. Nat.*

Icon. ⸺

Inhabits the shores of the Caspian.

Obs. Pallas found but three cheek-teeth on each side in the upper jaw in this species; if this were not accidental, it should constitute a specific character.

619. 5. *D. Maximus* (Great Jerboa.) Bright gray above; a black line over each eye, uniting on the forehead; white underneath; four toes to the fore-feet, and three to those behind. About the size of a moderate Rabbit.

Dipus Maximus, *Blainville. Desmarest, Nouv. Dict. d'Hist. Nat.*

Icon. ——

Habitat uncertain.

Obs. We have inserted the several species considered as such by M. Desmarest, some of which, it will be seen, Dr. Pallas treats as varieties.

GERBILLUS. Desmarest. *Incisive teeth* $\frac{2}{2}$; *canines* $\frac{0}{0}\frac{0}{0}$; *cheek-teeth* $\frac{3}{3}\frac{3}{3} = 16$; *cheek-teeth tuberculous, the first with three, the second with two, and the third with one tubercle. Ears moderate; fore-feet short, with four toes and the rudiment of a thumb; the hind legs long, or very long, terminated by five toes, with nails, each with a distinct metatarsus; tail long, covered with fur.*

Obs. M. Desmarest separated a number of species, heretofore inserted in several different subdivisions of the Rodentia, into one genus, under the name Gerbillus, without, we believe, having examined the teeth of any of the animals he refers to it, except those of the Egyptian species. The Baron admits also the Indian species into Desmarest's genus; but of those then remaining, he says, that having examined the teeth of some of them, he finds some distinct, and others he refers to new groups.

Frederick Cuvier has restricted the genus, and gives the

character of the teeth in his work on the Teeth of Quadrupeds.

620. 1. *G. Tamaricinus* (Tamarisk Gerbil.) Yellowishgray above, white underneath ; tail about as long as the body, annulated gray and brown ; body about seven inches long.

Mus Tamaricinus, *Pallas, Glires*, 322.

Sciurus Tamaricinus, *Erxleb.* Dipus Tamaricinus, *Gm.* Myoxus Tamaricinus, *Desmarest, Nouv. Dict. d'Hist. Nat. 1st ed.* Gerbillus Tamaricinus, *ejusd.* 2d ed., *et Ency. Méthod. Art. Mammalogie, Sp.* 513.

Tamarisk Rat, *Pen. Quad.* II. 175. Tamarisk Jerboa, *Shaw, Zool.* II. *p.* 1. 191.

Icon. *Pallas, l. c. pl.* 19. *Schreb. tab.* 232.

Inhabits the vicinity of the Caspian.

621. 2. *G. Indicus* (Indian Gerbil.) Red-brown above, sprinkled with small brown spots, disposed in longitudinal lines; white underneath ; tail a little longer than the body, brown, terminated by a tuft, about seven inches long.

Yerbua Indica, *Hardwicke, Trans. Lin. Soc. t.* VIII. 279.

Icon. *Nouv. Bull. Soc. Philom.* v. 35. 121. *pl.* 1. *f.* 1.

Inhabits Hindostan.

622. 3. *G. Meridianus,* (Torrid Gerbil.) Grayishyellow above, white underneath, with a central line of redbrown ; limbs white; about five inches long; tail between three and four.

Mus Meridianus, *Pallas, Glires*, 314.

Le Jird, *Vicq. d'Azyr. Syst. des. Anim. t.* II. 413.

Icon. *Pallas, l. c.* pl. 18, *B.*

Inhabits the deserts near the Volga.

239

Obs. Considered by some authors as the Mus Longipes
of Linnæus.

623. 4. *G. Ægyptius* (Egyptian Gerbil.) Upper part
of the body bright yellow, under part pure white ; tail a
little longer than the body, brown, and terminated with a
few long hairs ; hind legs very long ; about the size of a
Mouse.

Dipus Gerbillus, *Olivier, Bull. de la Soc. Philom. n.* 40.
ejusd. Voyage dans l'Empire Ottom. t. iii. 157. Mus
Longipes, *Lin.* Dipus Pyramidum, *Geoff.* Gerbillus
Ægyptius, *Desmarest, Ency. Méthod. Mam. Sp.* 516.

Icon. *Oliv. Voy. dans l'Empire Ottom. pl.* 28. *f. A, B, C.*

Inhabits the vicinity of Memphis and the Pyramids of
Egypt.

624. 5. *G. Canadensis* (Canadian Gerbil.) Yellowish
above, white underneath ; ears short ; tail almost denuded, rather longer than the body, without tuft at the
end.

Dipus Canadensis, *Davies, Lin. Trans.* iv. 155. Gerbillus Daviesii, *Rafinesque Précis. des Découvertes Simio-
logique,* 14.

Icon. *Lin. Trans.* iv.

Inhabits the vicinity of Quebec

625. 7. *G. Labradorius* (Labrador Gerbil.) Fur brown
above, beneath white, without a dividing line ; toes four
before, five behind ; tail more than half the length of the
body.

Mus Labradorius, *Sab. Append. Frank. Voy.* 661.
Gerbillus Labradorius, *Harlan, Faun. Amer.* 157.

Inhabits Labrador.

M. Raffinesque has named others as distinct species of

the Gerbil; but in the present uncertain state of species belonging to this subdivision, we shall not enumerate them as species otherwise than by this notice. These are the *Gerbillus Soricinus.* Gray-brown, with a longitudinal red line on the flanks. *Gerbillus Megalops. (American Monthly Mag.* 1818, *p.* 446.) Black; tail longer than the body, terminated by a white tuft; eyes large and black; body three inches long. *Gerbillus Conurus. (Id.)* Uniformly yellow; eyes small; tail as long as the body, black, terminated by a yellow tuft. *Gerbillus Hudsonius. Dipus Gerbillus, Zimmerman.*

ASPALAX. Incisive-teeth, $\frac{2}{2}$, yellow, large, square at the top and bottom, those below twice the length of those above; canines, $\frac{0}{0}\frac{0}{0}$; check-teeth, $\frac{3}{3}\frac{3}{3}$, with tuberculous crowns; body long, cylindrical; eyes very small, entirely covered by the skin; no external ears; paws short, pentadactylous; tail naked; fur short, and soft.

626. 1. *A. Typhlus* (the Spalax.) Fur blackish ash-colour at the base, reddish toward the point; head large, and thick, and the whole animal cylindrical; eyes merely rudimentary; about as big as a Rat.

Mus Typhlus, *Pallas, Glires,* 154.

Spalax Microphthalmus, *Guldenst.* Spalax Major et Glis Zumui, *Erxleb.* Ασπαλαξ or Mole of the *Greeks.*

Icon. *Pallas, l. c. pl.* 8. *Schreb. tab.* 206.

Inhabits Asia Minor, Syria, Mesopotamia, Persia, and Southern Russia.

Var. A. With large irregular white spots.

BATHYERGUS. Incisors, $\frac{2}{2}$; canines, $\frac{0}{0}\frac{0}{0}$; cheek-teeth, $\frac{4}{4}\frac{4}{4}$, or $\frac{3}{3}\frac{3}{3}$, according to F. Cuvier, = 20. Incisors very long, large, and square; cheek-teeth slightly tuberculous, indented on the edges ; body thick and cylindrical ; head thick, muzzle truncated ; eyes small ; 1 o external ears ; feet short ; toes five, with digging nails ; tail very short.

627. 1. *B. Maritimus* (the Coast Bathyergus.) Whitish gray ; tail flat, covered with rough hairs ; body above a foot long ; tail about three inches.

Mus Maritimus, *Gm.* Bathyergus, *Ill. Prod.* Arctomys Africana, *Lamark, Voy. de Thunberg.* 1. 188. and 11. 475.

Taupe du Cap, *Lacaille, Journ.* 299. Grande Taupe du Cap, *Buff. Supp. t.* vi. Taupe des Dunes, *Allamand, Supp. t.* v. 24. Zand Mole, *Cape Colonists.*

Icon. *Buff. Supp.* vi. *tab:* 38. *Allamand, l. c. tab.* 10. *Lamark, l. c. t.* 11. *pl.* 1. *Schreber, tab.* 204, *B.*

Inhabits the Cape of Good Hope.

628. 2. *B. Capensis* (the Cape Bathyergus.) Brown, with white round each eye and ear, and on the top of the head, and end of the muzzle ; about six inches long.

Mus Capensis, *Pallas, Glires,* 172.

Georychus, *Illiger, Prod.*

Taupe du Cap de Bonne Esperance, *Buff. Supp.* ii.

Icon. *Pallas, l. c. pl.* 7. *Buff. l. c. tab.* 36. *Schreber,* 204. *Thunberg, t.* ii. *pl.* 2.

Inhabits the Cape of Good Hope.

M. Desmarest treats this as distinct, and not as a smaller variety ; and Illiger even refers this and the last to two genera.

Raffinesque, in the Annals of Nature, describes *Spalax Vittata*, which is yellow on the upper part, with three brown longitudinal bands, and white underneath ; about seven inches long, and refers it to this sub-genus.

629. 3. PEDETES. Incisors, $\frac{2}{2}$; canines, $\frac{0.0}{0.0}$; cheek-

teeth, $\frac{44}{44}$; lower incisors, cut obliquely, and not pointed; cheek-teeth formed, of two elliptical parts, united at their internal extremity, and separated above by a deep furrow; head short, large, and flat; muzzle obtuse, terminated by small nostrils at right angles; ears long, narrow, pointed; eyes large; no cheek-pouches; large whiskers; anterior feet with five toes, and long narrow digging nails; posterior feet with four toes, the external very small, the intermediate of the other three much the longest, the rest being equal, all furnished with thick strong nails; tail long, thick. An abdominal pouch in the females like that of the Didelphes, but not enclosing the teats.

630. 1. *P. Capensis* (Cape Pedetes.) Bright fulvous, varied with black on the upper part; white underneath, with a line of the same colour in the folds of the arms; legs brown; tail thin, reddish above, near the insertion, gray below, and black at the end.

Yerbua Capensis, *Sparman, Acta. Stockholm.*, 1778, and *Travels in Africa.* Mus Cafer, *Pallas, Glires*, 87. Dipus Cafer, *Gm.* Gerboa Major, *Allam. Monog.* 1776. Helamys Marmot, Helamys Cafer, *F. Cuvier, Dict. des Sciences Nat. t.* 20, 344.

Cape Jerboa, *Penn.* ii. Leaping Hare, *Cape Colonists.*
Inhabits the Cape of Good Hope.

ARCTOMYS. Incisors, $\frac{2}{2}$; canines, $\frac{0}{0}\frac{0}{0}$; cheek-teeth, $\frac{55}{44} =$ 22. Incisors very strong; anterior surface rounded; cheek-teeth, with the upper surface furnished with ridges and tubercles; body thick and heavy; head large; eyes large; ears short; paws strong, anterior with four toes, and the rudiment of a thumb; the posterior with five toes; nails strong, compressed; tail generally short.

M. F Cuvier has divided this genus into two, viz.

243

I. Arctomys. *No cheek pouches. Habits social.*

631. 1. *A. Marmotta* (the Marmot.) Yellowish-gray, with an ashy tint on the head; top of the head, and end of the tail, black; under part yellowish-white; nearly eighteen inches long.

Mus Alpinus, *Pliny, lib.* viii. 37. Glis Marmotta, *Klein. Quad.* 56. Mus Marmotta, *Lin. Sys. Nat.* 81. Arctomys Marmotta, *Gm.*

Marmotte, *Buff. Hist. Nat.* 8. Alpine Marmot, *Pen.* ii. 128.

Icon. *Perrault, Hist. des Anim.* iii. *f.* 7. *Shreb. tab.* 207. *Buff. l. c. pl.* 28.

Inhabits the Alps, Pyrenees, and the other high mountains of Europe and Asia.

632. 2. *A. Bobac* (the Bobac.) Yellowish-gray, with red tint near the head; under part of the body reddish; about eighteen inches long.

Mus Arctomys, *Pallas, Glires,* 97. Glis Polonica, *Brisson.* Glis Marmotta, *Erxleb.* Arctomys Bobac, *Gm.* Mus Arctomys, *Boddaert.* Mus Marmotta, *Forster, Phil. Trans.* 57.

Bobak, ou Marmotte de Pologne, *Buff.* t. xiii. 1.

Icon. *Pallas, l. c. tab.* 5, and 9. *f.* 1, 2, 3. *Buff. l. c. pl.* 18. *Schreb.* 209.

Inhabits Poland, and Northern Russia.

633. 3. *A. Monax* (Maryland Marmot.) Brown above, paler on the sides, and belly; muzzle bluish-gray and black; tail half as long as the body, black; nearly eighteen inches long.

Mus Monax, *Lin. Sys.* 81. Glis Monax, *Erxleb.* Cuniculus Bahamensis, *Catesby, Carolina,* ii. 79.

Maryland Marmot, *Pen. Quad.* ii. 130. Monax, ou Marmotte de Canada, *Buff. Supp. III.* Wood-chuck, or Ground Hog of the United States.

Icon. *Schreb.* 208. *Buff. l. c. pl.* 28. *Catesby, pl.* 79.

Inhabits North America.

634. 4. *A. Empetra* (Quebec Marmot.) Blackish-brown, dotted with white-red underneath; tail short, black at the end.

Mus Empetra, *Pallas, Glires,* 75. Arctomys Empetra, *Gmelin.*

Quebec Marmot, *Pen. Quad.* ii. 130.

Icon. *Schreb. tab.* 210. *Pen. l. c. pl.* 74. *f.* 1.

Inhabits Canada, and the shores of Hudson's Bay.

635. 5. *A. Brachyura* (Short-tailed Marmot). Cinereous brown above, light red beneath; tail flat, reddish, two-rowed, one-seventh the whole length.

Anisonys Brachyura, *Raffinesque, Amer. Mag.* vi. 45.

Burrowing Squirrel, *Lewis* and *Clarke, Exped.* vi. 173.

Icon. ——

Inhabits Missouri.

636. 6. *A. Rufa* (Red Marmot.) Reddish-brown; fur short, thick, and silky; ears short, thin, and pointed, covered with hair.

Arctomys Rufa, *Harlan, Amer. Faun.* 308. Anisonys Rufa, *Raffinesque, Amer. Month. Mag.*

Sewellel, *Lewis* and *Clarke, Exp.* vi. 176.

Icon. ——

Inhab. ——

Obs. Raffinesque formed these two animals into a genus, under the name of *Anisonys,* the character of the nails being unequal; but Lewis and Clarke do not mention their being so.

II. Spermophilus. *With large cheek-pouches. Habits solitary.*

637. 7. *A. Citillus* (the Souslik.) Yellowish-brown, waved or spotted with white in transverse stripes; white underneath; with cheek-pouches; external ears scarcely visible.

Mus Noricus, aut Citillus, *Agricola, An. Subter.* 485. Mus Citillus, *Pallas, Glires*, 549. Mus Suslica, *Guldens. Nouv. Can. Petrop.* xiv. 389. Glis Citillus, *Erxleb.* Mus Marmota, *Forster, Phil. Trans.* 57, 343.

Zizel, *Buff. t.* 15, 139. Souslik, *ejusd. t.* 15, 144. Jèvraschka, ou Marmotte de Siberia, *ejusd. Sup. t.* 3, 191. Earless Marmot, *Pen.* ii. 135.

Icon. *Pallas, l. c. tab.* 21. *Guldens. l. c. tab.* 7. *Schreber, tab.* 211. *A. B. Buff. Sup.* iii. *tab.* 31.

Inhabits parts of Germany, and Russia in Europe and Asia.

Obs. There are three varieties of this species, distinguished, 1st. by the wavy disposition of the colour of the back; 2dly, by white spots instead of waves; 3dly, uniform.

The following species discovered by our late intrepid navigators, seem referable to this division of pouched Marmots.

638. 8. *A. Franklinii* (Franklin's Marmot.) Head broad; ears small; snout very blunt; tail elongated; body variegated, fuscous.

Arctomys Franklinii, *Sabine, Lin. Trans.* xiii. 587.

Icon. *Lin. Trans.* xiii. *t.* 27.

Inhabits Canada.

639. 9. *A. Richardsonii* (Richardson Marmot). Ears short; snout acute; tail moderate; body fuscous.

Arctomys Richardsonii, *Sabine, Lin. Trans.* xiii. 589.
Icon. *Lin. Trans.* xiii. *t.* 28.
Inhabits Canada.

640. 10. *A. Parryii* (Parry's Marmot.) Snout very blunt ;
ears very short; tail elongate, tip black ; body above marbled with confluent white and black spots ; beneath ferruginous. Length of head and body one foot ; tail four inches.

Arctomys Alpina, *Parry, Nar.* 2d *Voy.* 61. Arctomys
Parryii, *Richardson, Append. Franklin. Voy.*

Ground Squirrel, *Hearne, Jour.* 141. Quebec Marmot,
Forster, Phil. Trans. lxxii. 378.

Inhabits Canada.

641. 11. *A. Tridecim-lineata* (Wood's Marmot.) Fur deep
chestnut above, striped with six white lines, alternating
with an equal number of longitudinal rows of white spots ;
white beneath.

Arctomys Woodii, *Sab. Trans. Lin. Soc.* xiii. 599. Sciurus Tridecim-lineatus, *Mitchell, Med. Repos.* vi. Arctomys
Tridecim-lineata, *Harlan, Faun. Amer.* 164.

Striped and Spotted Ground Squirrel, *Say. Exped.
Rocky Mount.* ii. 171. Federation Squirrel, *Mitchell.*

Inhabits ——

Obs. This species, at first referred to the Squirrels, seems
allied to M. F. Cuvier's division of Spermophilus.

The two following species are referred conditionally to
their present situation.

642. 12. *A. Ludoviciana* (Prairie Marmot.) Fur light,
dirty reddish brown above, intermixed with some gray, also
a few black hairs ; the hair next the skin bluish-white, then
light reddish, tips gray; below dirty white. Length of
head and body one foot four inches ; tail two inches and
three-quarters.

Arctomys Ludoviciani, *Ord. Guthrie, Geog.* v. 303: Arctomys Missouriensis, *Warden, U. States,* v. 6. vi.

Prairie Dog, *Lewis* and *Clarke, Exp. Missouri.*

Icon. Nobis.

Inhabits North America.

643. 13. *A. Latrans* (Barking Marmot.) Uniform brick-red, lighter beneath ; cheeks furnished with pouches ; a few long hairs are inserted on each jaw, and directly over the eye.

Arctomys Latrans, *Harlan, Faun. Amer.* 306.

Barking Squirrel, *Lewis and Clarke, Exped.* vi. 175.

Icon. ——

Inhabits Missouri.

Obs. Perhaps a *Spermophilus,* but it lives in society.

Several other Marmots have been mentioned by different writers ; but as their great leading character of dentition is not noticed, they have been rejected from the genus until better authenticated: of these are the Hoary Marmot of *Pen. Quad.* ii. 130; the Marmot Gundi, of the same, *p.* 137 ; the Tailless Marmot, of the same, *p.* 137 ; the Mus Maulinus, of *Molina Chili,* 268; the Glis Tscherkessicus, of *Erxleb.;* and the Orctomys Missouriensis, of *Warden's United States,* t. v. 627.

Among these, also, we shall insert the species, or variety, we have noticed at *p.* 170, of Vol. iii, and figured under the name of Marmot Diana.

SciURUS. Incisors $\frac{2}{2}$; canines $\frac{0\,0}{0\,0}$; cheek-teeth $\frac{5\,5}{4\,4}=22$; upper incisors flat in front, and wedge-shaped at the extremity ; the lower pointed and compressed laterally ; cheek-teeth tubercular, the fives in the upper jaw found only in the young state ; body small ; ears erect ; head small ; eyes large ; anterior feet with four long toes, and a tubercle instead of a thumb ; the posterior with five long toes, all furnished with long crooked nails. Tail long, often very villose ; two pectoral teats, and six ventral.

This genus has been divided into groups distinguished by a flat cylindrical tail, and by the presence or absence of cheek-pouches.

644. 1. *S. Vulgaris* (Common Squirrel.) Bright red ; ears terminated by a pencil of hairs.

Sciurus *of the Ancients.* Sciurus Vulgaris, *Linn. Syst. Nat.* 86.

Icon. *Schreb. tab.* 212 ; and most naturalists. Skull, *Fisch. Ade. Zool. f.* 6, *f.* 3.

Inhabits Europe and the north of Asia.

Obs. Several varieties of the Common Squirrel have been noticed ; indeed the colours are various, from red through different shades of cinereous, even to black. In high latitudes they vary with the season, and become bluish ash-colour in winter.

645. 2. *S. Alpinus* (Alpine Squirrel.) Deep brown varied with yellowish-white on the back ; beneath white ; feet yellow, with a yellow band separating the white of the neck and the grey of the outside of the limbs from the brown of the back ; hairs of the tail very long, black, and rugged, with yellow at the base ; ears ending in a tuft of hairs.

Sciurus Alpinus, *F. Cuv. Mam. Lithog.*
Icon. *F. Cuv. l. c.*
Inhab. Pyrenees.

646. 3. *S. Maximus* (Great Squirrel.) Upper part of the head, flanks, and legs, purpurescent reddish-brown, with a transverse stripe on the shoulders ; lower part of the back, loins, and tail, black ; under part of body and interior of limbs pale yellow ; nearly as big as a Cat.

Sciurus Maximus, *Gm.*

Grand Ecureuil de la Côte de Malabar, *Sonnerat, Voyage,* *t.* 2, 139.

Icon. *Schreb. tab.* 217, *B. Sonnerat, l. c. pl.* 87.

The Baron identifies this specifically with the S. Macrourus of Gmel. Desmarest treats them as distinct, and gives as synonyms to that species the Long-tailed Squirrel of *Pen. Indian Zool.* and the Ceylon Squirrel, *ejusd. Quad.,* the Sciurus Ceilonicus and Zeylonicus of *Boddaert* and *Ray,* and with *Schreber's figure,* 217.

647. 4. *S. Madagascariensis* (Madagascar Squirrel.) Upper part of the body deep black; throat yellowish-white; belly yellowish-brown; tail black, longer than the body; body from eighteen to twenty inches long.

Sciurus Madagascariensis, *Shaw's Zoology* II. *part* 1, 128. Ecureuil de Madagascar, *Buff. Hist. Nat. Sup.* 7.

Icon. *Buff. l. c. pl.* 63.

Inhabits Madagascar.

648. 5. *S. Ceylonensis* (Ceylon Squirrel.) Fur above black; below yellow; tail gray.

Sciurus Ceylonicus, *Rai. Quad.* 215. S. Ceylonensis, *Wodd. Elench.* 117. S. Macrourus, *Gmel.*

Long-tailed Squirrel, *Penn. Ind. Zool.* Ceylon Squirrel, *Penn.*

Icon. *Penn. l. c. t.* 1. *Schreb. t.* 217. *Ency. Méthod.* *t.* 75, *f.* 4.

Inhabits Ceylon.

Obs. Cuvier is of opinion that this is a mere variety of the S. Maximus.

649. 5. *S. Prevostii* (Prévost's Squirrel.) Black above, yellow on the flanks, and reddish-brown underneath; tail brown. About the size of the Common Squirrel.

Sciurus Prevostii, *Desmarest, Ency. Méthod. Mammalogie,*
sp. 537.

Icon. ——— ?

Inhabits India.

650. 6. *S.Leschenaultii* (Jeralang, or Leschenault's Squirrel.) Ochery-brown above ; head, throat, belly, and anterior part and internal sides of the limbs, yellowish-white ;
brown above, and yellowish underneath. Body rather
more than a foot in length ; tail about the same.

Sciurus_Albiceps, *Geoff. Collect. du Mus.* S. Leschenaultii, *Desm. Mam.* 335.

Icon. *Horsfield, Zool. Journal.*

Inhabits Java.

Var. A, darker.

651. 7. *S. Bicolor* (Two-coloured Squirrel.) Fur deep
brown or blackish above ; bright yellow beneath ; eyes surrounded by a black circle ; ears not bearded. Three feet
from nose to end of tail.

Sciurus Bicolor, *Sparman, Act. Soc. Goth.* S. Javanensis, *Schreb.* Javan Squirrel, *Pennant, Quad.*

Icon. *Schreb. f.* 216.

Inhabits Java.

652. 8. *S. Bilineatus* (Two-rayed Squirrel.) Fur above
gray, with a longitudinal white line on each side ; below
yellowish ; tail rather shorter than the body.

Sciurus Bilineatus, *Geoff. Col. Mus.* S. Notatus, *Bodd.*
Elench. i. 119. S. Platani, Var. *Ingerman.*

Plantain Squirrel, *Pen. Hist. Quad.*

Icon. *Horsf. Java, t.*

Inhabits Java.

653. 9. *S. Affinis* (Allied Squirrel.) Fur ash-gray above ;

beneath nearly white, with a reddish brown line on each side.

Sciurus Affinis, *Raffl. Lin. Trans.* XIII.

Icon. ――

Inhabits Sumatra.

654. 11. *S. Nigrovittatus* (Black-banded Squirrel.) Fur foxy-gray varied with brown above ; edges of the abdomen and circle round the eyes paler; beneath gray, with a lateral black line. Tail longer than the body, ringed with black.

Sciurus Nigrovittatus, *Horsf. Zool. Java.*

Icon. ――

Inhabits Java.

655. 12. *S. Tenuis* (Slender Squirrel.) Fur above finely variegated with deep gray and black ; lateral edge foxy ; beneath yellowish-gray ; tail foxy, banded with black.

Sciurus Tenuis, *Horsf. Zool. Java.*

Icon. ――

Inhabits Singapore.

656. 13. *S. Finlaysonii* (Finlayson Squirrel.) Fur milk white; back yellowish; eyes, whiskers, and soles of the feet, black ; tail with scattered black hairs.

Sciurus Finlaysonii, *Horsfield, Zool. Java.*

Ecureuil Blanc de Siam, *Buff. Hist. Nat.* VI. 256.

Icon. ――

Inhab. Java.

657. 14. *S. Palmarum* (Palm Squirrel.) Upper part of the body gray-brown, marked with three longitudinal bands of a pale white, the two lateral terminating at the eyes, under part white; tail reddish above, whitish underneath. Length of the body about six inches.

252

Sciurus Palmarum, *Gm.* 149. S. Pencillata ?, *Leach.*
Palmiste, *Buff. Hist. Nat.* 10, 126. Rat Palmiste, *Brisson, Règ. Anim.* 156.

Icon. *Buffon, l. c. p.* 26. *Leach, Zool. Misc. t.*
Inhabits India, Africa.

658. 15. *S. Getulus* (Barbary Squirrel.) Brown above, with longitudinal white lines reaching to the tail; body about five inches long; tail the same.

Sciurus Getulus, *Gm.*

Barbarian Squirrel, *Edwards,* 198. White-striped Squirrel, *Pen. Glean. Quad.* ii. 150. Baresque, *Buff. t.* 10.

Icon. *Edwards, l. c. tab.* 198. *Schreber,* 221. *Buff. l. c. pl.* 27. *Ency. Méthod. t.* 76. *f.* 3.

Inhabits Northern Africa.

659. 16. *S. Capistratus* (the Masked Squirrel, or Capistrate). Body ashy; head black; muzzle, ears, and belly, white. Larger than the species of Europe.

Sciurus Capistratus, *Bosc. Ann. Mus. t.* 1. 281.

Ecureuil à Masque, *Cuvier, Règne Animal.* i. 205.

Icon. *Ency. Méthod. Supp. t.* ii. *f.* 2. *Schreb. t.* 313, B. *Brown, Illus. t.* 47.

Inhabits South Carolina.

Obs. There is a black variety of this species, (Brown's Illustration, pl. 47,) and a gray variety, with a black belly, (Desmarest,) Ency. Méthod. Mammalogie, 333.

M. F. Cuvier also considers the Coquallin of Buffon, t. 13, Sciurus Variegatus, of Gmelin, to a variety of this species.

There is still a black Squirrel of North America, which M. Desmarest considers different from the black variety of the Capistrate by its smaller size, the softness of the fur, and because the nose and ears are not regularly white, and are different from the black variety of the Gray Squirrel and the tail shorter. He identifies this with the Black Squirrel of Catesby, Carol. t. 273, and ,Barham's Travels in North America, II. 31, and with Schreber, fig. 215.

660. 17. *S. Cinereus* (Gray or Carolina Squirrel). Larger than the European species generally; ash-coloured, with a white belly, a yellowish line on the belly.

Sciurus Cinereus Virginianus Major, *Ray. Syn. Quad.* 215. Sciurus Carolinensis, et Cinereus, *Gm.*

Petit Gris, *Buff. t.* 10. Gray Squirrel, *Pen. Quad.* II. 144.

Icon. *Buff. l. c. pl.* 25. *F. Cuvier, Mam. Lithog.*

Inhabits North America.

Var. B. Rubrolineatus. Fur grayish on the sides, with a red line on the middle of the back; belly white.

Sciurus Rubrolineatus, *Desm. Mam.* 333.

Ecureuil Rouge, *Warden, Descrip.* v. 630.

661. 18. *S. Rufiventer* (Red-bellied Squirrel.) Fur gray-above, bright red beneath; feet brown; tail shorter than the body; base gray-brown and yellow.

Sciurus Rufiventer, *Geoff. Col. Mus. Desm. Dict. Hist. Nat.* x. 103.

Icon: ——

Inhabits North America.

662. 19. *S. Ludoviciana* (Red River Squirrel.) Body, and upper part of the tail, dark gray; the belly, inside of the legs, and thighs, and under part of tail, reddish-brown; ears not bearded; tail longer than the body, very broad.

Sciurus Ludoviciana,·*Curtis. Barton's Med. Phys. Journ.* VI. 47.

Icon. ——

Inhabits the shores of the Red River, in America.

663. 20. *S. Grammurus* (Lined-tail Squirrel.) Body cinereous; fur very coarse; three black lines on each side of the tail.

Sciurus Grammurus, *Say, Long's Exped.* vi. 72.

Icon. ——

Inhabits the Rocky Mountains, America.

664. 21. *S. Lateralis* (Side-marked Squirrel.) Above brownish, cinereous; each side of the back marked with dull yellowish stripes, white dilated line, broader before.

Sciurus Lateralis, *Say, Long's Exped. Rocky Mount.* vi. 46.

Icon. ——

Inhabits the Rocky Mountains, America.

665. 22. *S. Quadrivittatus* (Four-banded Squirrel.) Head brownish, intermixed with fulvous, marked with four white lines; sides fulvous ; beneath whitish.

Sciurus Quadrivittatus, *Say, Long's Exped. Rocky Mount.* vi. 45.

Icon. ——

Inhabits the Rocky Mountains, America.

666. 23. *S. Magnicaudatus* (Large-tailed Squirrel.) Body above, and each side, mixed with gray and black ; sides of the head and orbits pale ferrugineous; cheeks, under the eyes, and ears, dusky.

Sciurus Macrourus, *Say, Long's Exped. Rocky Mount.* vi. 115. S. Magnicaudatus, *Harlan, Faun. Amer.* 178.

Inhabits Canada.

Length of head and body one foot seven inches; tail ten inches.

667. 24. *S. Clarkii* (Clarke's Squirrel.) Silvery gray above; shoulders, flanks, belly, and insides of limbs, white,

with a slight ochery tint; tail flat, widest in the middle, and terminating in a point.

Icon. Hamilton Smith, *MSS.*

Inhabits near the Missouri, in North America.

668. 25. *S. Æstuans* (the Guerlinguet Squirrel.) Olive-gray, mixed with red above, pale red underneath; tail round, longer than the body, slightly annulated, brown, black and yellow; paws, the colour of the body; body about eight inches long.

Sciurus Æstuans, *Gm.* Myoxus Guerlinguet, *Shaw, Zool.* II. *part* I. 171. Le Grand Guerlinguet, *Buff. Sup. t.* 7.

Icon. *Buff. l. c. pl.* 65. *Ency. Méthod. t.* 77. *f.* 1.

Inhabits Guiana and Brazil.

669. 26. *S. Insignis* (Bokol Squirrel.) Fur, gray brown above, with three black lines; head gray; outer side of the limbs and sides red; chine, neck, and belly, white; tail cylindrical, brown.

S. Insignis, *Des. Mam.* 544. Larog. *F. Cuv. Mam. Lithog.* Bokkol, *of Java.*

Icon. *F. Cuv. l. c. t. Horsf. Java, t.* 3.

Inhabits Sumatra

670. 27. *S. Pusillus* (the Little Guerlinguet.) Above part of the body gray-brown and olive, mixed; lower parts of the same colour, but a lighter tint; muzzle yellow; tail round, shorter than the body, covered with brown and yellow hairs, intermixed.

Sciurus Pusillus, *Geoff. Col. Mus.* Le Petit Guerlinguet, *Buff. Supp.* VII. 263. The Wood-Rat of Cayenne.

Icon. *Buff. l. c. pl.* 46. *Ency. Méthod. t.* 77. *f.* 2.

Inhabits Cayenne.

672. 28. *S. Ginginianus* (Gingi Squirrel.) Upper part

of the body blackish sprinkled with white, with a white stripe on each side; under part white; tail round at its insertion, divided near the end, and varied with black and white; nails very long, compressed, and slightly arched; body above a foot in length; tail two-thirds as long.

Sciurus Dschinschicus, *Sonnerat, Voy.* ii. 140. Sciurus Ginginianus, *Shaw's Zool.* ii. *p.* i. 147. Sciurus Erythropus, *Geoff. Collect. Mus.* Sciurus Albovittatus, *Desmarest, Nouv. Dict. d'Hist. Nat.* x. 110. Sciurus Brazilinsis, *Brisson, Règne Anim.* 154?

Icon. *Sonnerat, l. c. tab:* 89.

Inhabits Cape of Good Hope.

Var. a. Brownish-gray above, lighter underneath; tail appearing black, though white hairs are intermixed.

671. 29. *S. Annulatus* (Annulated Squirrel.) Greenish-gray above, without lateral bands, white underneath; tail longer than the body, round, annulated, black and white; size of the Palm Squirrel.

Sciurus Annulatus, *Desmarest, Ency. Méthod. Mammalogie, Sp.* 546, *from the Paris Museum.*

Lewis's Squirrel, nobis?

Icon. Nobis?

Habitat unknown.

671. 30. *S. Bivittatus* (Two-banded Squirrel.) Fur black-brown, picked with yellowish above, bright red beneath, with a white upper and a black lower line on each side; tail round, black, brown, end red.

S. Bivittatus, *Desm. Mam.* 543. S. Vittatus. *Horsf. Java.*

Ecureuil Toupays, *F. Cuv. Mam. Lithog.* Tupai, *Raffles.*

Icon. *F. Cuv. l. c. t.*

Inhabits Sumatra.

Var. b. With the upper lateral line and belly yellow.

Several other species have been named as such by different zoologists, but which, for want of more detailed observation, remain doubtful. Of these are,

1. *S. Persicus, Gm.* 148, *Voy. t.* 40. Dusky yellow underneath, with white sides; beardless ears; and blackish-gray tail, with a white band. Inhabits Persia.

2. *S. Anomalus, Gm.* 148 (Georgian Squirrel, *Shaw, Zool.*)
Dusky ferruginous, with tail and under parts fulvous,
and rounded beardless ears. Inhabits Georgia. *Ency.
Méthod. t.* 75. *f.* 2. *Schreb. f.* 215.

3. *S. Erythræus* (Ruddy Squirrel of *Pennant* and *Shaw.*)
Yellowish-brown, with the under parts and tail red-
dish-ferruginous, and ciliated ears. Inhabits North
America? *Horsf. Java, N.* 10.

4. *S. Abyssinicus.* Rusty black above; belly and fore-
feet gray. Three times larger than S. Vulgaris.

5. *S. Indicus.* Purple-brown, yellow underneath; tip of
the tail orange-coloured. Inhabits the vicinity of
Bombay. Sixteen inches long; tail seventeen.

6. *Plantain Squirrel, Pen. Quad.* ii. 151. Lighter coloured than the Common Squirrel, with a yellow line
along the side; resembles the common species.

7. *Mexican Squirrel* of *Seba, Thes.* i. *p.* 76, *f.* 2. Ashy-
brown, with five or seven longitudinal white stripes.
Desmarest thinks this is a factitious species.

8. *S. Flavus.* Yellow, with roundish ears without pen-
cils. Less than half the size of the Common Squirrel.
From Guzarat, in India, according to Pennant; but
from South America, according to Linnæus.

Raffinesque, in the Annals of Nature, has described five

North American species of Squirrels ; S. Ruber (not Warden,) S. Felenus, S. Phaiopus, S. Melanotus, S. Lateralis.

PTEROMYS. *Dentition similar to that of genus Sciurus.*
Head round ; ears round ; eyes large ; anterior feet with four
elongated toes, furnished with compressed sharp talons, with
the rudiment of a thumb, having an obtuse nail; posterior
feet with five long toes, much divided, and fitted for seizing;
tail long, villose; skin of the sides extended from the anterior
to the posterior extremities, forming a sort of parachute.

675. 1. *P. Sibiricus* (Common Flying Squirrel, or Pola-
touch.) Ashy-gray above ; white underneath ; tail half
the length of the body ; about seven inches long.

Mus Ponticus, aut Scythicus Volans, *Gesner.* Sciurus
Volans, *Lin. Faun. Suec.* i. 13.

Ecureuil Volant de Sibère, *Briss. Rêg. Anim.* 159. Eu-
ropean Flying Squirrel, *Pen. Quad.* 155. Common Flying
Squirrel, *Shaw, Zool.* ii. *p.* 1, 151. Polatouche Sapan,
Desm. Nouv. Dict. d'Hist. Nat. t. 27, 404, and Pteromys
Sibiricus, *ejusd. Ency. Méthod. Mammalogie, sp.* 553.

Icon. *Schreb. tab.* 223. *Shaw, l. c. tab.* 149.

Inhabits Finland, Lapland, &c., in Europe and in Siberia.
Type of F. Cuvier's genus Sciuropterus.

673. 2. *P. Volucella* (Assapan.) Grayish-brown above ;
white underneath ; tail nearly as long as the body ; body
about five inches long.

Sciurus Volucella, *Pallas, Glires,* 353.

Polatouche, *Buff. t.* 10. Flying Squirrel, *Catesby, Carolina,* ii. Assapan, *F. Cuvier, Mam. Lithog.*

Icon. *Schreber, tab.* 222. *Buff. l. c. tab.* 41. *Catesby,
Carolina, pl.* 76. *Edwards, Birds, pl.* 191. *Ency. Méthod.
t.* 77, *f.* 4. *F. Cuv. Mam. Lithog.*

Inhabits Canada and the United States.

674. 3. *P. Genibarbis* (the Kechubu.) Hoary on the upper part, with a yellowish dorsal line ; tail oblong, obtuse, flat, and distichous ; numerous vibrissæ on the cheeks.

Pteromy Genibarbis, *Horsfield, Zool. Java, No.* 4.

Icon. *Horsfield, l. c.*

Inhabits Java.

675. 4. *P. Nitidus* (the Bright Pteromys.) Deep chestnut above ; bright red underneath ; tail deep brown, particularly near the end, and cylindrical. Body about sixteen inches long.

Pteromys Nitidus, *Desmarest, Ency. Méthod. Mammalogie, sp.* 551.

Ecureuil Eclatant, *Geoff. Collect. du Mus.*

Inhabits Java.

676: 5. *P. Sagitta* (the Barbed Polatouche.) Deep brown above ; white underneath ; tail bright brown. Length of body about six inches, of tail nearly the same.

Pteromys Sagitta, *Cuv. Règ. Anim.* 207. Sciurus Sagitta, *Penn.* Polatouche Flêche, *Geoff. Collect. du Mus.*

Icon. ——

Inhabits Java.

677. 6. *P. Petaurista* (Sailing Squirrel) Chestnut colour, with the hairs tipped with white on the shoulders ; whitish-gray underneath ; thighs red ; feet brown ; tail blackish and cylindrical.

Sciurus Petaurista, *Pallas, Miscel.* 54. Tagnan, *Buff. Hist. Nat. Sup.* iii. Sailing Squirrel, *Pennant, Quad.* ii. 152.

Icon. *Pallas, l. c. p.* 6, *f.* 1 and 2. *Buff. l. c.* iii. *tab.* 21 and 22 *bis,* and *Sup. tab.* 67. *Schreb. tab.* 224. *Ency. Méthod. t.* 77, *f.* 5, 6.

Inhabits India and the Islands.

260

678. 7. *P. Genibarbis* (the Kechubu of Java.) Gray on the upper part; white underneath; vibrissæ on the cheeks and side of the head.

Pteromys Genibarbis, *Horsfield's Zoological Researches.*
Icon. *Horsfield, l. c.*
Habitat Java.

679. 8. *P. Lepidus.* Fur blackish-brown; beneath white; head and middle of the back gray; tail longer than the body, oblong, flat; ears oblong, simple, naked; vibrissæ very large.

Pteromys Lepidus, *Horsf. Java.*
Icon. *Horsf. l. c.*
Inhab. Java.

Obs. For the Cheiromys or Aye Aye, placed after Pteromys in the text, see p. 51 of Table.

SECTION II. *With imperfect clavicles, or none.*

Genus III.—HYSTRIX.

Incisors $\frac{2}{2}$; canines $\frac{0.0}{0.0}$; cheek-teeth $\frac{44}{44} = 20$; cheek-teeth, with the tops flat, but furnished with ridges of enamel. Head strong; muzzle thick; ears short and long; tongue furnished with spiny scales; anterior feet with four toes, and the rudiment of a thumb; posterior with five; nails strong; spines on the body, sometimes intermixed with hair; tail more or less long, sometimes prehensile.

Obs. M. F. Cuvier has given us some remarks on the teeth and characters of this genus, and has divided it into five genera; it had before been usually divided into two sections according to the tail. See *Memoires du Museum d'Histoire Naturelle*, tom. IX. p. 413.

680. 1. *H. Cristata* (Crested Porcupine.) Very long spines on the back, annulated black and white; a mane of

261   2 T

long stiff hairs on the head and neck ; tail short; length of body upwards of two feet; of tail about three inches,

Hystrix Cristata, *Lin. Sys. Nat.* 76. Hystrix Dorsata, *Gm.*

Porc-epic, *Buff. His. Nat.* 12. Crested Porcupine, *Pen. Quad.* ii. 122.

Icon. *Buff.* 12, *tab.* 51, 52. *Schreber, tab.* 169. *Ency. Méth.* 64, *f.* 3.

Inhabits Africa, and naturalized in Southern Europe.

Obs. Type of M. F. Cuvier's genus *Hystrix.*

681. 2. *H. Fasciculata,* (Pencillated-tail Porcupine.) Spines like strips of parchment ; those on the body flat, black. Less than the Common Porcupine.

Histrix Fasciculata, *Lin.* Mus Fasciculatus, *Desmarest, Mammal.* 308. Acanthion Daubentonii, *F. Cuv.*

Porc. Epic. de Malacca, *Buff. Sup.* 67. Brush-tailed Porcupine, *Shaw, Zool.* ii. 11.

Icon. *Buffon, l. c. pl.* 77. *Shaw, Gen. Zool.*

Inhabits India.

Obs. Type of F. Cuvier's genus *Acanthion.*

682. 3. *H. Longicauda.* (Long-tailed, or Marsden's Porcupine.) Like the last but tail shorter, notwithstanding the name given by Marsden.

Histrix Longicauda, *Marsden, Sumatra.* Acanthion Javanicum, *F. Cuvier.*

Landak of Java.

Icon. *Marsden, l. c. t.* 17. *F. Cuv. Mem. Mus.* x.

Inhabits Sumatra.

683. 4. *H. Macroura* (Rice-tailed Porcupine.) Very like the Cöendou, but the tail having a bundle of spines at the end formed like grains of rice.

Histrix Macroura, *Gm. S. N.* 77. Histrix Orientalis,

Brisson, Quad. 131. Porcus Aculeatus Sylvestris, *Seba,*
Thes. t. 12. Mus Macrourus, *Desmarest, Mammal.* 309.

Rice-tailed Porcupine and Long-tailed Porcupine, *Pen.*
Iridescent Porcupine, *Shaw, Gen. Zool.* vol. II.

Icon. *Seba, l. c. pl.* 52. *Shaw, l. c.* 124. *Pen. Quad. t.* 72.
Inhabits India.

684. 5. *H. Couiy* (the Couiy of D'Azara.) Body covered
with numerous short spines, yellowish at their base and
point, and brown in the middle; tail thick, shortish; the
latter half naked, and prehensile; nearly two feet long;
tail nine or ten inches.

Hystrix Prehensilis, var. 7, *Gm.* Hystrix aculeis appa-
rantibus caudâ brevi, *Briss. Règ. Anim.* 127. Erethizon
Buffonii, *F. Cuv. Mem. Mus.* Eucritus, *Fischer, Zool.* VII.
102.

Couiy *D'Azara, Quad. du Paraguay, t.* II. 105. Coëndou,
Buff. t. 12. (This name is probably referrible to the next
species.) Hoitzlacuatzin seu Flacuatzin, *Hernandez.*
Mexican Porcupine, *Pennant.*

Icon. *Buff. l. c. tab.* 54.
Inhabits South America.

685. 6. *H. Dorsata* (Canada Porcupine.) Hair long;
prickles short; male deep brown; female lighter brown;
tail long; bristles of the head and neck long.

Histrix Hudsonius, *Briss.* H. Dorsata, *Gmel.* H. Pilo-
sus, *Casteby, Car. App.* 30. Cavia Hudsonius, *Klein.* Erethizon Dorsatum, *F. Cuv.* Urson, *Buff.* XII.

Icon. *Schreb. t.* 169. *Buff.* XII. *t.* 55. *Ency. Méthod.*
t. 65, *f.* 1.
Inhabits Canada.

Obs. Type of M. F. Cuvier's genus *Erethizon.*

686. 7. *H. Cuandu* (the Couendou.) Body covered with

short spines, annulated black and white, without any mixture of hair on the upper part; tail two-thirds the length of the body, pointed and prehensile.

Hystrix Cuandu, *Desmarest, Mamm.* 346.

Hystrix Prehensilis, var. β. *Gm.* 76. Hystrix Americanus Major, *Brisson, Règ. Anim.* 130. Sinœthere Prehensilis, *F. Cuv.*

Coëndou à longue queue, *Buff. Sup.* vii. Brazilian Porcupine, *Pennant.*

Icon. *Buffon, l. c. tab.* 78. *Johnston, tab.* 10. *Shaw, Zool.* vi. *t.* 123. *Pennant, Quad. t.* 73.

Inhabits South America.

Obs. Type of M. F. Cuvier's genus *Sinœthere.*

687. 8. *H. Spinosa* (Spiny Sphiggurus.) Spines rather long, dark at the end; tail beneath naked.

Sphiggurus Spinosus, *F. Cuv. Mem. Mus.* v.

Le Coni, *D'Azara.*

Inhabits Paragua.

Obs. Type of M. F. Cuvier's genus *Sphiggurus.*

688. 9. *H. Villosa* (Hairy Sphiggurus.) Spines hid in the long thick hairs.

Sphiggurus Villosus, *F. Cuv.*

Orico, *Brazilians.*

Inhabits Brazils.

Genus IV.—Lepus.

Incisors $\frac{4}{2}$; canines $\frac{0.0}{0.0}$; cheek-teeth $\frac{6.6}{5.5} = 20$. Upper incisors in pairs, two in front, and two immediately behind them; the former large and cuneiformed, with a longitudinal furrow down the front, the latter small; the lower incisors square; cheek-teeth with flat crowns with transverse laminæ of enamel; ears and eyes large; five toes to fore-feet, and four to those behind, with nails slightly arched ; interior of the mouth, and soles of the feet to the nails, covered with hair ; tail short ; mammæ from six to ten ; cæcum very large.

689. 1. *L. Timidus* (the Common Hare.) Brownish red-gray ; chin and belly white ; ears black at the point ; tail white underneath, black above. About two feet in length.

Λαγως, *Ælian.* L. Timidus, *Lin. S. N.* Lepus, *Plin.*
Common Hare. Hare *of Authors.*
Icon. *Schreb.* 2?3 *A. Buff. l. c.* 38. *Pennant, l. c. tab.*
Inhabits Europe, northern and temperate parts of the Old World.

690. 2. *L. Variabilis* (Variable Hare.) Yellow-gray in summer, white in winter ; ears shorter than the head, and black at the tip at all times ; tail white in winter, gray in summer. Larger than the Common Hare.

Lepus Variabilis, *Pallas, Glires,* 40. Lepus Albus, *Bris. Rêg. Anim.* 139.
Varying Hare, and Alpine Hare, *Penn, Br. Zool.* i. *n.* 20.
Icon. *Schreb. tab.* 235, *B. Penn. Quad. t.* 96, *f.* 1.
Inhabits Northern parts of Europe, Asia, and America.
Obs. The Lepus Hybridus of Pallas, the Spurious of *Pennant,* is probably a variety of this species.

691. 3. *L. Glacialis* (Snowy Hare.) White ears, black at the tip, longer than the head ; nails strong, broad, and depressed. Larger than *L. Variabilis.*

Lepus Glacialis, *Sabine, Suppl. Parry's Voy.*
Lievre du Grœnlandon Rekalek, *Desmarest,* 349.
Icon ——
Inhab. within the Arctic Circle. Mus. Brit.

692. 4. *L. Virginianus* (Virginian Hare.) Grayish-

brown in summer, white in winter ; the orbits of the eyes surrounded by a reddish fawn colour at all times ; ears and head of nearly equal length ; tail very short.

Lepus Virginianus, *Harlan, Amer. Fauna,* 196.

Varying Hare, *Warden's Descrip. Unit. States, V. B.* 2.

American Hare, *Pennant, Quad. ?* Hare or Hedge Coney, *Lawson,* 122. ?

Icon ———

Inhab. Virginia.

Var? Plumbeous above; white beneath during summer, of a pure white in winter ; tips of the ears black or reddish-brown at all seasons; body covered with fine close fur ; tail round, bluntly pointed,

Varying Hare, *Lewis* and *Clark's Exped.* vi. 179.

693 5. *L. Cuniculus* (the Rabbit.) Gray and yellow mixed ; reddish about the neck, throat, and belly ; white tail ; brown on the upper side about seventeen or eighteen inches long.

Δασυπους, *Aristotle.* Dasypus, *Pliny.* Cuniculus, *Johnston.* Lepus Cuniculus, *Lin. Sys. Nat.* 1. 77. Lepusculus, *Klein.*

Rabbit or Coney *of Authors.*

Icon. *Schreb.* 236, *A. Buff. t.* 6. *tab.* 50. *Ency. Méthod. t.* 62, *f.* 3, *t.* 63, *f.* 1, and *t.* 62, *f.* 4.

Habitat by transportation almost all parts of the world, except the north of Asia ; said to have been originally from Africa.

In domestication the Rabbit varies without end. The most remarkable are—1. The Angora. 2. The Russian Rabbit, figured by *Pennant* from *Edwards, t.* 69, *f* 2.

694. 6. *L. Tolai* (Baikal Hare.) Gray mixed with brown and yellow ; belly white ; neck yellowish ; white above, yellowish underneath; paws yellow; ears of the

female shorter than those of the male. Larger than the Common Hare.

Lepus Tolai, *Pallas, Glires,* 17. Lepus Dauricus, *Erxleben.*

Tolai, *Buff.* 15. 138. Lapin de Sibèrie, *Cuv. Règ. Anim.* i. 211. Baikal Hare, *Pennant, Quad.* 104.

Icon. *Schreb. tab.* 234. *Pallas, l. c. tab.* 4. *f.* 2.

Inhabits Mongolia and Tartary.

695. 7. *L. Americanus* (American Hare.) Yellow gray, varied with brown; neck yellow; throat and belly white; ears shorter than the head, without black tips; about as big as a rabbit.

Lepus Americanus, *Erxleb.* Lepus Hudsonius, *Pallas, Glires, pa.* 30. Lepus Nanus, *Schreber.*

American Hare, *Forster, Phil. Trans.* 62, *pa.* 376.

Icon. *Schreber,* 234, *B.*

Inhabits North America.

696. 8. *L. Africanus,* (Cape Hare.) Ears one fifth larger than the head; size and colour of the Common Hare, but the legs are ferruginous, and a little larger.

Lepus Capensis, *Lin. S. N. ?* Lepus Ægyptius, *Geoff. Mem. d'Egypt.*

Lièvre d'Afrique, *Cuv. Règ. Anim.* 211.

Icon. *Geoff. l. c. tab.*

Inhabits the whole of Africa.

Obs. The *Lepus Capensis* has been treated as distinct from the *Lepus Ægyptius,* but the Baron places them together.

697. 9. *L Brasiliensis* (Tapiti, or Brasilian Hare.) Varied with brown and yellowish above; a white half collar on the throat; ears much shorter than the head; tail very short.

Lepus Brasiliensis, *Lin. S. N.* 78. Lepus Tapeti, *Boddaert.* Tapeti Brasiliensibus, *Marcgr. Brasil.* Lepus Ecaudatus, *Brisson, Quad.* 97.

Tapéti, *Marcg. Bras.* 22. *Pison, Ind.* 102. Lièvre Tapeti, *D'Azara, Quad. of Paraguay,* II. 57.

Brazilian Hare, *Pen. Quad.* II. 107. Collared Rabbit, *Wafer.*

Icon. *Marcgrave, l. c.* 223. *Pison, l. c.* 102.

Inhabits South America.

The *L. Viscaccior* of Gmelin, 160, is said to have four toes on the fore-feet and three behind, with a long tail. It seems probable that an animal of some other species was intended.

698. 10. *L. Nigricollis* (the Moussel.) Top of head sprinkled yellow ; red sides ; gray chin, and throat white ; grayish-white band from the muzzle to the ear ; upper part and sides of neck and shoulder bright black ; size of a rabbit.

Lepus Nigricollis, *F. Cuvier, Dict. des Sciences,* XXVI.

Moussel, *Id.*

Icon. ——

Inhabits Malabar and Java.

Obs. Brought by MM. Leschenault, Diard, and Duvaucel from India.

699. 11. *L. Saxatilis* (Rock Hare.) Reddish-gray ; under parts white ; ear red behind ; black-brown at tips.

Lepus Saxatilis and Lièvre des Roches, *F. Cuvier, Dict. de Sciences Nat.* XXVI.

Icon ——

Inhabits the Cape.

Obs. Brought by M. de Leland from the Cape.

Sub-genus LAGOMYS. *Teeth and toes similar to those*

of Lepus; ears moderate; eyes round; hind legs not much larger than those before; fur under the feet; tail none; mammæ four or six; clavicles nearly perfect.

700. 1. *L. Alpinus* (the Pika.) Reddish-yellow ; ears, and palm of the feet dark-brown. About ten inches long.

Lepus Alpinus, *Pallas, Glires*, 45. Lagomys Pika, *Geoff.*

Pika or Picka, of the inhabitants of the shore of Lake Baikal. Alpine Hare, *Pen.* ii. 107.

Icon. *Pallas, l. c. tab.* 2. *Schreber,* 238. *Pennant, Quad. t.* 70, *f.* 2.

Inhabits the Northern Mountains of the Old World.

701. 2. *L. Ogotoma* (the Ogotone, or Gray Pika.) Pale brownish-gray ; feet yellowish ; ears oval, of the same colour as the body. About seven inches long.

Lepus Ogotona, *Pallas, Glires*, 59. Lagomys Ogotona, *Desm. Mamm.* 353. Lepus Alpinus, *Erxleb.*

Ogotome *of the Mongole Tartars.* Ogotoma Hare, *Pen. Quad.* ii. 109.

Icon. *l. c. tab.* 3. *Schreb. tab.* 239. *Pennant, Quad. t.* 70, *f.* 3.

Inhabits Mongolian Tartary

702. 3. *L. Pusillus* (Calling Hare of Pennant.) Gray-brown ; ears nearly triangular, edged with white. About six inches long.

Lepus Pusillus, *Pallas, Glires,* 31. and *Nov. Com. Petrop. t.* 13. 534. Lagomys Pusillus, *Desm. Mamm.* 353.

Calling Hare, *Pen. Quad.* ii, iii.

Icon. *Pallas, Glires, tab.* 1. *Com. Petrop. tab.* 14. *Schreber,* 237. Sulgam, *Vicq. d'Azyr. Syst. des Anim. Pen. Quad. t.* 70, *f.* 1.

Inhabits South-eastern parts of Russia.

Genus V.—Hydrochærus.

Incisors $\frac{2}{2}$, without longitudinal furrow, the lower compressed, and sharp ; canines $\frac{00}{00}$; cheek-teeth $\frac{44}{44}$, laminous. Muzzle compressed ; eyes large; ears moderate ; round anterior feet with four palmated toes ; posterior with three. No tail. Two mammæ. Hair scattered and bristly.

703. 1. *H. Capybara* (the Capybara.) Colour dingy, deepest above; head very large; nostrils distant. Length nearly three feet.

Capybara Brasiliensibus *Marcgrave, Bras.* 230. Sus Maximus Palustris, *Barrère.* Hippopotamus Ecaudatus, *Hill. Anim.* 569. Le Cabiai Hydrochærus, *Briss. Règ. An.* 117. Mus Hydrochærus, *Lin.* Sus Hydrochærus, *ejusd.* 103. 12*th Ed.* Cavia Capybara, *Gm.* Hydrochærus Capybara, *Erxleb.*

Cochon d'Eau *Desmarchais, Voy. t.* 3. 298. Cabiai, *Buff.* 12. 384. Capward, *Troger's Voy.* 122. River Hog, *Wafer in Damp.* III. 400. Irabubos, *Gumil*, 22 or III. 238· Capygona, *D'Azara, Quad. of Praguay*, II. 12. Thick-nosed Tapir, *Pen. Synops.* 83. Capybara Cavy, *Pen. Quad.* II. 88.

Icon. *Marcgrave, l. c. Buff. l. c. tab.* 49. *Schreb.* 174.

Inhabits the shores of the great rivers of South America.

Genus VI.—Cobaya.

Incisors $\frac{2}{2}$; canines $\frac{00}{00}$; cheek-teeth $\frac{44}{44}$ = 20. Body thick ; muzzle short, compressed ; eyes large; ears round; legs short; four toes on the fore-feet, and three only on those behind; not palmated. No tail; two teats ventral.

704. 1. *C. Cobaya* (the Cobaya or Guinea Pig.) Wild Var. Reddish-gray, or like a hare on the upper parts.

Tame Var. Varied with large patches, black, yellow, and white. Length nearly one foot.

Aperea Brasiliensibus, *Marcgr. Bras. Ind. Pison.* Amæna,

F. Cuvier. Cabaya, *G. Cuvier.* Cuniculus Brasiliensis, *Briss. Rêg. Anim.* 149. Cavia Aperea, *Erxleb.* Aperea, *D'Azara, Quad. of Paraguay*, ii. 6. Cavia Cobai Brasiliensibus, *Marcg. Bra.* 224. Porcellus Indicus, *Johnston, Quad.* Cavia Cobaya, *Pison, Erxleb.* Mus seu Cuniculus Americanus et Guineensis, *Ray. Syn.* 223. Mus Brasiliensis, *Lin.* Mus Porcellus, *Lin. Syst. Nat. ed.* 12. 79. Cuniculus Indicus, *Briss. Rêg. Anim.* 146.

Cochon d'Inde, *Buff. Hist. Nat.* viii. Restless Cavy, *Pen. Quad.* ii. 89. Variegated Cavy, *Shaw, Zool.* ii. *pl.* 1. 17. Rock Cavy, *Pen. Quad.*

Icon. *Marcg.* 224. *Buff. l. c. pl.* 1. *Schreber, tab.* 173.

Inhabits Brazil, Paraguay, &c.; the domesticated variety has been transported to almost all the temperate parts of the world,

Genus VII.—Dasyprocta.

Incisors $\frac{2}{2}$; canines $\frac{0\,0}{0\,0}$; cheek-teeth $\frac{4\,4}{4\,4} = 20$. Head rather elongated; forehead flat; muzzle thick; eyes large; fore-paws with four toes, and a tubercle for a thumb; hind-legs longer than those before, with three toes, and long strong nails; sole of the foot naked and callous.

705. 1. *D. Acuti* (the Agouti.) Brown, sprinkled with yellow or reddish; orange on the crupper; ears short; tail rudimentary. Nearly two feet long.

Mus Sylvestris Americanus, *Ray, Syn.* 226. Cavia Aguti, *Gmel.* Cuniculus Americanus, *Brisson, Rêg. Anim.* 143. Dasyprocta Acuti, *Illiger.* Chloromys, *F. Cuv.* Platypyga, *Illiger ?* Long-nosed Cavy, *Penn. Quad.* vi. 94. Long-nosed Rabbit, *Wafer.* Small Indian Coney, *Brown, Jam.* 484.

Acutis, *Johnston, Quad.* Agouti, *Buff. t.* viii.

Icon. *Marcgrave, Bras. Johnston, Quad. tab.* 63. *Seba, tab.* 41. *f.* 2. *Buff. l. c. tab.* 50. *Schreb. tab.* 172.

Inhabits South America.

706. 2. *D. Cristata* (the Crested Agouti.) Blackish, sprinkled, with red hair on the occiput ; crupper very long ; ears and tail short.

Cavia Cristata, *Geoff. Coll. du Mus.* Dasyprocta Cristata, *Desmarest.*

Agouti a Crête, *F. Cuvier, Dict. des Sci. t.* 6.

Icon. *Menag. Nation, No. 5, pl.* 3.

Inhabits Surinam.

707. 3. *D. Acuschy* (the Akouchy.) Brown, spotted with yellow ; crupper blackish, and belly red.

Cavia Acuschy, *Gm.* Cavia Acuschy, *Erxleb.*

Olive Cavy, *Penn. Quad.* Akouchy, *Buff. t.* 15.

Icon. *Burrere Fr. equinox, pl.* 153. *Buff. Supp.* III. 36. *Schreb. tab.* 171.

Inhabits the West Indies.

708. 4. *D. Patagonica* (the Patagonian Cavy.) Brownish-gray ; dotted on the back, darker on the crupper, white on the thighs and belly ; yellowish on sides ; ears long ; tail short, about thirty inches long, and the average height seventeen or eighteen.

Cavia Patachonica, *Shaw, Gen. Zool.* vol. II.

Lievre Pampa, *D'Azara, Quad. du Paraguay,* Patagonian Hare, *Byron's Voyage.* Patagonian Cavy, *Penn. Quad.*

Icon. *Pennant, Quad. pl.* 68. *Shaw, l. c.* 165.

Inhabits Patagonia.

Obs. This species seems nearly allied to *Lepus.*

709. 5. *D. Viscacha* (the Viscache.) Dirty white ; sides of the head black ; moustache seven inches long ; body and neck thick, large, and cylindrical ; tail nine inches long, naked at tip, but with bristly hairs on the upper part of the remainder ; anterior feet with four toes, and digging nails posterior, with three. As big as a hare.

Dolichotes Vischacha, *Desmarest, Jour. de Phys.* Lepus

Viscaccia, *Molina Chili*, 272.   Viscacha, *Niremberg, Hist. Nat.* 161.   Viscache, *D'Azara, Quad. du Paraguay.*

Icon. ——

Inhabits Brasil Chili.

Obs. This species appears intermediate between the genera *Lepus* and *Dasyprocta.*

Genus VIII.—Cælogenus.

Incisors $\frac{2}{2}$; canines $\frac{0.0}{0.0}$; cheek-teeth $\frac{4.4}{4.4}$. Toes five on all the feet; the interior toes of the fore-feet, and interior and exterior toes of those behind, very small.

710. 1. *C. Subniger* (the Brown Paca.)   Dingy-brown, spotted with white; head large; neck short; body thick; ears round; fur short and harsh.

Paca Brasiliensibus, *Marcgrave, Brazil. lib.* 6. 224. Cælogenus Subniger, *F. Cuvier.*

Cottie, *Johnston, Quad.* III.   Pag or Pague, *Lery. Hist. d'un Voy. à Brasil.* 138.   Paca, *Male, Buffon, Sup.* III. 35. Pay, *D'Azara, Quad. du Paragua,* II. 20.

Icon. *Johnston, l. c. tab.* 63. *Buff. l. c. pl.* 35. *F. Cuvier, Anim. du Mus.* x. *pl.* 9. and *Mam. Lithog.*

Habitat South America.

711. 2. *C. Fulvus* (the Yellow Paca.)   Like the last, only with the ground colour yellow.

Cuniculus Paca, *Briss. Règ. Anim.* 145.   Cælogenus Fuscus, *F Cuvier.* Paca Femelle, *Buff.* x.

Icon. *Buff. l. c. pl.* 43. *Annales du Mus. t.* x. *pl.* 9. *Ency. Méthod. t.* 65, *f.* 4.

Inhabits South America.

Order V.—EDENTATA.

No incisive teeth in general; canines in some, but not in all; some of the genera with cheek-teeth only, and some perfectly edentatous ; toes varying in number, and generally armed with great nails ; orbits and temporal fossæ united.

Food various, vegetable for some of the genera, insects and flesh for others.

Habits various, generally inactive.

Inhabits South America, Central Africa, the Indian Islands, and New Holland.

SECTION I. *E. Tardigrada* or *Sloths.*
Genus I.—BRADYPUS.

Incisors $\frac{0}{0}$; canines $\frac{1}{1}$; cheek-teeth $\frac{4}{3}\frac{4}{3} = 18$; canines larger than the cheek-teeth, pyramidical, and pointed; cheek-teeth cylindrical ; head small ; muzzle truncated ; neck short ; nostrils at the extremity of muzzle ; anterior extremities longer than the posterior, with two or three toes armed with strong nails ; fur harsh and long ; intestines short ; no cæcum ; arteries of limbs commence by infinite ramifications, which finally unite.

712. 1. *B. Tridactylus* (the Three-toed Sloth.) Three nails to all the feet; has nine cervical vertebræ ; gray, generally spotted on the back, brownish, or white ; soles of the feet hairy ; fur harsh.

Ignavus Arcthopithecus, *Gesner. Quad.* 869. Papio, 2. *Johnston.* Bradypus Tridactylus, *Lin.* Tardigradus, *Briss. Rêg. Anim.* 34. Cuaikare, *Barrere.*

Ai, *Marcgrave, Brazil,* 221. Sloth, *Edward's Av.* 220.

Icon. *Johnston, tab.* 61. *Buffon,* XIII. 5 and 6. *Edwards, Shaw,* &c. *Ency. Méthod. t.* 25, *f.* 1. *Penn. Quad. t.* 91.

Inhabits all South America.

The type of F. Cuvier's genus *Acheus,* and considered as the true Bradypus by Illiger, and most succeeding authors.

Obs. The Ai seems to vary considerably as the Spotless Ai, the Yellow-faced Ai, the Collard Ai (treated as a distinct species by Desmarest), the Ash-coloured Ai, and others mentioned in the Supplement, and figured in this work.

713. 2. *B. Didactylus* (the Unau.) Two long nails only on the fore feet ; lower jaw rather pointed ; generally of a brownish-gray colour; and from two to three feet long; soles of the feet naked.

Tardigradus Ceylonicus Catulus, *Seba Thes.* i. 54. Bradypus Didactylus, *Lin.* Simia Personala, *Klein Quad.* 42. Chælopus, *Illig. Prod.*

Unau, *Buffon*, xiii. Two-toed Sloth, *Pen. Quad.* 242. Icon. *Seba, l. c. Buff. l. c. pl.* 1. *Schreb. tab.* 65. Inhabits South America.

Obs. For notice of the Little Unau of Buffon, and other varieties, see Supplement. This is the type of Illiger's genus *Chælopus*, and *Bradypus* of F. Cuvier, and the Unaus of Gray, in the *Medical Repository.*

SECTION II. *E. Effodientia, or Digging Edentata.*
Genus II.—DASYPUS.

Incisive teeth $\frac{0}{0}$; or $\frac{2}{4}$ canines $\frac{00}{00}$; cheek-teeth varying in the several species in all from 28 to 68 : these teeth cylindrical, separate, and without enamel on the inner side; head long, mouth small, tongue partially extensible ; body altogether covered with a shell or plate armour; five toes to the hind-feet, four or five to the fore-feet, with long nails for digging ; mammæ two or four; tail rather long, round; stomach simple, intestines without cæca.

Living in woods on roots and putrid animals, rolling themselves up, for protection. Confined to the warm parts of South America.

714. 1. *D. Apar* (the Apara, or Three-banded Armadillo.) Cheek-teeth $\frac{8 8}{8 8}$; generally with three moveable transverse bands to the body; tail short and flat; five toes on all the feet.

Dasypus Tricinctus, *Lin.* Armadillo Orientalis, *Briss.* *Rêg. Anim.* 38. D. Trachyurus *Fischer. Zoon.* Toly peutes, *Illiger, Prod.* Tatusies, *F. Cuv.*

Tatu Apara, *Marc. Brasil,* 232. Tatou Apar, *Buff. t.* x. Tatou Mataco, or 8th Tatou, *D'Azara, Quad of Paraguay* II. 202. Three-banded Armadillo, *Pennant, Quad.* 246. Icon. *Schreb. tab.* 71.

Inhabits South America about Brazil and Paraguay.

The type of F. Cuvier genus *Tatuises* and Illiger's *Toly-pentes*, characterized by their having no teeth in the inter-maxillary bone.

The *Dasypus Quadricinctus* of Linnæus, the *Armadillo Indicus* of Brisson, and the *Cheloniscus*, of Colon, appears to be allied to, if not identified with, this species.

715. 2. *Dasypus Peba* (the Peba.) Cheek-teeth $\frac{8.8}{8.8}$; tail round, with rings nearly its whole length, and almost as long as the body; body with seven, eight, or nine mobile bands; plates of the shield small, rounded; those of the bands rectangular; ears very long; teats four. About two feet long.

Dasypus Peba, *Desmarest. Mam.* 368. D. Septemcinctus, D. Octocinctus, and D. Novemcinctus, *Lin.* D. Serratus, *Fischer. Zoonom.* Armadillo Brasilianus, *Briss.* 40. A. Mexicanus, *Briss.* 41. A. Guyanensis, *Briss. Rêg. Anim.* 42.

Cachicame, *Buff. Hist. Nat.* x. Tatou noir, or 5th. *D'Azara Parag.* II, 175. Tatouhou, *Guaranis.* Tatu Peba. *Marc. Brazil.* 231. Nine, eight or seven-banded Armadillo, *Penn. Quad.* Pigheaded Armadillo, *Grew.*

Icon. *Ency. Méthod. t.* 27, *f.* 2, *t.* 27, *f.* 1. *Buffon Hist. Nat.* x. *t.* 37. *Schreb. t.* 72, 73, 74, 76.

Inhabits Brazil.

716. 3. *Dasypus Hybridus*, (Mule Armadillo.) Cheek-

teeth $\frac{8.8}{8.8}$?; tail round, nearly half as long as the body; nose long; ears large; legs short; shield with six or seven moveable bands.

Dasypus Hybridus, *Desmarest, Nov. Dict. Hist. Nat.* XXXII. 492. *Fischer, Zoog.*

Tatou Mulet, *Azara, Paragua*, II. 288.

Icon ——

Inhabits Brazil.

717. 4. *Dasypus Giganteus* (Giant Tatou.) Cheek-teeth $\frac{17}{17}\frac{17}{17}$; tail round, half as long as the body, covered with plates; shield with twelve or thirteen bands, composed of long scales; ears small; head rather broad; muzzle long; claws very strong.

Dasypus Gigas, *Cuv. Règ. Anim.* I. 221. Dasypus Giganteus, *Desmarest, Mamm.* 269. *Fischer, Zoog.* Dasypus Maximus, *Gmelin.*

Grand Tatou, *Azara, Parag.* II. 132. Giant Armadillo, or Greatest Armadillo.

Icon. *Buffon Hist. Nat.* x. *t.* 41.

Inhabits Paragua.

This animal is the type of F. Cuvier's genus *Priodontes*, characterized by the very great number of its teeth, &c.

718. 5. *Dasypus Tatouay* (the Twelve-banded Armadillo.) Cheek-teeth $\frac{9.8}{7.7}$; tail round, less than half the length of the body, covered with scattered tubercles; shield, with twelve or thirteen moveable bands, formed of broad rectangular plates; ears large; head rather convex; muzzle long. About two feet three or four inches long.

Dasypus Unicinctus, *Lin. Sys. Nat.* D. Duodecemcinctus, *Gmelin.* D. Dasycerus, *Fischer, Zoogn.* Armadillo Africanus, *Briss. Règ. Anim.* 43. Tatu Mastelinus, *Ray,* 235.

Kabasson, *Buffon, Hist. Nat.* x.? Tatou Tatouay, *Azara,*

Parag. II. 155. Weasel-headed Armadillo, *Grew.* Eighteen and twelve-banded Armadillo, *Penn. Quad.*

Icon. *Buffon, Hist. Nat.* x. *t.* 40. *Ency. Méthod. t.* 27, *f.* 3. *Seba. t.* 30. *Schreb. t.* 75. *Shaw, Zool.* I. *t.* 83. *Penn. Quad. t.* 93.

Inhabits Brazil.

719. 6. *D. Sexcinctus* (Six-banded Armadillo.) Incisive teeth $\frac{2}{4}$, canines $\frac{0.0}{0.0}$; cheek-teeth $\frac{8.8}{8.8}$; six or seven moveable transverse bands; tail round, half as long as body; five toes to all the feet.

Dasypus Sexeinctus, and D. Octodecemcinctus, *Lin. S. N.* D. Flavipes, *Fisher's Zoogn.*

Encoubert, *Buff. Hist. Nat.* x. Tatou, *Belon, Obs.* 211. Tatou Poyou, *Azara, Hist. Parag.* II. 142. Weasel-headed Armadillo, *Grew, Mus. Gresh.* Cirquineou, *Buffon, Hist. Nat.* x. Six-banded and twelve-banded Armadillo, *Penn. Quad.* 249.

Icon. *Buffon, Hist. Nat.* x. *t.* 42. *Suppl.* 11. *t.* 57. *Ency. Méthod. t.* 26, *f.* 4. *F. Cuv. Mam. Lithog.* VI. *t.* *Penn. Quad. t.* 93.

Inhabits Paragua.

The type of F. Cuvier's genus *Tatous*, or Armadillos with incisive teeth.

720. 7. *Dasypus Villosus* (Hairy Armadillo) Cheek-teeth $\frac{8.8}{8.8}$; tail rather more than half as long as the body, ringed at the base; shield edged with serrated scales, furnished with six or seven moveable bands, formed of rectangular plates; ears moderate; frontal plate forming irregular scales, with the edge between the ears and the eye acute and prominent; hair very abundant, long, and brown.

Dasypus Villosus, *Desmarest, Nouv. Dict. Hist. Nat.* XXXII. 489.

Tatou Velu, *Azara, Paragua*, II. 164.

Icon. —— ?

Inhabits south side of the river Plata.

721. 8. *D. Minutus* (the Pichiy.) Cheek-teeth? tail round, ringed at the base, nearly half as long as the animal; shield tooth-edged, with six or seven bands, formed of rectangular plates; ears very small; sharp frontal plates, formed of smooth irregular scales, cut in on the sides over the eyes, but not over the ears; hair very abundant on the lower part of the shell.

Dasypus Minutus, *Desmarest, Mam.* 371.

Tatou Pichiy, *Azara, Paragua*, II. 192.

Icon. *F. Cuv. Mam. Lithog.* I.

Inhabits Buenos Ayres.

Obs. D'Azara observes that there is another kind, which he could not get a specimen of, found in Paragua.

Genus III.—ORYCTEROPUS.

Incisive teeth $\frac{0}{0}$; canines $\frac{0\cdot0}{0\cdot0}$; cheek-teeth $\frac{5\cdot5}{5\cdot5}$, separate, formed of bony substance, traversed longitudinally by a number of parallel tubes; head elongated; four toes before, and five behind, with the hind-feet plantigrade; nails very thick and like hoofs. The tarsi and metatarsi very like those of the Pachydermata.

722. 1. *O. Capensis* (Cape Orycteropus or Ant-eater.) Pale gray, inclining to red on the flanks; feet deep brown; general appearance pig-like. As big as a badger.

Myrmecophaga Afra, *Pal. Miscel.* VI. 64. Myrmecophaga Capensis, *Gm.* Orycteropus Capensis, *Illiger, Cuv. &c.*

Cochon de Terre, *Kolbe, Descript. de Cap.* Cape Anteater, *Pen.* Hardvark or Ground Hog *of the Colonists.*

Icon. *Buff. t.* 6, *pl.* 3. *Allamand, Sup.* v. *f.* 11.

Inhabits South Africa, near the Cape.

2 U 2

Genus IV.—MYRMECOPHAGA.

Perfectly toothless ; head elongated ; muzzle tapering to a point ; tongue protractile ; all the toes united to the root of the nails, four before and five behind, or two before and four behind, armed with strong digging nails; mammæ two pectoral, or two pectoral and two ventral ; tail sometimes prehensile.

723. 1. *M. Jubata* (Great or Maned Ant-eater.) Four toes before, five behind; tail furnished with long flowing hair; with an oblique black line on the shoulders; muzzle formed like a trumpet. Upwards of four feet long.

Myrmecophaga Tridactyla, *Lin. ed.* 10. M. Jubata, *ejusdem ed.* 12. Tamandua Gangu Brasiliensibus, *Johnston, Quad.* 136. Bear Ant-eater, *Dampier, Voy.* Tamanoir, *Buff. t.* x. Gnouroumy or Yogouy, *D'Azara, Paraguay.*

Icon. *Marcg. Brasil. Johnston, l. c. tab.* 62. *Buff. l. c.* and *Supp.* iii. *f.* 45. *pl.* 19. *Schreb. tab.* 67. *Shaw, vol.* i. *pl.* 19.

Inhabits South America.

724, 2. *M. Tamandua* (the Tamandua.) Four toes before, five behind ; tail round, naked toward the point, prehensile; varying much in colour, but most commonly pale gray, with a band on the shoulders. About two feet long.

Tamandua Brasiliensibus *Marcg. Brasil.* Myrmecophaga Tetradactyla et Tridactyla, *Lin.* M. Nigra, *D'Azara, Voy. to Paraguay.*

Tamandua, *Buff. t.* x. Cagouré, *D'Azara, Quad. of Paraguay.* Little Bear Ant-eater *of the Spanish Americans.*

Icon. *Schreb. tab.* 66. *D'Azara, Voy. Marcgr. Bras.* 225.
Inhabits South America.

Var. A. Yellowish-gray; transverse band, visible only in certain directions.

Var. B. Like the last, but with black before the eyes.

Var. C. Yellow, with an oblique line on each shoulder.

Var. D. Yellow, with the crupper, flanks, belly, and shoulders, bare.

Var. E. Uniformly yellow.

Var. F. Black.

Var. G. Pale yellow, with a brown mantle, figured in this Work under the name of the Ursine Ant-eater.

Var. H. With an annulated tail, figured in this Work under the name of Tamandua; annulated var. It is probably the *M. Annulata* of Desmarest, from Krusensten's Voyage.

Var. I. With a triangular brown spot about the eyes, and tail annulated.

725. 3. *M. Didactyla* (Little or Two-toed Ant-eater.) Only two nails on the fore-feet, one of which is very large ; four on the hind-feet; tail long and prehensile, naked underneath at the extremity ; fur wholly yellow, with a dorsal deeper stripe; length of body seven or eight inches.

Myrmecophaga Didactyla, *Lin.* M. Minima, *Brisson, Règn. Anim.* 28.

Little Ant-eater, *Edwards, Glean.*

Icon. *Buff. t.* x. *pl.* 30. *Edwards, l. c. pl.* 200. *Shaw, Zool. vol.* i. *pl.* 52. *Schreb. tab.* 66.

Inhabits South America.

Var. A. Without the dorsal stripe. The M. Unicolor *of Geoffroy's MSS.*

Obs. The Tamandua *of Buff. Supp.* iii. 36, and of *Shaw* and *Boddaert,* is a factitious species.

Genus V.—MANIS.

Toothless; body elongated, and reptile-like; muzzle

pointed; tongue protractile; feet with five toes formed for digging; tail long; body covered with hard scales or plates; and capable of being rolled up into a sperical shape.

726. 1. *M. Crassicaudata* (the Short-tailed, or Indian Manis.) Tail shorter than the body, thick at the base; scales forming eleven longitudinal series, about one foot ten inches long in the body; tail one foot five inches.

Phattager, *Ælian.* Lacertus Indicus Squamomus. *Bontius Ind.* 60. Tatu Mustelinus, *Klein, Quad.* 47. Manis Pentadactyla, *Lin.* Manis Brachyura, *Erxleben.* Manis Pangolinus, *Bodd.* Manis Acroura, *Desmarest, Ency. Méthod.*

Grand Lezard Ecaillè, *Perrault, Anim. t.* III. 87. Pangolin, *Buff.* x. 34. Pangolin a queue courte, *Cuv.* Broad-tailed Manis, *Pennant? Quad.* 254.

Icon. *Pennant, l. c. tab.* 87. *Schreb.* 54. *Buff.* 34. *Schreb. tab.* 69. *Seba, Thes.* I. 54. *Bontius,* 60. *Perrault, Anim.* III. *f.* 17.

Inhabits Bengal and the Indian Islands.

Obs. The specimen of the Manis in the Paris Museum, which Desmarest has described under the name of *M. Javanica,* accords with this, except in having seventeen ranges of scales.

727. 2. *M. Longicaudata* (Long-tailed, or African Manis.) Tail twice the length of the body, turned upward, compressed; scales armed with three points on their edge, smaller than the preceding.

Lacertus Squamosus Peregrinus, *Clus. Exot.* 374. Manis Tetradactyla, *Lin.* Manis Macroura, *Erxleb.* Manis Phatagus, *Bodd.* Pholidolus Longicaudatus, *Briss. Rég. Anim.* 31. Manis Longicaudata, *Geoff.*

Scaly Lizard, *Grew.* Lezard de Clusuis, *Perrault,* III. 89. Phatagin, *Buff.* x.

Icon. *Buffon, pl.* 35. *Schreb. tab.* 70. *Pennant, Quad.*
f. 94.

Inhabits Central Africa.

Obs. This or a similar species is figured in *Marsden's*
Sumatra, t. 18.

Genus VI.—CHLAMYPHORUS.

Incisors $\frac{0}{0}$; canines $\frac{0.0}{0.0}$; cheek-teeth $\frac{8.8}{8.8}$; the two first
pointed, the rest flat at top, and cylindrical in form; shell
composed of a series of transverse plates; toes five before,
and five behind, with long laterally-compressed nails;
tail short and turned downward; lower-jaw articulated
almost in the manner of the Ruminantia and Pachydermata.

728. 1. *C. Truncatus.* Body covered with a leather-like
shell, abruptly truncated behind, white silky hair underneath; tail short, and bent under the abdomen. Length
about five inches.

Chlamyphorus Truncatus, *Harlan, New York Lyceum of*
Nat. Hist.

Icon. *Harlan, l. c.* and *Zoological Journal, vol.* II.

Inhabits North America.

SECTION III.—*Monotrema,* or *Monotremes* *.

Genus VII.—ECHIDNA.

Toothless, but the palate aculeated; muzzle flat, narrow,
and small; tongue protractile; eyes small; external ears,
none; paws short, and five toes; a moveable sharp pointed
spur on the inner side of the hind-legs, through which an
acrid secretion is ejected; tail short; body covered with
spines; large marsupial bones. Body capable of a spherical
shape.

* The location of these two anomalous genera, in the present class,
places the systematist in the dilemma of admitting Mammalia without teats,
which have not as yet at least been discovered.

729. 1. *E. Hystrix* (the Spiny Echidna, or Porcupine Ornithorynchus.) Upper part of body covered with thick spines, without hairs ; about the size of the Hedgehog ; under part, with bristly hair ; deep brown spines, tipped black.

Ornithorynchus Hystrix, *Home, Phil. Trans.* 1802. Myrmecophaga Aculeata, *Shaw, Gen. Zool. t.* 175.

Porcupine Ant-eater, *Naturalist's Miscel.* 1792. Aculeated Ant-eater, *Pennant, Nat. Mis. f.* 109.

Icon. *Home, l. c. Shaw, l. c. pl.* 54.

Inhabits New Holland.

730. 2. *E. Setosus* (Bristly Echidna, or Ornithorinchus.) Body covered with stiff hairs, among which on the back, are to be found, on close inspection, some short spines.

Alter Ornithorynchus Hystrix, or O. Setosus, *Home, Phil. Trans.* 1802. *Shaw, Gen. Zool.*

Icon. *Home, l. c. pl.* 13. *B. Bull. Soc. Phil.* iii. *f.* 15.

Inhabits New Holland.

Genus VIII.—ORNITHORYNCHUS.

Incisors $\frac{0}{0}$; canines $\frac{0\cdot0}{0\cdot0}$; cheek-teeth $\frac{2\cdot2}{2\cdot2}$, which are merely fibrous, and are not fixed in any bone, but only in the gum ; a sort of horny beak, resembling a duck's bill ; nostrils contiguous, opening at the end of the upper beak, or mandible ; cheek-pouches ; paws pentadactylous, formed for swimming, and united behind by a web, with a spur, behind in the male, as in the last genus.

731. 1. *O. Rufus* (the Red Ornithorinchus.) Uniformly reddish-brown above, lighter underneath.

Platypus Anatinus, *Shaw, Nat. Miscel.* Ornithorynchus Paradoxus, *Blumem. Manuel, t.* 165. O. Rufus, *Peron* and *Leseur Voy.*

Duck-billed Platypus, *Shaw, Gen. Zool.* i. 229. Water Mole *of the Colonists.*

Icon. *Nat. Miscel.* and *Gen. Zool. Blumem. l. c. pl.* 14. *Peron, l. c. pl.* 34.

Inhabits New Holland.

732. 2, *O. Fuscus* (Brown Onithornchus). Fur flat, crisp, and blackish-brown above.

Ornithorynchus Fuscus, *Peron* and *Leseur Voy. Teras Aust.* O. Crispatus, *Wernerian Trans.*

Icon. *Peron Atlas, t.* 34, *f.* 1, 5, 6. *Leach, Zool. Misc. t.* 111.

Inhab. New Holland.

Order VI.—PACHYDERMATA.

Skin very thick, whence the Order is named. Some genera partially edentatous, others with the three sorts of teeth; quadrupedal, generally with hoofs, and the toes varying in number; stomach simple; without clavicles.

Herbivorous or omnivorous.

Habits various.

Inhabits the temperate and torrid zones.

Obs. Three families or sections of this Order have been marked:—1. The Proboscidiana, including the Elephants. 2. The Pachydermata, including all the remaining genera in the Order, except 3. The Solidungula, or Horses.

There are several extinct genera belonging to this Order known only by their fossil remains.

Genus I. Elephas.

Incisives enormously elongated, and called tusks $\frac{2}{0}$; canines $\frac{0.0}{0.0}$, cheek-teeth $\frac{2.2}{2.2} = 10$. Incisives slightly arched toward their extremity, composed of ivory, incased with a crust of enamel; cheek-teeth composed of vertical and transverse lamina, springing up from the bottom of the jaw obliquely forward; five toes on all the feet; nose greatly elongated, forming a long cylindrical proboscis, moveable with admirable precision in all directions with a sort of finger or organ of tact and holding, at the end; body very large and massive; head very large; tail rather short, pencillated at the end; mammæ two; nasal fossæ greatly elevated.

733. 1. *E. Indicus* (Indian Elephant.) The head oblong, forehead concave; ears large, but less than those of the African species; four hoofs on the hind feet; crown of cheek-teeth marked by transverse undulating bands of enamel; ordinary height about ten feet.

Ελεφας, *Aristot. Hist. Anim.*

Elephas Maximus, *Lin.* E. Indicus, *Cuv. Mem. de l'Inst. t. 2.*

Icon. *Cuv. Menag. du Mus. Encyclopedia Metropolitana.*

Inhabits all Southern Asia and the large Islands.

734. 2. *E. Africanus* (African Elephant.) Head round; forehead convex; ears very large; three hoofs to the hind-feet; crown of cheek-teeth marked by lozenge-shaped ridges of enamel. Less than the Asiatic species.

Elephas Maximus, *Lin.* E. Capensis, *Cuv. Mem de l'Instit.* E. Africanus, *ejusdem, Règn. Anim.*

Icon. *Gesner, Quad.*

Inhabits Africa.

Obs. This is probably the Elephant of the Greeks and Romans. The distinctness of the two species was discovered by the Baron Cuvier.

Genus II. Hippopotamus.

Incisors $\frac{4}{4}$; canines $\frac{1\cdot1}{1\cdot1}$; cheek-teeth $\frac{7\cdot7}{7\cdot7}$, $= 40$; upper incisors thick, short, conical, bent inward, the lower cylindrical, directed obliquely forward, the intermediate being the strongest ; the canines greatly developed, forming strong tusks ; the three or four first cheek-teeth conical and simple ; head thick and square ; muzzle very large; eyes and ears small; body very thick and heavy ; legs short, terminated with four toes; tail short ; mammæ two, ventral ; skin without hair, except at the extremity of the tail.

735. 1. *H. Amphibius* (the Hippopotamus.) Dark dirty-brown, body very heavy, and low on the legs ; ears far back ; end of the jaw very wide to accommodate the enormous teeth.

Ποταμος *of the Greeks.* Hippopotamus *of the Moderns ;* H. Amphibius, *Lin.*

Icon. *Prosper, Alpin Egypt,* i. 22 and 23. *Buff. t.* xii. 3 and 6. *Sup.* iii. 28, and vi. 4 and 5.

Inhabits nearly the whole of Africa.

Obs. Desmoulins has divided this species into two distinguished by the character of the skull of specimens from different parts of Africa. These he names *H. Capensis,* and *H. Senegalensis.*

Genus III. Sus.

Incisors $\frac{4}{6}$ or $\frac{6}{6}$; canines $\frac{1\cdot1}{1\cdot1}$; cheek-teeth $\frac{7\cdot7}{7\cdot7}$. The lower incisors directed obliquely forward ; the upper conical ; the canines increasing during the whole life of the animal, growing out of the mouth, and frequently bending toward the end ; cheek-teeth simple and tuberculous ; four toes on all the feet, the two middle ones only touching the ground ; nose elongated, cartilaginous, and furnished with a particular bone to the snout ; mammæ twelve ; body covered with a thick skin, furnished with stiff hair.

736. 1. *S. Scropha* (the Hog.) When wild, generally of a blackish-gray, striped with bands during nonage ;

tusks strong, triangular, and directed almost laterally. Varying infinitely in a domestic state.

Καπρος, *Aristot. Anim.* II.   Sus Ferus, and Porcus, *Plin. Hist. Nat.* VIII. ch. 51. Sus Aper, *Briss.* Sus Scropha, *Lin.*

Le Sauglier and Morcassin, *Buff.* v.

Icon. *Buff. l. c. pl.* 14 and 17.

Inhabits almost all the habitable world.

Obs. The principal of the many varieties of this species are noticed in our supplementary observations upon it. One is peculiar for being solidungulous.

737. 2. *S. Babyrussa* (the Babyroussa.)   Tusks not so thick as in the other species, but more elongated and curled, particularly those of the upper jaw ; legs long.

Υς τετραχερως, *Ælian, Ani.*   Babyroussa, *Bontius, Ind. Orient.* Sus Babyrussa, *Lin.*   Hog-deer *of Travellers.*

Icon. *Bontius, l. c. Buff.* XII. 48, and *Sup.* III. 12.

Inhabits the Indian Islands.

738. 3. *S. Larvatus* (Masked Boar.)   Tusks moderate, angular, and directed laterally ; a fleshy tubercle on each cheek.

Sus Larvatus, *F. Cuvier.*   Sus Africanus, *Schreb.*

Sanglier de Madagascar, *Daubenton, Description du Cabinet du Roi.* No. 1885. Sanglier a Masque, *Cuvier, Règn. Anim.*

Icon. *Schreb. tab.* 327.   *Daniel's African Scenery. f.* 22.

Inhabits Madagascar and the neighbouring parts of Africa.

Obs. The *S. Koiropotamus* and the *S. Papuensis,* or Pig of New Guinea, have very lately been noticed. The former by a figure inserted by *M. Desmoulin* in the Classical Dictionary of Natural History, and the latter by *Lesson* and *Garnot* in Captain Trecinet's Voyage.

Genus IV.—PHASCOCHÆRUS.

Incisors $\frac{2}{6}$; canines $\frac{1\cdot1}{1\cdot1}$; cheek-teeth $\frac{4\cdot4}{4\cdot4}$; the two intermediate lower incisors smaller than the rest, and apart from each other; canines or tusks of enormous size, like horns, those in the upper jaw the longest; cheek-teeth formed of clustered cylinders, first small, the rest very large in the upper jaw; the three first in the lower jaw small and apart, the rest very large; toes like Sus, large fleshy excrescencies on the cheeks.

739. 1. *P. Africanus* (the Æthopian Boar.) Tusks round, thick, directed sideways and upwards; a large fleshy lobe on each cheek.

Aper Æthiopicus, *Pall. Misc.* and *Spic. Zool.* ii. Sus Æthiopicus, *Gm.* Sus Angalla, *Bodd.*

Emgalo or Engalo, *Barbot, Guin.* 487. Sauglier du Cap, Vert or Sanglier d'Afrique, *Buff. Hist. Nat.* xv. 148, and xiv. 409, and *Supp.* iii.

Icon. *Pallas, Misc. tab.* 2, and *Spic. Zool.* 11—1. *Buff. Sup.* iii. *pl.* 11.

Inhabits Africa.

Obs. The French Zoologists make but one species of the African Boar; but see our observation at page 409, and the figure.

Genus V.—DICOTYLES.

Incisors $\frac{4}{6}$; canines $\frac{1\cdot1}{1\cdot1}$; cheek-teeth $\frac{6\cdot6}{6\cdot6}$. Canines or tusks not projecting out of the mouth; other teeth like those of Sus; four toes before, three behind; the external little toe of the hind feet of the swine wanting in this species; an opening on the back, from which is extracted a fetid humour, secreted within; tail a mere tubercle.

740. 1. *Dicotyles Torquatus* (Collared Peccary.) Hair

of the fur annulated black and white; a large blackish oblique band descending from the shoulders to the ribs.

Sus Tajassu, *Lin.* Dicotyles Torquatus, *F. Cuv. Dict. de Sciences Naturelles,* ix. 568.

Pecari, *Buff.* x. Pecari or Tajassou, *Daubenton Descrip. Anatom.* Taytetou, *D'Azara, Quad. du Paraguay,* i. 31.

Icon. *Buff. l. c. pl.* 3. F. Cuvier, *Mam. Lithog.*

Inhabits eastern side of South America.

741. 2. *D. Labiatus* (White-lipped Pecary.) Fur uniformly blackish-brown, with white round the mouth.

Sus Tajassu, *Lin.* Dicotyles Labiatus, *F. Cuv. Dict. des Sci. Nat.* ix. 519. Taguicati, *D'Azara, Quad. of Paraguay,* i. 25.

Icon. *F. Cuv. Mam. Lithog.*

Inhabits Paraguay, and probably other parts of South America.

Genus. VI.—Rhinoceros

Incisors $\frac{0}{0}$, or $\frac{2}{2}$, or $\frac{4}{4}$; canines $\frac{0\cdot0}{0\cdot0}$; cheek-teeth, $\frac{7\cdot7}{7\cdot7}$ or $\frac{6\cdot6}{6\cdot6}$. The incisors unequal among themselves when they exist; the anterior cheek-teeth small; the posterior increasing progressively; the eyes small, lateral, and placed far back, like the ears; one or two horns placed on the nose; three toes on all the feet; tail short, laterally compressed near the end; mammæ two; inguinal skin very thick, nearly without hair, and forming, in some species, thick and heavy folds.

742. 1. *R. Indicus* (Indian Rhinoceros.) Two incisors in each jaw, with a small tooth on each side of them in the upper jaw; one horn on the nose; skin forming several deep folds or plaits; length upwards of ten feet; height about five feet.

Rhinoceros, *Pliny, t.* iii. *ch.* 20, and xviii. *ch.* 1. R. Unicornis, *Lin.* R. Indicus, *Cuv. Menag.*

Icon. *Buffon, t.* xi *pl.* 7. *Parson's Phil. Trans. Edwards's Gleanings, pl.* 221. *F. Cuv. Mam: Lithog. Thomas's Phil. Trans.* 1800.

Inhabits India, especially the banks of the Ganges.

Obs. Camper has described a rhinoceros with two incisors in each jaw, as distinct from this. M. Cuvier thinks it the same species, but M. de Blainville otherwise. He has called it *R. Camperis.*

743. 2. *R. Africanus* (African Rhinoceros.) No incisors in either jaw ; two horns placed longitudinally on the nose ; skin without folds or plaits. About the size of the Asiatic species.

Rhinoceros Bicornis, *Lin.* Africanus, *Cuv.*

Icon. *Buff. Sup.* vi. *pl.* 6. *Facycis Essai de Geologie, t.* i. *pl.* 9 and 10.

Inhabits South Africa.

744. 3. *R. Bicornis Sumatrensis* (Sumatran Two-horned Rhinoceros.) Four incisors, two large and two small in each jaw, and cheek-teeth $\frac{6.6}{6.6}$. ; two horns on the nose ; skin with slight indications of folds, and one large one on the shoulders.

Sumatran Rhinoceros, *Bell, Phil. Trans.* 1793.

Icon. *Bell, l. c. Shaw, Gen. Zool.* i. *pl.* 62.

Inhabits Sumatra.

745. 4. *R. Sondaicus.* Teeth ; one horn ; body lighter than *R. Indicus ;* skin with slight folds, and covered with occasional short stiff hairs.

Rhinoceros Sondicus, *Cuv.* R. Sumatranus, *Raffles, Lin. Trans.*

Icon. Horsfield's Java.

Inhabits Sumatra.

746. *5. R. Camus.* Teeth ? undescribed horns two; muzzle truncated ; skin without folds. Nearly double the size of the common two-horned species of Africa.

Rhinoceros Simus, *Burchell, Journal de Phys.*, June, 1817, and *African Travels*, ii. 75.

Icon. *Journal de Phys. l. c. Burchell's Travels.*

Inhabits Southern Africa.

Obs. A more complete description of this species is promised by Mr. Burchell.

Colonel Gordon indicated a species as new, which Allamand edited in his edition of Buffon. Blainville thinks it probably the Simus of Burchell.

Genus VII.—HYRAX.

Incisors $\frac{2}{4}$; canines $\frac{0 \cdot 0}{0 \cdot 0}$; cheek-teeth $\frac{7 \cdot 7}{6 \cdot 6}$, $= 32$. Incisors large and bent, with a void space between them and the cheek-teeth ; anterior cheek-teeth in the upper jaw, with flat triangular crowns, the others with the crown slightly concave ; the posterior lower cheek-teeth with a transverse ridge dividing the middle of the crown ; toes before, four or three, behind four ; head large, with a slight muzzle ; nostrils oblique ; eyes small, with a large membrane ; upper lip cleft ; ears short, large, round ; no tail ; fur of two sorts, short and woolly, and long and silky ; mammæ six, two pectoral, and four ventral.

747. 1. *H. Capensis* (Cape Hyrax.) Toes four on all the feet ; grayish-brown above, whiter underneath ; inside of ears white ; length about two feet six inches, height about eight inches.

Cavia Capensis, *Pall. Misc.* 34. Hyrax Capensis, *Gm.*

Daman and Marmotte du Cap. *Buff. Sup. t.* vi. Klipdaas or Cape Badger *of the Colonists.*

Icon. *Pallas, l. c. pl.* 3. *Buff. t.* vi. *pl.* 42 and 43 ; and
iii. *pl.* 39.

Inhabits the Cape of Good Hope.

748. 2. *H. Syriacus* (Syrian Hyrax.) Differing from the
South African species principally in having only three toes
on the anterior feet, and long bristles or hairs dispersed
over the upper part of the body.

Hyrax Syriacus, *Gmel.* Askhkoko, *Bruce's Travels.*
Bristly Cavy, *Pen.* Daman Israel, *Buff. Sup.* iv. 276.

Icon. *Bruce,* v. *f.* 29, *Pen. Quad.* 68. *A. Schreb.* 211. *B.
Buff. Sup.* vi. 63.

Inhabits Syria and Abyssinia.

Genus VIII.—Tapir.

Incisors $\frac{6}{6}$; canines $\frac{1 \cdot 1}{1 \cdot 1}$; cheek-teeth $\frac{7 \cdot 7}{7 \cdot 7}$; intermediate incisors shorter than the exterior, which appear like canines ;
canines moderate, a void space between them and the cheek-
teeth, the crowns of which have two transverse ridges ;
fore-feet with four toes, the posterior with three, each toe
with a short round hoof ; nose elongated, forming a small
moveable probosis, but not prehensile like that of the Elephant ; eyes small, ears long and mobile ; tail short ; mamm-
mæ two, inguinal.

749. 1. *T. Americanus* (American Tapir.) Head laterally
compressed ; a ridge from between the shoulders along the
neck to between the eyes, which has a slight mane in the
male ; colour dirty-brown ; length upwards of six feet, height
about five feet.

Tapurete Brasiliensibus, *Marcg. Brazil,* 229. Sus Aquaticus Multisulcus, Tapir Mapouri, *Barrère, pa.* 160. Hippopotamus Terrestris, *Lin.* Hydrochærus Tapir, *Erxleb.*
Tapir Americanus, *Gm.*

Tapiluies, *Thevet. Cosmog.* II. 987. Danta, *Nieremb. Hist. Nat.* 187. Antes, *Menh. Brasil.* 23. Mountain Cow, *Dampier.* Anta *in Brasil.* Niborèbi, *in Paraguay.*

Icon. *Marcg. l. c. Dampier, l. c. Tapir, Buff.* XI. 43.

Inhabits South America very generally.

750. 2. *T. Malayanus* (Malay Tapir.) Black or dirty-brown, with a large white patch on the posterior part; when young, black, spotted, and striped with fawn-colour, or white.

Tapir Malayanus, *Raffles, Lin. Trans.*. T. Sumatrensis, *Gray's Med. Repository.* T. Indicus, *Desmarest.*

Mariba, *F. Cuv. Mam. Lithog.*

Icon. Horsfield, *Zool. Researches, F. Cuv. l. c.*

Inhabits Sumatra.

Genus IX.—Equus.

Incisors $\frac{6}{6}$; canines $\frac{1\cdot1}{1\cdot1}$, or $\frac{0\cdot0}{0\cdot0}$, in the females of some species; cheek-teeth $\frac{6\cdot6}{6\cdot6}$. Cheek-teeth furrowed on each side with flat crowns and several ridges of enamel. Between the canines and cheek-teeth is a void space; upper lip capable of considerable motion; eyes large; ears rather large, pointed, and erect; feet with a single apparent toe, covered with a thick hoof; tail with long hair, or with a tuft at the extremity; mammæ two, inguinal; stomach simple and membranaceous; intestines and cæcum very large.

751. 1. *E. Caballus* (the Horse.) Not known in its pristine state. May be characterized specifically by its long tail with long hair all over, long mane, and want of the humeral stripe.

῞Ιππος, *Aristot. Hist.* Equus, *Pliny.*

Icon. Most Zoological Works, with figures.

Inhabits the temperate climates of the Old World.

752. 2. E. Hemionus (the Dziggtai). Light-bay in summer, redder in winter, mane and dorsal line black ; tail terminated by a black tuft. As big as a moderate horse.

Equus Hemionus Dziggtai Dictus, *Pallas, Nov. Can. Petrop.* vii. 394. Probably the Wild Mule of antiquity.

Icon. *Pallas, l. c. pl.* 7.

Inhabits the Deserts of Mangolia.

753. 3. E. Asinus (the Ass.) The Wild Ass (which is presumed to have sprung from emancipated tame individuals) is as big as a moderate-sized horse, with ears not quite so long as in the domesticated race. Gray or brownish-yellow, with a brown dorsal band, and one on each shoulder. The domesticated races vary but little in colour, being generally gray, with a black humeral stripe, and long hair at the end only of the tail.

"Ονος, *Aristotle.* Onager, *Pliny.* Equus Asinus, *Lin.* Kaidon *of Southern Russia.*

Common Ass *of Authors.*

Icon. *Johnston's Quad.* 16, *Buff.* iv. 11, &c.

Inhabits

754. 4. E. Quagga (the Couagga). Head and neck dark-brown, with transverse grayish-white stripes; the under part and legs whiter ; tail terminated with long hairs. About four feet high at the withers.

Equus Quagga. *Gm.* Asinus Quagga, *Gray's Zool. Journ.* Quaccha, *Pen. Quad.* Couagga, *Buff. Sup.* vii. Female Zebra, *Edwards's Gleanings.* Opeacha, *Masson.*

Icon. *Buff. l. c. pl.* 7. Young, *Edwards, l. c. pl.* 223. *Cuv. Menag. du Mus.*

Inhabits Southern Africa.

755. 5. E. Zebra (the Zebra.) White, with numerous brownish-black bands of more or less intensity, and lighter down the middle of each band. As big as a moderate horse.

Equus Zebra, *Lin.* Zebra, *Ray.* Zebra Indica, *John.* Equus Brasiliensis, *Jacob. Mus. Règn.* Hippotigre, *Dion. lib.* 77.

Icon. *Jacob. l. c. pl.* 3. *Buffon, t.* xii. *pl.* 1, and 2. *Cuv. Menag. du Mus.*

Inhabits Southern, and probably nearly the whole of Africa.

756. 6. *E. Montanus* (the Dauw, Mountain, or Berg Paart.) Covered with pure single black and white stripes down to the hoofs.

Equus Montanus, *Burchell's Travels. Gray, Zool. Journ.*

Icon. *Gray, l. c. pl.*

Inhabits Southern Africa.

ORDER VII.—RUMINANTIA. *Pecora,* Lin.

By Charles Hamilton Smith, Esq., F.R.S., *&c. &c.*

TEETH of three sorts; incisors in the lower-jaw only, usually eight in number, opposed to a callosity in the upper-jaw; canines in some species in the upper-jaw, in others, in both, in most none; cheek-teeth or molars almost always six on each side, in both jaws; articulation of the jaw disposed for a grinding motion; no clavicles; extremities disposed for walking; the toes externally, two anterior, rudimentally in most, two posterior, all unguiculated, excepting the posterior of some. Single metacarpal and metatarsal bones to each foot; organs of digestion disposed for chewing the cud; four stomachs; intestines long; mammæ two or four, always inguinal; horny or osseous horns in the males, and often the females, of most species.

Food invariably vegetable.

Manners peaceable, residing in pairs, families or herds, in forests, or on the plains.

Inhabit nearly the whole earth, New Holland and Terra del Fuego, and smaller islands excepted.

The Order is divided into five tribes,

Camelidæ. Cervidæ. Giraffidæ. Capridæ. Bovidæ.

Tribe I.—*Camelidæ.*

No horns; no succentorial hoofs; no muzzle; nostrils slit; upper-lip divided, separately moveable, and extensible; horny soles to the feet; toes covered with crooked unguicular claws or nails; canines in both sexes; neck long; limbs long; lower abdomen drawn up under the pelvis, retromingent.

Genus I.—Camelus, Lin.

Incisors $\frac{2}{8}$; canines $\frac{1\cdot1}{1\cdot1}$; false molars $\frac{1\cdot1}{1}$; molars $\frac{5\cdot5}{5\cdot5} = 36$. Inferior incisors in trenchant quoins, the superior lateral and cuneiform; canines conical, straight, robust; false molars on each side, separated from the other teeth; in the diastema, and uncinated; head long; chaffron convex; no sinus under the eyes; nostrils slit obliquely, and closing at pleasure; eyes prominent; ears small; pores at the back of the head; feet with toes only free, the rest united; neck bent; one or two hunches on the back much developed; callosities on the sternum, and flexures of the extremities; tail reaching to the tarsus; mammæ four; hair woolly; the ventriculus with membranous cells, one of which is very large to contain water; male organs slender, reversed in a state of repose; scaphoid and cuboid bones of the tarsus separated; stature very large; belong to the old continent.

757. 1. *C. Bactrianus* (the Bactrian Camel.) Two hunches on the back; colour generally brown.

Κάμηλος Βαχτριανος, *Arist.* Camelus Bactriæ, *Plin.* C. Bactrianus, *Lin.* and *Auctor.* Chameau, *Briss. Buff.* G. and *F. Cuv.* Bactrian Camel, *Pent., Shaw.* Ditylus, *of the Lower Empire,* Werbljud, *of the Sclavonic.* Deva Deve, *of the Hunnic.* Tjuja Tue, *of Tartar Nations.* Tong, *Chinese.*

Icon. *Buff. Pent. G.* and *F. Cuv.*

Inhabits Tartary, Persia, Turkey, China, domesticated.

758. 2. *C. Dromedarius* (the Arabian Camel.) One hunch on the back, colours pale brown, whitish and fawn.

Κάμηλος Αραβιος, *Arist.* Camelus Arabiæ, *Pliny.* C. Dromedarius, *Lin.* and *Auctor.* Dromedaire, *Buff.* G. and *F. Cuv.* Gemal Gemel, in the East Arabia, *&c.* Oont, India, *Shuttur,* Persia. *Geldowesi,* Turkish. The several races, Mahairy, Ashaary, *&c.* In Morocco Egin, female Nago.

Icon. *Buff. F. Cuv. An. Lithog.*

Inhabits Arabia, Turkey, Northern Africa, India, *&c.* ; domesticated.

Obs. There appears to be a species distinct from the Bactrian and Arabian Camels, in the possession of the Ruguere.

Genus II.—AUCHENIA, Illig.

Incisors $\frac{0}{8}$; canines $\frac{1 \cdot 1}{0 \cdot 0}$; false molars $\frac{1 \cdot 1}{0 \cdot 0}$; molars $\frac{4 \cdot 5}{5 \cdot 5} = 32.$ Teeth in general resembling those of the Camel; nose slightly turned; no sinus at the back of the head; eyes large, clear; neck slender, vertical; ears long, pointed, moveable; toes protected with small hoofs, more free than in the Camels; sole of the foot shorter; no hunches on the back; tail short; two mammæ; callosities on sternum, and knees developed; male organs reversed; no vesicular appendices to the ventriculus; generic tone of colours pale purplish brown; belong to the New World.

759. 1. *A. Glama* (the Lama.) Head long ; chaffron slightly arched, joining the forehead, without sensible interruption ; back rather straight ; fur composed of long soft hair, very abundant, variously coloured, but mostly brown, with a cast of purple and white ; tail not much elevated.

Elaphocamelus, *Marcg.* Chameau du Perou, *Briss* Camelus Glama, and C. Lama, *Auctor.* Lama, *Buff. G.* and *F. Cuv.* Allocamelus, *Gesner.* C. Huanacus, *Schreb.* Guanaco, *Shaw.*

Icon. *Buff. F. Cuv. Nobis* from life.

Inhabits Peru and Southern Andes ; domesticated and wild.

760. * 2. *A. Huanaca* (the Guanaco.) Head more pointed than the preceding ; nose slightly arched ; forehead covered with woolly hairs ; lips less turned ; ears longer ; back slightly arched ; tail erect, or reversed on back ; abdomen more drawn up ; fur short, coloured pale purplish-brown and buff, not so abundant ; four feet at the shoulder ; neck vertical. Confounded with Lama, of which perhaps it is only a variety.

Cervo Camelus, *Johnst.* Pennich-Cat. *Hern. Mex.* C. Huanacuo, *of Schreb.* and *Shaw,* is a true Lama.

Icon. *Nobis* from life.

Inhabit the High Andes ? Mexico ; domesticated.

761. 3. *A. Paco* (the Paco.) Smaller by a fourth than the last ; no callosities on sternum or joints ; hair long and soft, abundant, mostly fulvous brown and gray ; lips tumid; neck rather short.

Paco, *Lact.* Paco Alpaco, *Molina.* Pacos, *Pent.* Ca-

* An asterisk before the number of the species of this Order, designates such as are not positively determined.

melus Paco, Alpaca, *F. Cuv.* Guanaco of *Stewart Trail,* Mem. of *Wern. Soc.* appears to be a variety of this.

Icon. *F. Cuv. Mamm. Lithog.*

Habitat. Peruvian and Chilian Andes ; domesticated and wild.

762. 4. *A. Vicugna* (the Vicugna.) Still smaller than the last, not three feet at shoulder ; lighter in form ; head shorter ; eyes large ; lips tumid ; body covered with very fine woolly hair ; colours pale vinous brown and buff.

Camelus Laniger, *Klein.* Vicognes, *Frezier.* La Vigogne, *Buff.* Vicunna, *Pent.* C. Vicugna, *Lin.* and *Auctor.*

Icon. *Buff. Nobis* from life.

Inhabit Peruvian Andes ; not reclaimed.

763. 5 * *A. Araucana* (the Chilihuque.) Snout curved or arched ; pendulous ears and tail ; colours various ; probably only a variety of the true Lama ; said to be the oldest domesticated race ; reduced by the Caciques of Chili.

Camelus Araucanus. Chilihuque, *Molina.*

Icon. —

Inhabit Chili.

764. 6. * *A. Huemel* (the Huemel.) Size of an ass ; colour ashy ; voice neighing ; a very doubtful species.

Equus Bisulcus. Huemel, *Molina.*

Icon. ——

Inhabit the Chilian Andes, to the Strait of Magellan. The natives on that coast are often dressed in Auchenia skins ; Can they be of this animal ?

TRIBE II.—*Cervidæ.*

No horns, or deciduous horns; feet truly bisulcated;

structure elegant, slender, mostly with muzzle, suborbital sinus, and with canines in the upper-jaws of the males; succentorial hoofs.

Genus I.—Moschus.

Incisors $\frac{0}{8}$; canines $\frac{1\cdot1}{0\cdot0}$ in the males; molars $\frac{6\cdot6}{6\cdot6} = 34$; therefore two more than in the females. Incisors and molars as in other ruminants; two or four inguinal mammæ; form of the body gathered up; the hind-quarters more elevated than the anterior; slight appearance of callosity on the breast of some; general colouring, gray-brown, with white and black, in streaks about the throat; white in the young of some on the body, and even on adults; no horns.

765. 1. *M. Moschiferus* (the Thibetan Musk.) Size of a roebuck; hair very coarse; brittle, gray-brown; a pouch on the abdomen, before the prepuce of the male containing an odoriferous unctuous substance (the musk); canine teeth long, curved back, edged; very long succentorial hoofs.

Var? Slaty blue, small, often with white on the throat; a blackish streak downwards intervening; also white from albinism.

Moschi Capreolus, *Gesn.* Animal Moschiferum, *Nieremberg, Ray.* Capra Moschi, *Aldr.* Tragus Moschiferus, *Klein.* Kabarga, *Gmel.* Le Musc, *Buff.* The Thibetan Musk, *Pent., Shaw.* M. Moschiferus, *Auctor.* Xe *of the Chinese.*

Icon. *Buff.*, and in our possession, ditto of variety.

Inhabits China, Tartary, Mountains of Thibet, and Northern India.

766. 2. *Memina* (the Memina.) Size of a rabbit; fur

SYNOPSIS OF THE

olivacious ash above, white beneath; sides and back marked with irregular white spots; no musk-bag.

Memina, *Knox, Ceylon.* Tragulus Memina, *Boddaert.* Indian Musk, *Pent.* Chevrotain a peau marqué, de taches blanches, *Buff.* Mos Memina, *Auctor.*

Icon. *Buff., Pent.* In our possession from the living specimen.

Inhabits Ceylon.

767. 3. *M. Javanicus.* (the Kantchil.) Size of a rabbit; deep red-brown on back, bay on the sides, white below; three white streaks under throat; canines long, edged, curved back; no musk-bag; very active.

Chevrotain de Java, *Buffon.* Kantchil, *Raffles.*

Icon. ——

Inhabits Java in the deep forests.

768. 4. *M. Napu* (the Napu). Size of a hare; ferruginous-gray above; whitish-gray on the sides; five white stripes under throat, divided by black stripes; canines short, straight, obtuse; no musk-bag.

Syn. M. Javanicus. Napu, *Raffles, F. Cuvier.*

Icon. *F. Cuvier, Mam. Lithog.* We are tempted to consider a drawing in our possession, from a living specimen smaller than the above, and more active, with streaks less regular, as the true Napu.

Inhabit Java, in the bushes near the sea-shore and human habitations.

769. 5. *M. Pelandoc* (the Pelandok.) Resembling the former, but with a heavier body and larger eyes.

Var.? Gray-yellow; only three streaks on the throat, white? *Lev. Mus.*

Syn. Nupu, *F. Cuv.*

Icon. *F. Cuvier, Mam. Lith.?*

Inhabit Java, in the same places as the former.

302

770. 6. * *M. Pygmeus* (Pigmy Musk.) A very doubtful species, said to be without succentorial hoofs, and to have short canines. The Antilope Pygmea, usually mistaken for it. Left here on the authority of M. Desmarests.

The Pigmy Musk of Sumatra, figured in this Work, may however be considered as a species. It is the size of the Kant-chil ; ferruginous-gray above ; three white stripes beneath the throat; legs buff. The specimen at Exeter 'Change shewed no canines externally protruding, but the muzzle was long and pointed.

Obs. All the American species are supposed to be fawns of Deer.

Genus II.—CERVUS, Lin.

Incisors $\frac{0}{8}$; canines $\frac{0\cdot0}{0\cdot0}$, or $\frac{1\cdot1}{0\cdot0}$; molars $\frac{6\cdot6}{6\cdot6} = 32$, or 34. The canines in some males compressed and bent back ; head long, terminated in most by a muzzle ; ears large ; pupils elongated ; suborbital sinus in most ; tongue soft ; no gall bladders ; four inguinal mammæ. Horns solid, deciduous ; existing in the males only, in the females with one exception none, palmated, branched or simple ; the horn consisting in a burr, or rose-shaped foot, a beam and branches, or antlers ; succentorial hoofs in all.

Sub-genus I.—ALCE. *Horns united into one blade or palm, more or less indented ; no muzzle ; no canines ; tail, very short.*

771. 1. *C. Alces* (the Elk.) Horns spreading into a broad palm, with exterior snags ; no separate branches ; the snout very tumid, overhanging ; ears long ; neck short ; legs very long ; stature considerable ; colours dark ashy-brown, sometimes white.

Var. With the basal part of the palm very deeply indented, almost separated, and generally bifurcate. It appears also that there is a fossil var. or species not as yet clearly established.

Syn. Alces, *Plin. Aldrov. Gesner.* Cervus Alces, *Auctor.* Elan, *Buff., C.* and *F. Cuv.* Orignal, *Chartevoix.* Moose-Deer, *Dudley, &c.* Elend and Elch *of the Germans.* Los *of the Sclavonic Nations.* Bulan *of the Tartaric.* Elk *of the British.* Moose *of North Americans,* and *Algonquins.* Mongsoo *of the Cree Indians.* Kistu *of the Cluches,* and Moluck *of the Columbia River Indians.*

Icon. *Fred. Cuvier, Mam. Lithog. Nobis,* from the life.

Inhabits a zone south of the arctic, from the sixty-fifth to the thirty-fifth degrees of north latitude, on both continents.

772. 2. * *C. Coronatus* (Crowned Elk.) Known only from a pair of horns about one foot long, with seven processes on each; described by Baron Cuvier as belonging to the following; but perhaps an intermediate animal between the two sub-genera. They are in the Paris Museum.

Syn. C. Coronatus, *Geoff.* Cerf Couronni, *F. Cuvier, Schreb.*

Icon. *Nobis, G. Cuv. Oss. Fossils.*

Inhabits

Sub-genus II—RANGIFER. *Horns in both sexes; palmated or pointed at brow, and bezantlers, and at top; incipient muzzle; canines in both sexes.*

773. 3. *Tarandus* (the Rein-deer.) Horns varying greatly, but in complete adults with a palm on the brow and bezantler; the beam forming a concave bend, terminated with a third palm, or with snags; smaller in the females; the

muzzle only a naked triangular spot; colours white, with intermixture of brown, or various; tail very short.

Var. It seems that the American varieties have the horns shorter, more robust, straighter; palms narrower, and fewer processes, but occupying more of the horn: 1. Caribou des Bois *. 2. Great Caribou of the Rocky Mountains. 3. Labrador, or Polar Caribou.

Tarandus, *Pliny, Ælian, Aldrov.* Rangifer, *Gesner.* Caribou, *Charlevoix.* C. Grœnlandicus, *Briss.* C. Rangifer, *ejusd.* C. Tarandus, *Auctor.* Renne, *Buff. G.* and *F. Cuv.* Rein-deer, *Pent., Shaw.* C. Mirabilis, C. Palmatus, *Johnst.* Rennthies *of the Germans.* Olen *of Sclavonic.* Juscha, Putsche, Sægau, &c., *of the North East of Asia.* The Attenk *of the Cree Indians in the Labrador Caribou.*

Icon. *Buff. Schreb. Fred. Cuv. Mam. Lithog.* Siberian American specimens. *Nobis* from Mr. Temminck's Museum and Plymouth.

Inhabit the arctic circle of both continents; in Europe, never south of the Baltic, nor in America, south of the St. Lawrence.

† *C. Guetardi* (Fossil Rein-deer.) Small, slender, almost filiform fragments of horns; belonging to an animal not larger than a fallow-deer; found near Etampes in France.

---

* A specimen, conjectured to be this variety, measured about three feet six inches at the shoulder, six feet six in length; the head one foot eight inches; each horn three feet four inches; the brow palms meeting on the forehead, the second spreading each of five snags, and one foot three, and one foot four inches long; the terminal tip developed; one snag to the rear; all very robust; general colour dark chocolate-brown, whitish intermixed; no naked triangular space between the nostrils; face very flat; ears four inches long; gray outside.

† In the former part of this Synopsis, the fossil non-existing species are not inserted. In this catalogue of Ruminants by Major Hamilton Smith, they are inserted in his respective divisions of the Order, omitting the consecutive numerals.—ED.

Syn. Bois de Cerf, trouvés a Etampes. *Guetard* Mem. *G. Cuv. Oss. Foss.* Cerf d'Etampes, Cervus Guetardi, *Desmarests.*

Sub-genus III.—DAMA. *Horns round, with brow and bezantler pointed; summit palmated, lengthways; no canines; a muzzle.*

774. 4. *C. Dama* (the Fallow-deer.) Horns in the male only, round, with brow and bezantlers pointed; the summit palmed lengthways, indented above and below; colours brown with white spots; tail long, black above, white below; no canines; a muzzle; black streak on buttocks, behind which the rest is white.

Var. Brown, and in the north, one nearly black.

Platyceros and Dama, *Pliny.* Dama Vulgaris, *Gesn. Aldrov.* Platogna, *Belon.* Daim, *Buffon,* G. and F. *Cuv.* C. Dama, *Auctor.* Fallow-deer, *Pent., Shaw.*

Icon. *Buff. Schreb. F. Cuv. Mam. Lithog.*

Inhabits Europe; Western Asia.

C. Giganteus (Fossil Dama of Ireland.) Horns of very large dimensions; broad palm with snags on both borders, fewer than in the Elk, brow and bezantler; first snag of the palm longest; skeleton resembling the Stags, but approaching in size that of the Elk.

Syn. Irish Elk. Fossil Elk of Ireland, *Molineux, Pent. &c.* C. Giganteus, *G. Cuv.* C. Hibernus, *Desmarests*; perhaps the *Machlis* of *Pliny.* Euryceros *of Oppian.* Segh *of the Britons,* and Schelch *of the Ancient Germans.*

Icon. Heads in *Phil. Transactions.* Skeleton, *Encyclop. Britain Supl. Idem, Cuv. Oss. Foss.*

Inhabit. Found in the peat bogs of Ireland, in Germany, near Worms on the Rhine, and in England, France, &c. The skeleton in the Isle of Man.

C. Paleodama (Fossil Dama of Scania.) Horns resembling the living Fallow-deer, but with only one antler on the beam, which is much more curved ; the palm narrower in proportion ; its anterior border without snags, forming a considerable segment of a circle ; the extremity to the front, and downwards ; length forty-seven inches.

C. Paleodama, *Retsius, Mem. Acad. Stockholm. G. Cuv. Oss. Foss. Desmarests.*

Inhabit. Found in peat ground near Svedala in Scania, Sweden.

* *C. Somonensis* (Fossil Dama of Abbeville.) Horns resembling the Fallow-deer, rising from the head, without pedicles, larger, wanting an anterior antler.

Syn. Daim d'une Grande Taille, *Cuv.* Cerf d'Abbeville. Cervus Somonensis, *Desm.*

Icon. *Cuv. Oss. Foss.*

Inhabit. Found in the sands on the declivities of the Somme, near Abbeville, France, and in Germany.

Sub-genus IV.—ELAPHUS. *Horns round ; three antlers turned to the front ; summit terminating in a fork or in snags from a common centre ; suborbital sinus ; canines in the males ; a muzzle.*

775. 5. *C. Elaphus* (the Stag.) Horns with three anterior antlers, all curving upwards ; the summit forming a crown of snags from a common centre ; tail middle-sized ; lachrymary sinus ; muzzle ; canines in the males ; colour red-brown in summer, brown-gray in winter ; pale disk on buttocks.

Var. Barbary and Corsican Stag. Browner, smaller, lower ; horns terminating in forks ? This is the Bukr-al washi, and the female Fortass, or Broad Scalp, because without horns, of the Moors *.

Ελαφος, *Arist. Ælian.* Cervus, *Pliny.* Cerf, *Buff.* G. and *F. Cuv.* C. Elaphus, *Auctor.* Stag, *Pent.*, *Shaw.* Hirsch *of the Germans.* Olen, and Jelen, *Sclavon.* Buga *of Tartar.*

Icon. *Buff. F. Cuv. Schreb.*
Inhabits Europe, Western Asia, Barbary, Corsica.

776. 6. *C. Canadensis* (the Wapiti.) Horns very large, branching in serpentine curves, terminating in a fork; brow antler over the face; muzzle broad; suborbital opening wide; tail very short; disk on buttocks; summer colours dun-brown; winter, dark brown-gray; stature surpassing the Elaphus.

Var. The real *C. Canadensis*, somewhat smaller; antlers more bent up; termination of beam often trifurcate; colours darker.

C. Canadensis, *Briss.* E. Strongyloceros, *Schreb.* Stag of America, *Catesby.* C. Major, *Ord.* Wapiti, *Barton, Michell, Leach.* American Elk, *Bewick.* Le Wapiti, *G.* and *F. Cuv.* Elk. Round-horned Elk. Sometimes Reddeer *of the Americans.*

Icon. *Le Sueur. F. Cuv. Bewick. Nobis* mas. et. fem.
Inhabits Canada, Missouri, and Western States.

777. 7. * *C. Occidentalis.* Nob. (North-western Stag.) Horns with three antlers to the front; summit with one, two, and even three successive bifurcations; the forks parallel to the front; the medial or bezantler longest; colours dark; tail five or six inches long, with tuft at the end; ears long; size of the Stag of Europe?

C. Auritus, *Warden.* Mule-deer, *Le Raye.* Perhaps Wewaskish, or Wa-was-keesho *of Hearn and the Crees.*

* Bukr-goat, *Al-washi* of the Forests.

Icon. In our possession.

Inhabits remotest part of North Western America.

778. 8. *C. Wallichii* (Nepaul Stag.) Horns rather short, with two small antlers at base, pointing to the front; half way up the beam a small snag turned forward; large suborbital opening; colours yellowish brown-gray; large disk upon the croup; tail very short.

C. Wallichii, *G. Cuv.*

Icon. *F. Cuv. Mam. Lithog.*

Inhabits the Mountains of Nepaul, the only specimen known being brought from thence by Dr. Wallich.

* *C. Americanus* (Fossil Stag of America.) Fragments and part of a skull of a fossil species, allied to Canadensis, found with bones of Mastodon near the Falls of Opio.

C. Americanus. Fossil Elk *of the United States. Harlan Fauna Americana*, first noticed by *Dr. Wistar. Trans. Amer. Phil. Soc. New Series, vol.* 1.

Habitat. Fossil in North America.

Sub-genus V.—Rusa. *Horns trifurcate, with basal but no median antler; beam terminating in a perch, with one process or snag on the anterior or posterior side of the beam, and forming a fork; broad muzzle; deep suborbital slit; canines, sometimes even in the females; mane on neck; in most dark colours.*

779. 9. *C. Hippelaphus* (the Great Rusa.) Horns trifurcated; basal antler on the burr; beam reclining back and outwards, with a medial bifurcation, the branch being on the external anterior side; heavy mane and beard on the neck; large suborbital opening; tail long, terminated by a dark tuft; hair coarse fulvous-brown in summer, gray-brown in winter; no disk; large stature.

Great Axis *of Pennant.* L'Hippelaphe, *G. Cuv.* The Gaucohi are noticed by the Persian physicians as of three varities, probably these and the two following. Also named Gauzen and Gozen. In Arabic Iyyol or Uyyal. In the Indee, Barensing'ha.

Icon. *F. Cuv. Anim. Lithog. Nobis* from specimens in the Paris Museum.

Inhabit Java? Bengal, chiefly the Jungleterry district.

780. 10. *C. Unicolor,* Nob. (Gona Rusa.) Horns long, slender; antlers much developed; pedicle at base; rather elevated; basal antler on the burr, curving forward, upward, and inwards; half way up the beam, second antler short, directed inwards; ears broad, pointed; muzzle broad; throat covered with long bristly hair; shoulders higher than croup; tail rather short; colour entirely brown; size large.

We have applied Professor Schreber's distinctive name of Unicolor to this species, because his description does not positively determine the animal, and the Gona is entirely of one colour.

Icon. *Daniell's Scenery* in Ceylon, &c.

Inhabits the forests of Ceylon.

781. 11. *C. Aristotelis* (the Saumer, or Black Rusa of Bengal.) Horns short, robust, pointed; vertical antler on the burr; bifurcation by a branch pointing obliquely to the rear, inserted near the summit of the beam; heavy mane on neck and throat; no disk; colours black with dun points to the hair, and dark-brown; stature large.

C. Aristotelis, *G. Cuv.* Elk *of the British Indian Sportsmen.* Saumer *in Ramghur.*

Icon. Original drawing in the possession of M. F. Cuv. and also in our own.

Inhabit Bengal in the Prauss Jungles.

782. 12. *C. Equinus* (the Malayan Rusa.) Horns robust, pearled; basal antler on the burr; terminal bifurcation from the internal posterior side of the beam; points obtuse; suborbital opening very large, moveable, admitting air; orange-coloured disk on the buttocks; heavy mane; large stature; canines in both sexes.

C. Equinus, *G. Cuv.* Rusa, *Sir S. Raffles.* Mejangan Banjoe. Great Water Stag. Jamboe Stag. Elant *of the Dutch.*

Icon. *Nobis* from life, Exeter 'Change.

Inhabit Java, Sumatra, and probably other great islands of the Indian Archipelago. India?

783. 13. * *C. Peronii* (Rusa of Timor.) Horns rather slender, of a pale brown colour; anterior basal antler as before; second antler posterior, more equal in length with the terminal point of the beam; prominent longitudinal elevation of the cranium between the horns; posterior angle of the orbits much raised; canines; snout long and pointed.

C. Peronii, *G. Cuvier.*

Icon. *The horn, pl. v. fig.* 41. *Oss. Foss. vol.* iv. *G. Cuv.*

Inhabit Timor.

784. 14. *C.* ?(Rusa of Malacca.) Male unknown; female of the size of the Hind of Europe; of a brownish-black colour, with white border to the inside of the lips and base of the ears; inside of limbs white; legs white; edge of the buttocks ferruginous; a depression above each eye forming a sinus; hair hard and strong.

Female of C. Hippelaphus? *G. Cuv.* Female of Rusa. Etam? *of Raffles.*

Icon. *Fred. Cuv. Mam. Lithog.*

Inhabit Malacca.

785. 15. *C. Mariannus* (Rusa of the Mariannas.) Horns

heavy, robust, ashy-gray; basal antler nearly vertical, with a small process between it and the beam; near the summit a second antler, posterior and internal; the animal less than the Fallow-deer, with longitudinal eminence on the skull, and near the nose two remarkable convexities; no canines; colour dark brown.

C. Marianus, *Desmarest, G. Cuvier.*

Icon. *Nobis* from the specimen in the Paris Museum.

Inhabits the Marianna Islands.

Sub-genus VI.—Axis. *Horns similar to the former, but more slender; no canines; small, or no suborbital opening; generally spotted with white; no mane; tail down to the houghs; size middling or small.*

786. 16. *C. Axis* (the Axis.) Horns round, elongated, rather smooth; anterior antler near the burr; summits of beams converging; second antler medial, on the internal side of the beam turning to the rear; no canines, or suborbital sinus; colours bright fulvous, spotted with white; tail long, brown above, the end dark; spots on ridge generally oval; spot on forehead dark.

Var. Of Ceylon, browner with small white spots, irregular, none on the forehead; head more prolonged.

Common Hog-deer, smaller, more irregularly spotted; low on the legs; horns slender; antlers very short, the second near the summit.

Axis, *Pliny. Belon. Pen. Buffon.* Cervus Axis, *Auctor.* Axis, *G.* and *F. Cuv. Shaw.* Parrah *of Indostan,* and perhaps the Ruru *of the Institutes of Menu.*

Icon. *Buff. F. Cuv.* The varieties in our possession.

Inhabits the plains near Surput and the Jungles of Indostan, Ceylon, Java, Sumatra.

787. 17. *C. Porcinus* (the Brown Porcine Axis.) Horns

slender, with the antlers very little developed : the second near the summit ; ears round at tip ; pink coloured inside ; head short and ovine ; gray ; body brown ; legs short ; two feet high at the shoulder ; species rare.

C. Porcinus, *Pen.*? Porcine-deer.   Cerf Cochon, *Buff.*

Icon. Original drawing in the possession of F. Cuvier to whose kindness we owe a copy.

Inhabit India.

788. 18. * *C. Pumilio,* Nob. (Dwarf Axis.)   Fragment of a frontal not above three inches broad across the horns; pedicles low ; horns whitish, about two inches high ; small basal antler vertical, the beam flat and pointed, without bifurcation.   Specimen in Surgeons' College, London.

Icon. *Nobis* from the above specimen.

Inhabit.   Probably India.

Sub-genus VII—CAPREOLUS.   *Horns somewhat allied to the former ; a small antler to the front high upon the beam; the superior turned to the rear, forms a fork, somewhat flattened ; no canines, nor lachrymary sinus ; rudiment of tail.*

789. 19. * *C. Pygargus* (the Ahu, or Tartarian Roe.) Horns in the young male resembling the Common Roe ; in the old about fourteen inches long ; very robust, rugous, pearled, and denticulated ; first antler of the anterior part of the beam, vertical, with processes at base ; the beam spreading outwards ; the summit bilobed ; posterior antler horizontal, pointing to the opposite horn which is also bilobed ; colours brown, and brown-gray ; below yellowish ; large white disk ; only a rudiment of tail ; size equal to a stag.

Ahu, *G. Gmel.*   Cervus Pygargus, *Pallas* and *Auctor.* Chevreuil de Tartarie, *Cuv.*   Tartarian Roe, *Shaw.* Dikaja Kosa *of the Russians.*   Ahu, Saija.   *Tartar, Persian,* and *Bucharian.*

Icon. In our possession ; heads in Prague and Frankfort Museum.

Inhabits Mountains of Central Asia, descending in winter into the plains of Tartary, and probably of Northern India.

790. 20. *C. Capreolus* (the Roebuck.) Horns rather small, cylindrical; a small antler on the middle of the the beam pointing forward, a second high up, turned to the rear , tail very short , colours brown and reddish ; disk on the buttocks ; size below the middle.

Var. A blackish kind.

Caprea, *Pliny.* Capreolus Dorcas, *Gesner.* Cervus Capreolus, *Auctor.* Chevreuil, *Buff. Cuv.* Roebuck, *Pent. Shaw.* Rehe Rehbock *of the Germans.* Kosa Dikaja, (*i. e.* Goat) *of the Russ ;* more properly Jerna or Jaru, *in Sclavonic.* Ibec *of the Tartars.*

Icon. *Buff. Fred. Cuv. An. Lithog.*

Inhabits all Europe, and temperate parts of Asia, Scotland, Dorsetshire.

Fragments of jaws of a fossil roe have been found in fresh water calcareous strata at Montabusard, Dep du Loiret in France. The horns approach in character those of the Marianna Rusa, but the teeth of the upper Maxilla are different, especially the two anterior molars, which are simple, cutting and divided into three lobes, with only a collar at the base of the second, by which character the fossil species is distinguished from all known deer, and approaches the Musks. *See Oss. Fossils,* vol. iv. p. 105.

Sub-genus VIII.—MAZAMA. *Horns tending to flatten, bending into segments of a circle, the concave part to the front ; one anterior internal antler, the others posterior, and mostly vertical ; long tail ; suborbital pore forming a fold of the skin ; muzzle ; no canines.*

791. 21. *C. Virginianus* (Virginian Deer.) Horns middle-sized, tending to flatten, strongly bent back, and then forwards; a basal antler on the internal side, pointing backwards; several snags on the posterior edge, turned to the rear, and upwards; suborbital sinus making a fold, and small; muzzle; colours fulvous in summer, gray-brown in winter; no disk.

Var. Somewhat smaller; white coloured; triangle on the feet; black mark on the lower lip.

Fallow-deer, *Lawson, Catesby.* Cervus Virginianus, *Auctor.* Virginian Deer, *Pent. Shaw.* Cerf de Viginie, Cerf de la Louisiane, *Cuv.*

Icon. *Fred. Cuv. Mam. Lithog. Encyclop. Cerf de la Louisiane. Supl.* xiii. *fig.* 2. *Nobis.*

Inhabit North America, from Canada to Mexico.

792. 22. * *C. Mexicanus* (Mexican Deer.) Horns spreading outwards, curving to the front with extremities towards each other; an antler on the anterior face of the beam, pointing vertically, pointed or bifurcate, strongly denticulated; another at posterior part of the beam, divided into smaller snags; from the second antler the horns flatten into elongated palms, which in old specimens become very broad, with snags thrown off to the rear; the burr is replaced by large pearls, like incipient antlers; the beam tri-lateral; a muzzle; no canines.

Cervus Mexicanus, *Pent. Gmel.* Quantla Mazame, *Hernandes.* C. Ramosicornis, *Blain.*

Icon. *Pent. Buff.? Nobis.* Specimen in British Museum. Inhabit Mexico.

793. 23. * *C. Clavatus*, Nob. Horns deep yellow colour, very robust, pearled, extending horizontally, and then curving forwards and flattening; strong bifurcated vertical antler at base; on superior edge of beam, three bifurcate

315

snags, two others to the front at the summit, and one long heavy clavate and flattened branch hanging downwards from the inferior edge.

Icon. *Nobis* from the collection of Mr. Brooks.

Inhabits— ? probably America.

794. 24. *C. Macrotis* (Great-eared Deer.) Horns slightly grooved, and tuberculated at base; small antler on the internal anterior side of the beam; the beam less bent forwards, equally bifurcated at half its length, and each bifurcation again divided near the summit; the anterior snag of the posterior fork the longest; ears very long, (seven inches,) reaching to the forking of the horns; lateral teeth larger than in the Virginian: colour reddish-brown; dull cinereous about nose; tail reddish-cinereous, compressed; almost naked beneath; dark line on the neck near the head.

Great-eared Deer. Cervus Macrotis, *Say. Harlan. Major Long's Expedition.* Mule Deer? *Lewis* and *Clark.*

Icon. ——

Habitat the remotest north-western territories of the United States of America.

795. 25. * *C. Macrourus* (Long-tailed Deer.) Horns short, small, somewhat flattened; colour dark; belly white; tail nearly eighteen inches long, black above, edged with white, and held up erect when running; larger than the Red-deer, (C. Virginianus ?)

Long-tailed Deer. Cervus Macrourus. Black-tailed Deer, *Warden.* Deer with a large tail, *Lewis* and *Clark? Le Raye.*

Icon. ——

Inhabit about the River Kansas, Central North America.

796. 26. *C. Paludosus* (Guazupuco Deer.) Horns rather

316

large, cylindrical, terminated by a fork, with a branch above the burr, pointing forward and upwards, sometimes bifurcate; lachrymary sinus more developed; tail middling; red-bay above in summer, more gray in winter; white below; the hair of the inguinal parts, and under tail, long and white; spot on nose, and between eyes; white round eyes; size of a stag.

Goazoupouco, *D'Azara.* Cervus Paludosus, *F. Cuv. Desmarests.*

Icon. *Nobis* from specimens in London.

Inhabits Paraguay, in swampy places.

797. 27. *C. Campestris* (Guazuti Deer.) Horns middle-sized, rather slender, more or less rugous; beam sub-erect, with anterior antler pointing forward and upwards; behind one or two snags, turning with the point of the beam obliquely forward and inwards; small suborbital sinus; tail middle-sized; colours brown-bay.

Gouazouti, *D'Azara.* Cervus Campestris, *F. Cuv. Desmar.* Cervus Leucogaster, *Goldfus.*

Icon. *Nobis* from specimen at Exeter 'Change.

Inhabits the open plains of South America.

798. 28. *C. Nemoralis*, Nob. (Cariacou Deer.) Horns about eight inches long; sub-vertical, rugous at base; small anterior antler about the middle of the beam; posterior second antler, forming a fork with the summits of the horn, which flattens and turns inwards and forwards, making a hook; head roundish; black spot on nose, and one on each side of the mouth, and one on the lower-lip, all on a whitish ground; colour of back yellowish-brown-gray; twenty-eight inches high at shoulder, thirty at croup.

Cariacou *of Daubenton.* Squinaton, *Dobbs, Pent.?* Cariacou, *Laborde.* Jumping Deer, American Roebuck? Cerf Blanc ou des Paletuveirs *of Cayenne.*

Icon. *Nobis* from living specimens, Hospital of New York.

Inhabits Central America, round the Gulph of Mexico to Surinam.

Sub-genus IX.—SUBULO. *Horns small, simple, without branches or processes; small lachrymary sinus; muzzle widening to a glandular termination near the nostril.*

799. 29. *C. Rufus* (the Pita Brocket.) Somewhat higher than the Roebuck; head pointed; muzzle small above; small lachrymary sinus; canines in the male; horns about five inches long; colours lively reddish-bay; face and feet red-brown; lips and chin white; tail nearly nine inches long.

Cervus Rufus. Couassou, *Fred. Cuv.* Gouazu Pita, *D'Azara.* Biche de Barallou? *Laborde.* Antelope of Hondurus? confounded with an Aplocerus.

Icon. *Nobis* from specimen in the Paris Museum.

Inhabit. Gregariously in the forests of Eastern South America, Bay of Honduras and Paraguay.

800. 30. *C. Simplicicornis,* Nob. (the Apara Brocket.) About six inches lower at the shoulder; horns more pointed; no canines in the males; dark ring round the orbits, and spots about the corner of upper lip; colour bright fulvous; tail shorter, with longer hair of a red colour; head and neck ashy-brown in the young.

Guazu Apara of *Marcgrave.* Biche des Bois, *Laborde.* Indicated by *Baron Cuvier, Oss. Foss.*

Icon. *Nobis* from the specimens in the Museum of Prince Maxmilian of Wied.

Inhabits Brazil.

801. 31. *C. Nemorivagus* (the Bira Brocket.) About eighteen inches high at the shoulder, robust in structure; head ovine; ears rather round at tip; lachrymary sinus very small; horns reclined, solid, and pointed, two inches long; fur gray-brown above; much white beneath on belly, limbs, legs, and mouth; fore-arm convex.

Gouazoubira, *D'Azara.* C. Nemorivagus, *F. Cuv. Desmarests.*

Icon. *Nobis* from the specimen in the Museum of Frankfort.

Inhabits solitarily in the swampy forests, and near the sea of Eastern South America.

Sub-genus X.—STYLOCERUS. *Horns small, with only one anterior snag; standing upon elevated pedicles; long canines in most males; deep suborbital sinus; small muzzle.*

802. 32. *C. Muntjak* (the Kijang.) Horns upon pedicles from two and a half to three inches high, covered with the skin, flattened at the summit; the horns three or four inches long, points turned inwards; small antler at base, pointing forward; the pedicles prolonged in the form of ribs down to the nose; a fold of the skin between the ribs; body compact; legs slender; colour gray-brown, paler beneath; breast and inside of limbs white, this colour increasing with age; stature of a roebuck.

Var? A deep chestnut-brown colour, but probably only the sign of nonage.

Cervus Vaginalis, *Bodd.* Chevreuil des Indes, *Allam.* Muntjak, or Rib-faced Deer, *Pent.*

Icon. *Dr. Horsfield's* figure. Female, *Nobis* from Mr. Bullock's Museum.

Inhabit Java, Sumatra, Ceylon?

803. 33. *C. Philippinus* (Philippine Muntjak.) Specimen, without horns; pedicles low; ribs extending only

319

as far as the eyes; the face plane; forehead arched, dark-coloured, with a dirty-buff crescent between the eyes; fur brown sepia-gray; somewhat larger and heavier than the preceding.

Cervus Philippinus, *Desmarets.*

Icon. *Nobis* from the Paris Museum.

Inhabit the Philippine Islands.

804. 34. *C. Subcornutus* (Blainville's Muntjak.) Horns resembling the Kijang, but smaller, with regular burr, and small process in front; the point of the beam turned back, and not towards the opposite horn; pedicles short, strong, not much prolonged down the face; no canines?

C. Subcornutus, *Blainville.*

Icon. ——

Inhabit ? Skull in Surgeons' College, London.

805. 35. * *C. Aureus,* Nob. (the Ubi Muntjak.) Male unknown; female the size of Kijang; forehead square; snout tapering; two strong, hard, bristly, and curling spots of hair on the orbits, resembling dark eyebrows; lachrymary opening large; muzzle very small, not black, almost ovine; colour bright fulvous-yellow; ears broad, large, white inside; throat, belly, inside of limbs, and fetlocks, pure white.

Perhaps the Rusa Ubi of *Raffles.*

Icon. *Nobis* from specimen in Bullock's Museum.

Inhabit ? perhaps Malacca.

806. 36. * *C. Moschatus* (Nepaul Muntjak.) Male two feet eleven inches long, two feet high; horns upon elevated pedicles, slender, without branch, pointing backwards, simple; head seven inches long; hair rough, bristly, two inches long, dun-coloured all over the body; tail six inches and three-quarters long, dark.

Cervus Moschatus, *Blainv.?* Musk-deer of Nepaul, *Sir William Ouseley. Orient. Collections.*

Habitat. Nepaul.

TRIBE III.—*Giraffidæ.*

Frontal processes prolonged in the shape of horns, covered with hairy skin, which is continued from the scalp, and terminated by long hard bristles, in both sexes.

Genus I.—CAMELOPARDALIS, Lin.

Incisors $\frac{0}{8}$; canines $\frac{0.0}{0.0}$; molars $\frac{6.6}{6.6} = 32$. Head long prolonged with tuberculum on the chaffron ; osseous peduncles covered with skin, and hairy, terminated by a tuft of bristles ; no muzzle ; upper-lip entire ; no lachrymary sinus ; ears long ; tongue rough ; eyes large, soft, pupil elongated ; neck very long ; withers much elevated ; back oblique ; legs slender ; no succentorial hoofs ; callosity on the breast ; tail to the hough ; female four teats.

807. 1. *Camelopardalis Giraffa* (the Giraffa.) In stature the tallest of mammiferous animals ; coat of a dirty-white, marked with dark brown, or ferruginous spots or blotches, somewhat tending to symmetrical forms ; large and angular in their shapes ; short mane on neck and withers, in alternate parts of black and white ; tail terminated by a tuft of dark and long hair.

Camelopardalis, *Pliny, Oppian.* Heliodorus, *Gesner.* Anabula, Seraph. Alb the Great ; Gyraffa, quam Zurnapa, Græci et Latini Camelopardalus Nominant, *Bellon. Prosp. Albin.* Camelopardalus, *Lin.* Girafra, Camelus Indicus, *Johnst.* Giraffa Camelopardalis, *Briss.* Camelopardalis Giraffa, *Auctor.* Giraffe, *Buff.* Giraffa, *Shaw.* Zuraphate, *Arabic.* Seraphah, *Persian.* Jirataka Lin Amharic. Zomer, *Hebrew.* Deba, *Chaldaic, Æthiopic.* Nabis, *Pliny.* Naip *of the Hottentots.* Impatoo, *Bushmen.*

Icon. *Le Valliant, &c.* *Nobis* from male and female in the British Museum.

Inhabit Central Africa, from Caffraria, and the borders of the Gareep, across the deserts to Abyssinia *

TRIBE IV.—*Capridæ.*

Horns persistent, vaginating upon an osseous nucleus, totally or nearly solid; the horny sheath receiving its increase by annual ringlets at the base, which form in most species annuli, wrinkles, or knots; many striated longitudinally; the horns often compressed; angular, or subangular; animals in general of a light structure, calculated for springing or for swiftness; ears erect, funnel-shaped; pupils oblong; no canines in the mouth; vertebræ of the tail never descending below the hough; stature very various.

Genus I.—ANTILOPE.

Incisors $\frac{0}{8}$; canines $\frac{0\cdot0}{0\cdot0}$; molars $\frac{6\cdot6}{6\cdot6}$ = 32. Horns common to both sexes, or in the males only; bony core solid, without sinus or pores, round, or compressed, generally standing beneath the frontal crest; variously inflected, mostly distinguished by annuli, with longitudinal striæ between them; sometimes pearled and forked; the chaffron rather straight, with a muzzle, half muzzle, or simple nostrils; lachrymary sinus in most, and in some a suborbital pouch; eyes large, dark; ears in general long, pointed; inguinal pores; a gall-bladder.

Sub-genus I.—DICRANOCERUS. *Horns greatly compressed, rough, pearled, slightly striated, with an anterior*

---

* The fossil teeth of a large ruminating animal found in Siberia, indicate a lost genus, probably of this tribe. Mr. Bojanus has described them, and named the genus, *Merycotherium Sibiricum,* in the *Nov. Act. Acad. Cæs. Leop. Caræl, &c. tome* xii.

process, and the point uncinating backwards; dark coloured, placed upon the orbits, at right angles to the plane of the face, impending over the eyes; no suborbital sinus; no inguinal pores; no muzzle; facial line convex; tail very short; hair stiff, coarse, undulating, flattened, enclosing a sort of marrow. Female mammæ? horns? structure cervine. Confined to North America.

808. 1. *A. Furcifer* (Prong-horned Antelope.) Mixed resemblance between the Chamois and Roebuck; horns one foot in length, compressed, flat on the inner side, pearled and striated, with a compressed snag to the front; forking with the after part which forms a hook to the rear; eyes large, high in the head; nostrils ovine; colour foxy-dun, with a spot on the summit of the head; throat and disk on the buttocks, white; tail very short; stature about three feet at the shoulder.

Antilocapra Americana, *Ord.* Cervus Bifurcatus, *Raffinesque.* A. Furcifer, *Ham. Smith., Desmarests.* Cabree *of the Canadians.*

Icon. *Lin. Trans. vol.* xiii. *Nobis.*

Habitat. The borders of the Missouri, and plains of the North Western States, and along the Columbia.

809. 2. * *A. Palmata* (Palmated A.) Horns greatly compressed, with anterior and posterior edges, broad, dark, strongly pearled and striated; on the anterior edge near the base, a broad, flat, leaf-like, obtuse, and deflected process, forming a bifurcation with the posterior, which forms a curvilinear hook to the rear, and inwards; head shorter than the preceding; facial line nearly straight; fur softer, partially woolly, hoary without; a little white on the face, and on the croup; stature of a roebuck.

Mazame? *Hernandes,* lib. ix. cap. 14. Cervus Hamatus, *Blainv.* A. Palmata, *Ham. Smith.*

323

Icon. *Lin. Transactions, vol.* iii. *Nobis.*

Habitat. Baffin's Bay. Stony Mountains near the River Jaune ; may be only a variety of Furcifer.

Sub-genus II.—AIGOCERUS. *Horns very large, common to both sexes, pointed, simply bent back, annulated, placed above the orbits. Half muzzle ; no suborbital sinus ; no inguinal pores ; tail descending to the houghs ; mane reversed ; a white mark before the eyes ; throat and under-jaw somewhat bearded ; mammæ two ; stature large ; shoulders higher than the croup. Reside in Africa.*

810. 3. *A. Leucophæa* (Blue Antelope.) Four feet high at the shoulder ; horns slightly compressed, scimitar-shaped, about twenty-eight inches long, closely annulated, with twenty to thirty rings ; no striæ ; ears long ; colour silvery blue-gray ; spot before the eyes ; belly, and inside of the limbs, white ; short white mane turning towards the head ; hide black ; tail tufted at the end ; appearance of beard on the under-jaw.

A. Leucophæa, *Auctor.* Blauw-bock, *Kolbe.* Tzeiran, *Buffon, &c.*

Icon. *Buff. Allaman, Nobis.* Specimen in the Paris Museum.

Habitat. South Africa, rare. Last killed near Swellendam, Cape of Good Hope.

811. 4. *A. Equina* (Roan Antelope.) Four feet four inches at the shoulder ; horns very robust, about twenty-four inches long, strongly bent back, with seventeen to twenty-seven prominent rings, more remote from the orbits ; ears nine inches long ; hair coarse, undulating, loose, mixed red and white ; beneath the throat longer, whiter ; white spot round and before the eye, formed of a pencil of long hairs ; neck with short white reversed mane.

A. Equina. Antilope Ozanne, *Geoff. Cuv.* A. Aurita, *Burchell.*

Icon. *Dict. des Sciences Naturelles. Nobis.* Specimen, Paris and British Museum.

Habitat. South Africa, on the elevated ridge near the sources of the Gareep, &c.

812. 5. ** A. Grandicornis* (Long-horned Antilope.) Horns three feet and a half long, fifteen inches in circumference at base, curved like a scimitar, compressed, rounded behind, carinated, rough, with oblique wrinkles on the inner surface, furrows on the external.

A. Grandicornis, *Herman.* Empalanga? Empabunga? Empalunga? *Purchas. De Bry, Reg. Congo. p.* 22. Korooko *of the Bornouese?* El Bucher el Achmer *of the Arabs? Denham and Clapperton's Travels.*

Icon. If it be Empalanga, see De Bry in prima parte Iconum, *Ind. Orient. pars* 11.

Habitat. Central Africa? Bornou?

813. 6. * *A. Barbata* (the Takhaitze.) In size equal to the Equina, with a broad dark nose; white streak before the eye; horns scimitar-shaped, more erect and with fewer annuli; a considerable beard on the chin, and long flowing dark-coloured mane on the neck; colours blue-gray or rufous; no tuft to the tail.

Takhaitze *of Somerville* and *Daniell.*

Icon. *Daniell's African Scenery.*

Habitat. The parting ridge of the waters on the south-east coast of Africa.

Sub-genus III.—ORYX. *Horns common to both sexes; horizontal, very long, slender, without ridges, pointed, black, with annuli somewhat spirally twisted to half or two-thirds of their length; the animals large, with long ears, small or no suborbital sinus, ovine muzzle, darker coloured streak through the eyes, mane on the neck reversed; tail reaching to the houghs, and terminated by a tuft of long hairs: no tufts on the knees, nor inguinal pores? two mammæ. Stature large; general colours of the fur rufous or vinous gray upon a white ground.*

814. 7. *A. Oryx* (the Caffrarian Oryx.) Adult male three feet eight or three feet ten inches high, six feet six inches in length; horns three feet long, annulated, with twenty-eight to thirty-three rings, straight or very slightly bent, horizontal, diverging, and sharp at the points; eyes high in the head; black space round the base of the horns, descending in a streak down the forehead; another passing through the eyes, to the corner of the mouth, connected by a third which runs round the head over the nose. The rest of the head and ears white. General colour vinous buff; the breast, belly, and extremities white; a black list from the nape of the neck to the root of the tail; a broad bar of the same across the elbow, passing along the flank, and ending in a wide space on the thigh above the houghs. Black spot upon each leg beneath the joints.

A. Oryx, *Auctor.* Passan, *Buffon Resc. Sonnini.* Ceinse-bock *of the Dutch Colonists.*

Icon. *Daniell's Sketches. Nobis* male and female.

Habitat. Caffraria.

815. 8. *A. Leucoryx* (the White Oryx.) Adult male three feet seven inches high at the shoulder; head rather square, thick; neck short; body bulky; legs slender; horns three feet long, slender, horizontal, bent back, obliquely annulated, tips smooth; black spot at the base of the horns passing down the face, a second through the eyes towards the mouth, widening upon the cheek; a dark band upon the upper arm, passing down the fore-legs; lower part of

the thigh rufous, darkening into black about the hough, and upon the hind legs ; short dark mane, and end of the tuft of the tail black ; the rest of the body milk-white.

Oryx, *Oppian.* Antholops, *Eustathius.* Leucoryx, *Pent. and others.* El Walrush and Bukrus *of the Persians.* Ghau Bahrein *in India.* Jachmur and Yazmur *of the Arabs.*

Icon. *Pent. Dr. Flemming. Nobis.*

Habitat. Eastern Arabia, the Island and Province of Bahrein Mekran, Desert of Persia.

816. 9 * *A. Tao,* Nobis. (the Nubian Oryx.) May be a variety of the former. Near four feet at the shoulders ; seven feet in length ; horns three feet four inches long, more robust, very spirally annulated, equally curved backwards ; nose blunt ; the neck longer, the structure more elegant ; hoofs low and flat ; colour rufous and white, forming a gray on the nose, temples, cheeks, neck, upper arm, and lower part of the thigh ; more white over the shoulders, back, flanks, and croup ; a slight blackish mark above and beneath the eye, and a broad white streak passing before it to the corner of the mouth ; mane and tuft of tail white.

Tao *of the Hebrews and Egyptians.* Dante and Lout *of Congo? Leo Afric. de Bry.*

Icon. *Nobis,* from the superb specimen in the Frankfort Mus.

Habitat. Nubia, interior of North Africa.

817. 10. * *A. Besoastica* (the Algazel.) is perhaps a third variety. Three feet five inches high at the shoulder ; five feet two inches long ; horns three feet long, round, slender, bent back, with thirty-six annuli not spiral ; forehead narrow ; head long ; neck short ; body clumsy ; legs slender ; lachrymary sinus beneath the eye ; reversed ridge of short

white hair on the neck ; head white ; dark spot at the root of the horns passing down the face, another less distinct through the eyes ; body and neck fulvous-gray.

A. Algasel, *Fred. Cuvier.* A. Besoartria, *Licht. Pallas.* A. Eleotragus, *Schreber? Lichtenst?*

Icon. *Fred. Cuvier, An. Lithog. Nobis.* Specimen, Paris Mus.

Habitat. interior of Senegal. Seen by Major Denham south of the river Shary ?

Oryges passing into other Sub-genera.

818. 11. *A. Addax,* Nob. (the Addax.) Three feet seven inches high at the shoulders; three feet eight inches at the croup ; horns robust, black, round, divergent, with two and a half spiral turns, thirty-two to thirty-five annuli ; some dichotomous, extending three-fourths of the length ; two feet four inches long; no lachrymary sinus ; eyes large, dark ; dark-coloured mane on the neck ; tuft of long dark hair on the throat ; head thick ; forehead flat, covered with dark hairs, and surrounded by a narrow white line passing downwards before the eyes ; nose ovine ; chaffron, cheeks, and neck, liver-coloured gray, diluting on the shoulders, and the rest of the body milk-white ; hoofs flat, broad, round, and black ; tail and tuft white ; female two mammæ ; horns equally large. This species passes from the Orygine Sub-genus to the Damaline sub-genus Strepsiceros.

Strepsiceros and Addax *of Pliny and Caius in Gesner.* A. Addax, *Grætzmer.* El Bucher Abiad, *of Denham and Clapperton.*

Icon. *Nobis.* Male and female specimens, Frankfort Museum, head in Gesner.

Habitat. Nubia.

819. 12. * *A Kemas?* (the Chiru.) Total length, five feet eight inches. About three feet high at the shoulders;

neck long; croup more elevated than the withers; horns, twenty-one to twenty-six inches long, black, slender, striated, annulated, slightly lyrate, points turned forward and sharp ; no lachrymary sinus; nose ovine; ears short; body long ; tail eight inches without tuft; hair rough, thick, coarse, concealing a fine downy wool underneath; the face and legs, dark ; neck and back, blue-gray slate colour passing to rufous ; belly, inside of the limbs, and end of the tail, white. Female characters unknown.

Kemas ? *Ælian.* Chiru *of Bhootan,* pretended Unicorn *of the Natives.*

Icon. The horns ? *Nobis.*

Habitat. The Hymalaya Mountains.

Subgenus IV.—GASELLA. *Horns common to both sexes, placed nearer the orbits, more vertical, bending back, and the points forward, and also turned outwards, and again inwards, constituting a lyrate form: they are black, annulated and striated. These animals have small lachrymary sinus, inguinal pores, ovine nose ; mostly tufts on the knees, and dark-coloured bands on the flanks ; eyes very large and dark ; tail short and tufted ; mammæ two or four. Gregarious on open plains.*

820. 13. *A. Pygarga* (the White-faced A.) Adult male three feet eight inches at shoulder ; six feet long ; horns, twelve to fifteen inches long, seven inches circumference at base, black, very strong, with ten or twelve semi-annuli on their anterior side, and striated between. A patch of rufous hair at base of the horns, divided by a white streak, which passes down the face to the nose; ears long, reddish outside, sides of the head, neck, flanks and croup, deep purple-brown, the back hoary, bluish white, as if glazed; legs white ; no tufts on knees. Characters of the female unknown.

329

A Pygarga, *Auctor*. Nunni *of the Booshwanas*. A. Dorcas, *Pallas*. A Pygarga, *Ejusd.* A Pourpree, *Desmarets*, &c. Blessbock *of the Dutch*.

Icon. *Schreber.*

Habitat. Caffraria.

821. 14. *A Mytilopes*, Nob. (Broad-hoofed Antelope.) The male unknown ; the female two feet eight inches high ; four feet two inches long; head nine inches ; horns one foot, slender, round, sublyrate, black, with thirteen or fourteen obsolete rings, standing on a broad rufous spot ; ears six inches long; no lachrymary sinus perceptible ; incipient dark muzzle between the nostrils ; space between the eyes, mouth, under-jaw, breast, belly, croup and legs, white ; a bar across the nose, neck, shoulders and flanks lower part of buttock fulvous-ochre colour ; a space on the withers and back, of a glazed whitish-gray, as in the former; small callosities below the knees, and a dark brown spot at the spurious hoofs; hoofs broad, flat, rounded, black, muscle-shaped ; body rather heavy ; four mammæ.

A Naso Maculata, *Blainv.* A. Nez-tache. A. Mytilopes, *Nob. MS.*

Icon. *Nobis.* Specimen in British Mus.

Habitat. Western Africa.

822. 15. *A. Dama* (Swift Antelope.) Adult male three feet high at the shoulder, extremely light and elegant in structure ; head broad ; nose ovine, small ; horns black, one foot long, with twelve to sixteen annuli, lyrate, points turned forward and inwards ; small lachrymary sinus ; ears six inches long; tail short ; knees covered by two rows of bristly hairs, turned flat upon the joint, the points inwards ; the head white, with a spot of bright rufous hair at the base of each horn; ears six inches long, outside at the root

rufous, in the middle white, and tips black; the neck, shoulders, and back, whitish rufous ; a spot on the throat, the rest of the body, breast, limbs, and tail, white, with a rufous streak upon each of the fore-shanks. The female nearly equal in size to the male ; colours similar.

L. Nanguer, *Buffon.* A. Dama, *Auctor.* Swift Ant. *Pennant,* is the young animal. A. Ruficollis, *Grætzmer,* the adult. Engry ? *of the Bornouese.* Ngria ? *of the Byharmese.*

Icon. The adult, *Nobis* male and female. Le Nanguer, *Buffon.*

Habitat. The interior of North Africa from Nubia to Senegal.

823. 16. *A. Euchore* (Springer Antelope.) The adult male about twenty-two inches high at the shoulders, twenty-four inches at the croup ; head resembling a lamb's ; horns brown-black, lyrate, robust, with about twenty complete rings, tips turned inwards or forwards ; general colour of the fur pale-dun, with white about the head, limbs, belly, and croup, separated from the dun by a broad band along the flanks, another on the edges of the fold of the croup, and a dark streak through the eye ; females similar to the males ; horns more slender, with few distant annuli.

A. Euchore, *Forst. et Auctor.* Proukbock, *Vosmaer.* Springbock *of the Dutch.* A. Marsupialis, *Zieumer.* A. Pygarga, *Blumenb.* A. Dorsata, *Lacepède.* Tesbe *of the Caffres.*

Icon. *Sparraman. Nobis* male and female.

Habitat. Plains of South Africa.

824. 17. *A. Subgutturosa* (Persian Antelope.) Adult male about twenty-four inches high, by three feet six inches from nose to tail ; horns large, grayish-black, lyrate, annulated ; fur ashy-brown above, white beneath, with a

brown band on the flanks; the larynx tumescent; females smaller; no brushes on the knees.

A. Subgutturosa, *Guldenst. et Auctor.* Ahu *of Kæmpfer.* Tzeiran *of the Persians.* Jairou *of the Turks.*

Icon. *Schreber. Guldenstædt. Nobis.*

Habitat. Persia, Syria, Bucharia.

825. 18. *A. Dorcas* (the Barbary Antelope.) Adult male less than the Roebuck ; horns black, round, lyrated, thirteen inches long, annulated at base, semi-annulated in the middle, with twelve or thirteen bars, points slightly turned forwards, and the sides striated; facial line concave; face rufous, with black in the middle, and edged at the side with yellowish-white, which extends from the orbits to the nostrils ; a white and black streak from the eyes to the nose inside; ears streaked with black ; eyes large and black ; general colour pale fulvous; below white; tail short, tufted with black ; brushes on the knees; a broad brown band on the flanks; female with horns more slender, points turned inwards ; two mammæ.

Dorcas, *Ælian.* A. Dorcas, *Pallas et Auctor.* Gazal *of the Arabs.* Tzebi *of Scripture.* Gazalle, *Buffon.*

Icon. *Buff. Schreb. F. Cuvier. Nobis* male, female, and young.

Habitat. Northern Africa, Southern Syria, and Persia.

Var ? *A. Kevella* (the Kevel.) Adult male equal in size to the former; facial line straighter ; horns more robust, compressed at base, longer, with more decided flexures, with twelve to twenty annuli, points turned forwards ; orbits larger; eyes fuller, hazle colour; white space round the eyes, broader, and the same colour extending on the nether jaw; streak down the face fulvous ; below each eye fulvous-brown, without blackish intermixture ; general colour pale-fulvous, beneath white, and on the buttocks separated by a feint streak of brown ; the brown band on the flanks sometimes obliterated ; tufts on knees ; female resembling the former, and in a younger state, often mistaken for the Corinna.

A. Kevella, *Auctor.*   Le Kevel, *Buffon.*

Icon. *Buffon.*   *Nobis* male, female, and young.

Habitat. South-western Morocco, North Africa, between the Chain of Atlas and the Sahara.

Var ? *A. Corinna* (the Corinna.)   Adult male somewhat less than the Kevel ; horns black, more depressed at base, recumbent, and simply lyrate, slightly turmescent, about seven inches long, closely wrinkled beneath with obsolete small bars in the middle ; nose and mouth white ; chaffron and streak before the eyes bright fulvous ; forehead and general colour pale-fawn colour, mixed with gray on the flanks, beneath white ; a light chestnut band on the flanks ; small dark tufts on the knees.

A. Corinna, *Auctor.*   Corine ? *Buff.*   Korin *of the Negroes.*

Icon. *Nobis.*

Habitat. Central Africa.

Var ? *A. Cora,* * Nob. (the Cora.)   The male twenty inches high at the shoulder ; three feet two inches from nose to tail ; head round, with tapering small mouth, seven inches and a half long ; from nose to horns five inches ; the horns placed midway between the orbits, subvertical, about five inches long, round, slender, points turned backwards, smooth, without striæ, but one or two circular groves ; no perceptible suborbital sinus ; a black streak from near the base of the horns to the nostrils ; a second through the eyes towards the nose ; forehead and chaffron rufous ; occiput dark-brown ; mouth, nose, space between the streaks and region of the orbits white ; cheeks fawn-coloured ; general colour yellowish-rufous, beneath white ;

dark streak on the flanks; small callosity on the knees; tail five inches, tufted with black ; dark-brown tuft of hair on the anterior face of the pasterns to the division of the hoofs ; female resembled the male ; horns only four inches long.

Icon. *Nobis.*

Habitat. Shores of the Persian Gulf, Eastern Arabia.

Sub-genus V.—ANTILOPE. *Horns common to the males only, never truly lyrated, seated below the frontal crest, often sub-spiral or spiral ; suborbital sinus developed ; inguinal pores ; small bare space for a muzzle; two mammæ ; knees often tufted. Gregarious, or in families mostly on open plains.*

826. 19. *A. Melampus* (the Pallah.) The adult male above three feet high at the shoulder; nearly five feet in length. High on the legs; the horns black, about twenty inches, ascending obliquely upwards and outwards, and midway at an obtuse angle, obliquely inwards, rough and coarsely annulated at base, smooth at tip; ears seven inches long ; general colour fulvous ; brown on the back ; beneath and legs white, with a black spot round the spurious hoofs, and a dark streak sometimes double on the buttocks; tail white, eight inches long, without a tuft ; no brushes on the knees.

A. Melampus, *Lichtenstein, Desmar.* Pallah, *Daniell.* A. Pallah, *Cuv.* Pallah *of the Booshwanas.*

Icon. *Lichtenstein. Daniell. Nobis.*

Habitat. South Africa.

827. 20. *A. Forfex,* * Nob. (the Gambian Antelope ?) Male about twenty-five inches high at the shoulders, rather bulky in the carcass ; horns a foot long, black, close at base, slightly bent forwards, then opening laterally with their points again turned inwards, annulated with twelve

rings, the tips smooth; forehead broad; nose tapering with incipient black muzzle; ears large, open, with tufts of long hair hanging out of the conch; lengthened lachrymary opening; general colour fulvous dun; space round the orbits and inferior parts white; tail short, with black tuft; dark streak down the front of the legs, with spot on pastern joints; small dark brushes on knees; female smaller; two mammæ; no tufts on the ears.

Gambian Ant. *Pennant.*

Icon. Head, *Pennant.* Male and female, *Nobis.*

Habitat. Central and west coast of Africa.

828. 21. *A. Adenota*, Nob. (the Kob?) Male about twenty-six inches high at the shoulders; horns at base nearly vertical, spreading outwards, then bending back, tips slightly forward, nine inches and a half long, robust, black, striated, compressed at base, with ten semi-annuli on the anterior side, and the points smooth; head long, pointed, terminated with small black muzzle; general colour fulvous bay; space round orbits, lips, and under parts white; a small glandulous tubercle on the loins, from whence the hairs whirled in a circle over the body; a dark streak on the anterior face of the legs, with a band of the same colour at the fetlocks; a dark brush on the knees; tail short, wholly covered with long black hair; female resembling the male, but without horns.

Le Kob? *Buff.* A. Kob? *Desmarets.* Petite vache brune?

Icon. *Nobis* male and female.

Habitat. Central and Western Africa.

829. 22. *A. Colus* (the Saiga.) Male something less than the Fallow Deer; body bulky; head thick and heavy; horns distant, between spiral and lyrate, about ten inches long, erect, annulated, diaphanous, and yellowish in colour; the nose broad and cartilaginous; colour gray-dun, with

dark streak on the spine, beneath white ; in winter, although
hoary, small tufts on knees; the female hornless. In the
males there are sometimes three horns.

Κολος, *Strabo.* Colus *of Gesner* and *Johnston.* A. Saiga,
Pallas and *Auctor.* Sulok and Suhah *of the Poles.* Mar-
gatsch (the male), Saiga (the female), *in Russian.* Akoin
of the Turks.

Icon. *Pallas. Nobis.*

Habitat. South-eastern Poland, shores of the Danube,
Black Sea, to the Ural and Caspian.

830. 23. *A. Gutturosa* (the Dseren.) Adult male about
two feet six inches at the shoulder ; four feet eight inches
in length ; head thick, short ; horns about nine inches long,
annulated to near the tips, reclining backwards, and wavy
points turned inwards ; colour black or dark yellow ; nose
blunt, bristly ; larynx swelled externally, and surrounded
with long bristly hairs; glandulous bag on the abdomen
near the prepuce ; tufts on knees predominant ; colour in
summer yellowish-gray above, and white beneath ; in winter
almost white ; female smaller; no protruded larynx or
pouch on the belly.

Ant. Gutturosa, *Pallas* and *Auctor.* Le Dseren, *F. Cuv.*
Desmar. Hoang Yang or Yellow Goat *of the Chinese.*

Icon. *Pallas. Nobis.*

Habitat. The great desert of Cobi in Central Asia, and
Western China.

831. 24. *A. Cervicapra.* (the Common Antilope.) The
adult male about two feet six inches at the shoulders, and
four feet two from nose to tail ; horns black, round, annulated, twelve to twenty-two rings, spiral, or with three
flexures, and from twenty to twenty-four inches long ; the
head long, nose blunt, with incipient muzzle, or naked
space between the nostrils; general colour pale fulvous

336

above, white beneath, this colour darkening with age to nearly black, having part of the neck and thigh only fulvous; a white streak in younger animals runs along the side, in very old a black streak lower down on the flank; female hornless, paler in colour; tufts on knees.

A. Cervicapra, *Auctorum.* Antilope, *Buff.* Antilope, *des Indes Desmar.* Ena *of the Sanscrit.* Sasi or Sasin *of the Modern Hindoos.*

Icon. *Buff. Schreber. Nobis* in all states.

Habitat. India.

Var? An old male larger in size than the former, more robust in structure; horns eighteen inches long, very stout, spiral, with nine or ten semi-annuli close together at the base, fourteen to fifteen complete rings above, dark-brown tips, short; general colour deep rufous tawny; white spot round the eyes, on the cheek or throat, and beneath white; female nearly the same; a young male pale tawny, with darkish streak; the horns earthy brown.

Icon. *Nobis* male, female, and young.

Habitat. South-western Morocco ?

Sub-genus VI.—REDUNCA. *Horns in the males only, placed behind the orbits, black, reclining, tips bending forwards, annulated below, above smooth, short, slender; ears long, open, oval; imperfect suborbital opening; a small muzzle; inguinal pores; no tufts on knees; tail not longer than the buttocks; fur rather long, wavy; structure in general more robust; legs shorter; mammæ four; not gregarious; residing variously. Africa.*

832. 25. *A. Eleotragus* (the Rietbock.) Adult male two feet ten inches high at the shoulder, four feet six or eight inches long; ears six inches; tail nine or ten inches; horns ten or twelve inches long, recumbent below the plane of the face, divergent, regularly curved with the points

forward, wrinkled at base, and annulated with obsolete rings in the middle; general colour ashy-gray, tinged with ochre, beneath white; hair of the throat long, hanging down, and whitish; female smaller, in other respects resembling the male.

A. Eleotragus, *Schreb. Desmarets, &c.* A. Arundinum, *Bodd.* A. Arundinacea, *Shaw.* Rietbock, *Allam* and *the Dutch.*

Icon. *Allam. Supl. Buff. Nobis* male and female.

Habitat. Caffraria among the reeds of dried river courses.

833. 26. *A. Redunca* (the Nagor.) Adult male two feet eight inches high, four feet eight inches long; head nine inches; horns six inches long, approximating at base, a little compressed, not much divergent, sub-erect, bent forwards, with five obscure semi-annuli separated by striæ in front, points smooth, approximating; middle sized dark muzzle; ears long; head and neck tawny; back fulvous brown, with a cast of purple; the hair long, hard, loose, whirling in various directions; chin and lower parts white; the tail with much long hair, the base dark, the middle fulvous, and tip white; legs strong fulvous; the female marked in a similar manner.

The young entirely pale rufous, is the Nagor *of Buffon.*

A. Redunca, *Pallas* and *Auctor.* A. Reversa, *Pallas.* Le Nagor, *Buff.* A. Fulvo Rufula, *Afzel. Goldfus.* is the adult. A. Lalandiana, *Desmarets*, the female.

Icon. *Nagor. Buff.* The adult, *Daniell's Sketches of Africa. Nobis* male and female.

Habitat. Western Africa, Caffraria, lives among rocks.

834. 27. * *A. Isabellina* (Cream-coloured Antelope.) The male two feet six inches at the shoulders, four feet ten inches long; head ten inches; horns eleven inches, robust at base, approximating, parallel along the plane of the face,

the points turned forwards, round, shining, obliquely annulated, six or seven in front, eight or nine in rear, naked, triangular ; spot before the eye ; hair rather long, standing off, the shorter brown, the longer gray, forming a cream-colour, whirling in several places.

A. Isabellina, *Afzelius*.

Icon. ——

Habitat. Caffraria.

835. 28. *A. Villosa* (the Riet Rheebock.) Adult male two feet five inches at the shoulder, four feet six inches long ; head eight inches ; horns eight inches and a half long, straight, vertical, slightly inclining forwards, round, slender, with thirteen rings, sharp pointed ; black spot before the eyes ; suborbital sinus large beneath ; muzzle round the neck, long ; body very slender ; general colour whitish-gray, with a cast of buff ; beneath white ; hair very soft and villous ; tail five inches, gray, tipt with white ; female smaller, but similar in colours ; four mammæ.

A. Villosa, *Burchell, MS.* A. Capreolus, *Lichtenst, &c.* A. Lanata, *Dict. d'Hist. Nat.*

Icon. *Nobis* male and female.

Habitat. Deserts of South Africa, monogamous.

836. 29. *A. Scoparia* (the Orebi.) Adult male twenty-two to twenty-four inches high, four feet long ; head eight inches ; horns nearly vertical, slightly bent forwards, five inches long, with six or seven wrinkles at base, and five annuli above them, round black points, smooth ; lachrymary sinus well defined ; small muzzle ; tufts on the knees ; general colour of the face and back tawny, or pale fulvous ; a whitish arch over the eyes ; under parts white ; the throat and breast with loose white hairs ; tail short, blackish ; the hide sometimes black ; female the same, with brushes on knees ; no horns.

A. Scoparia, *Schreb. et Auctor.* Ourebi, *Pent. Suppl. Buff.*

Icon. *Buff. Schreb. Nobis.*

Habitat. The plains of Caffraria.

Sub-genus VII.—TRAGULUS. *Horns in the males only, placed near or upon the orbits, shorter than the ears, black, round, vertical, distant, parallel, straight, inclining slightly forward or backward, mostly without annuli or wrinkles, and without striæ; the ears long; the body in general slender; high on the legs; delicate; head round; black space before and about the eyes; a suborbital sinus; small black muzzle; tail very short; inguinal pores; two mammæ; no brushes; all monogamous or solitary in various situations. Africa.*

837. 30. *A. Oreotragus* (the Klipspringer.) Adult male twenty-one to twenty-two inches high, three feet seven inches long; forms robust; head short, round, and broad; horns about five inches long, distant, round, vertical, slightly inclined forwards, obscurely wrinkled at base, and annulated in the middle, tips smooth and pointed; legs robust; pasterns rigid; fur standing off, spirally-twisted, hard, ashy at base, brown in the middle, yellow at the tips, forming an agreeable olive.

A. Oreotragus, *Gmel.* and *Auctor.* A. Saltatrix, *Bodd.* Sauteur des Rochers, *Vosmaer, &c.*

Icon. *Goldfus. Nobis* male and female.

Habitat. The rocks and precipices of Caffraria.

838. 31. *A. Rupestris* (the Steenbock.) Adult male twenty inches at the shoulder, twenty-two at the croup, three feet six inches long; head oval; snout pointed; muzzle black, ending in a point upon the ridge of the nose; horns vertical, straight, parallel, round, slender, and pointed, one or two rudiments of wrinkles at base, not quite four

inches long; ears longer, open, pointed; general colour chocolate-rufous, below white; groin naked and black; tail not protruding beyond the hairs; pasterns short.

A. Tragulus Rupestris, *Forst. Lichten.* Tragulus, *Desmar.* A. Dama, *Cuv.* A. Ibex, *Afzel.* Steenbock *of the Dutch Colonists.*

Icon. *Nobis.*

Habitat. The bushes of high mountains in Caffraria.

839. 32. * *A. Rufescens* (the Vlackte Steenbock.) Male very high on the legs, two feet six inches from nose to tail; horns reclining slightly with the points turned upwards, round, smooth, without wrinkles or annuli, parallel, three inches and a half long, one inch and a half asunder at base, two inches from tip to tip; ears four inches and a half long; head squarer than the former, small black muzzle; general colour bright fulvous red, with a cast of crimson, beneath white; tail very short.

A. Rufescens, *Burchell, MS.* Vlackte Steenbock *of the Dutch.*

Icon. *Nobis* male and female.

Habitat. The open plains of Caffraria, very rare.

840. 33. * *A. Grisea* (the Grysbock.) Adult male nineteen and twenty inches high, three feet long; head oval, six inches long; horns four inches, smooth, round, vertical, slender, inclining forward, one inch and a quarter asunder at base, three inches from tip to tip; muzzle small and black; ears four inches and a quarter long, broad, open; colour deep chestnut-red, intermixed with numerous single white hairs; beneath rufous.

A. Grisea, *Cuv.* A. Melanotis, *Lichtenstein.* Grysbock *of the Dutch.*

Icon. *Nobis* male and female.

Habitat. Shrubby mountainous regions of Caffraria.

841. 34. *A. Palida* (the Bleekbock.) Adult male twenty-two to twenty-four inches high, three feet five inches long, very slender and light of form; head square; nose pointed; horns perfectly straight, inclining backwards, round, with an obsolete ridge in front, about four inches long, very pointed; black naked ring round the eyes; ears broader and shorter than the former; the tail near three inches long; general colour pale rufous fawn-colour above, and white beneath; females redder in colour; two mammæ.

A. Palida, *Lichtenstein.* A. Diotragus, *Afzelius.*

Icon. *Nobis* male and female.

Habitat. The plains of Caffraria, rare.

Sub-genus VIII.—RAPHICERUS. *Animals of diminutive stature; forehead narrow; horns without wrinkles, annuli, or striæ, black, slender, round, very sharp, subvertical, only known from the skulls. Asia.*

842. 35. * *A. Acuticornis* (Sharp-horned Antelope.) Horns three inches long, round, smooth, black, and pointed, about three-eighths of an inch in diameter at base, slightly bent outwards and slightly forwards, the frontal crest passing behind them, uniting with a broad parietal bone, terminated by a square ridge.

A. Acuticornis? *Blainville.*

Icon. *Blainville? Nobis* from the Royal College of Surgeons, London.

Habitat. The East Indies.

843. 36. *A. Subulata,* Nob. (Awl-horned Antelope.) Horns three-eighths of an inch in diameter, subvertical, round, smooth, four inches and a half long, bending outwards in the middle, the points slightly inwards, one inch two lines asunder at base, two inches in the middle, higher

on the frontals than the preceding, the sinciput broader, the parietal narrow.

Icon. *Nobis* from the Royal College of Surgeons, London.
Habitat. The East Indies.

Sub-genus IX.—Tetracerus. *Horns in the males only to the number of four, the upper or true horns rising on the frontal crest, straight, parallel, distant, without, or nearly without wrinkles, round, smooth, black, and pointed; the spurious or lower placed nearly between the orbits, conical, short, smooth, or slightly wrinkled at base; large suborbital sinus; tail short; monogamous. India.*

844. 37. *A. Chickara* (the Chickara.) Adult male twenty inches and a half high, two feet nine inches long; head seven inches and a half; superior horns, black, sub-ulate, round, without rings, smooth, erect, three inches long; spurious horns one inch four-tenths long, placed between, but rather above the middle line of the orbits, erect, stumpy, smooth; cylindrical three-fourths of an inch long; ears ovate, four inches three-fourths long; tail five inches; general colour bright-bay, beneath whitish; female paler; mammæ?

A. Chickara, *Hardwick.*
Icon. *Hardwick*, of male and female, *Trans. Lin. Soc.*
Habitat. Central India.

845. 38. * *A. Quadricornis* (Four-horned Antelope.) Skull seven inches long; superior horns longitudinally striated, transversely striolated, with rings at their bases; the spurious horns placed before the middle line of the orbits, sub-triagonal, yellowish on their inner surface, black on the outer: robust, vertical, one and two-thirds of an inch long, with three wrinkles at base; the general colour of the fur brownish, grayish beneath?

343   3 A 2

A. Quadricornis, *Blainville.* Tetracerus Quadricornis, *Leach.* Le Chikara? *Duvaucel. Fred. Cuvier.*

Icon. *Fred. Cuvier? Nobis?*

Habitat. Eastern side of the Burampootra, India, Nepaul?

Sub-genus X.—CEPHALOPHUS. *Horns in the male only, small, straight, or nearly straight, reclining, placed high on the forehead, black, with wrinkles or annuli; muzzle rather developed, black; hair of the forehead lengthened into more or less of tuft or spread; a pouch opening between the orbits and nostrils, by a puncture or a slit, independent of the lachrymary sinus, which in some is wanting; without tufts on knees, one only excepted; pasterns short; hinder shanks long; mammæ two or four; tail short, tufted; colours generally dark; stature middling or small; reside in covers or bushy plains. Solitary.*

846. 39. *A. Silvicultrix* (the Bush Antelope.) Adult male three feet, and three feet two inches high, five feet long; head ten inches; horns reclining, four inches long, straight, pointed, wrinkled at base, rugous higher up, smooth at tip, and slightly bent outwards; tail pendulous, with a brush; mammæ two; tuft between horns clear brown; general colour dark-brown above, with fawn-coloured longer hair over the spine and loins, grayish beneath; legs dark-chestnut; no tufts on knees.

A. Silvicultrix, *Afzel.* Bush Goat *of Sierra Leone.* Ant. des Buissons, *Desmar.*

Icon. *Nobis.* Mr. Landseer from living specimen, Exeter 'Change.

Habitat. The plains and bushes about the Pongas and Quia in Western Africa.

Var? *A. Platous*, Nob. (Broad-eared Antelope.) Specimen about equal in bulk to the former, but probably lower on the legs ; head long and pointed ; horns not five inches long, reclining, straight, divergent, irregularly annulated or rugous, pointed, and black ; ears very wide, pointed, longer than the horns, whitish within, dun-coloured at the back ; eyes large ; a black spot on the cheek, marking the opening of the sinus ; dark sepia streak on the chaffron, spreading in a coarser tuft about the horns ; general colour brown, and fawn-colour above, whitish-gray beneath ; no tufts on knees.

A. Platous, *Nobis MS.*

Icon. *Nobis.*

Habitat. The mountains on the west side of Caffraria.

847. 40. *A. Quadriscopa,* Nob. (Four-tufted Antelope.) Adult male about the size of a roebuck, lower on the legs ; head round ; nose tapering ; horns four inches long, reclining, straight, divergent, sharp at tip, with six or seven small annuli at base ; ears wide, longer than horns, two black striæ inside ; neck long ; darkish streak down the chaffron ; small lachrymary opening beneath the eye, and a naked line from thence towards the nose, indicating a second pouch on the cheek ; forehead covered with longish hair of a dark colour ; general colour brownish-yellow gray, beneath white, a feint lateral streak and several dark cross marks upon the arm ; legs slender, with tufts on the knees, and tufts on the upper anterior end of the posterior shanks ; pasterns short.

A. Quadriscopa, *Nobis MS.*

Icon. *Nobis.*

Habitat. West coast of Africa.

848. 41. * *A. Burchellii,* Nob. (Burchell's Antelope.) Adult male three feet five inches long, and about twenty-two inches in height ; head seven inches long ; ears six

inches; the horns five inches, slightly elevated above the plane of the face, approximated, parallel, the superior third part alone bent slightly outwards, and the points inwards and forwards; they are black, round, obtuse at the point, six to seven wrinkles at base, then striated, and above this again irregularly wrinkled, striated, and annulated; no external opening of the lachrymary sinus visible, and sub-orbital pouch not very evident; ears wide, long, and open, marked with three striæ; a space of long bright fulvous hairs upon the forehead; chaffron black; general colour brownish, rusty above, ashy beneath; the limbs robust, and fetlocks short and dark-coloured. It is possible that this is an old *A. Mergens* with the horns diseased, because the two are not exactly alike.

A. Burchelli, *Nobis MS.*

Icon. *Nobis.*

Habitat. Caffraria.

849. 42. *A. Mergens* (the Duiker Bock.) Adult male three feet two or six inches in length, twenty-one and twenty-three inches high; horns four inches long, more distant at base than in the former, more reclining, bending outwards, with a longitudinal ridge on the front, traversing four or five annuli of the middle, but not through the wrinkles at the base; forehead covered with a patch of bright fulvous coarse hair; ears five inches long, three dark striæ within; dark streak on the chaffron, and down the front of the legs; a suborbital slit on the side of the face; general colour light brown above, and white beneath; tail short, black, tipt with white.

A Mergens, *Blainv.* Cap. Merga, *Forster.* Duiker Bock *of the Dutch Colonists.* A. Mergens, *Desmar.*

Icon. *Nobis* male and female.

Habitat Southern and Western Africa, but principally Caffraria.

850. 43. *A. Ptoox* (the Dodger Antelope.) Male about twenty inches high, and three feet long. More delicately framed than the former; horns three inches long, with three annuli at base, round, bent outwards, reclined, without anterior ridge; a small pencil of vertical black hairs standing between the horns; rufous face and forehead; orbits prominent; lachrymary sinus a little prolonged; and further towards the nose a puncture, seeming to open in a second pouch; nose almost ovine; general colour pale dun above, beneath white; a black streak down the fore shanks, and a spot on the hinder pasterns; tail short, dun, and tipt with black. This may be a variety of the former in a junior state.

A. Grimmia, *Pallas.* A. Ptoox, *Lichtenstein.* Grimea A. *Pent.* The Grimm *of Leverian Museum in Shaw.*

Habitat. Southern and Western Africa, chiefly Guinea.

851. 44. *A. Grimmia* (the Grimm.) Adult male seventeen and eighteen inches high, twenty-seven inches long; structure very compact, more clumsily built than the former, head thick, terminated by a muzzle; horns very short, stout, reclining, almost concealed in the long dark hair of the forehead, which forms a kind of point between them; face dark; ears short and broad; a lengthened suborbital slit, containing an unctuous substance beneath the eye, but no lachrymary sinus; general colour fulvous fawn, with a dark ashy streak down the back; the inferior parts whitish, the legs dark, and tail longer than the preceding. Females darker.

A. Grimmia, *F. Cuvier.* Capra Silvestris Africana, *Grim.* Icon. *F. Cuvier, Mam. Lithog. Nobis* male and female. Habitat. Guinea and Western Africa.

852. * 45. *A. Maxwellii*, Nob. (Maxwell's Ant.) Adult

female about sixteen inches high, more slender in form than the last; ears longer; forehead square; nose more prolonged and pointed; a round muzzle; black spot beneath the eye, and on the cheek a puncture opening into the lower pouch; forehead and nose dark, a streak above the eyes resembling eyebrows; neck, back, and croup, dark-brown dun; beneath white; mammæ four yellowish, forming an udder; tail two inches long, black.

A. Maxwellii, *Nobis MS.*

Icon. *Nobis.*

Habitat. Sierra Leone.

853. 46. *A. Cærula*, Nob. (Slate-coloured Ant.) Adult male about thirteen inches at the shoulder, twenty-eight inches from nose to tail; head rather long, pointed, with small muzzle; no lachrymary opening, but suborbital pouch lower down, marked by a lengthened streak; horns one inch and a quarter long, recumbent tips turned upwards, black, pointed, with five semi-annuli; nearly concealed in the hair of the forehead; ears short, round, open; general colour slaty purplish-blue, beneath white; pasterns short, and legs buff; hoofs horn colour.

Blauwbockje, *of the Dutch Colonists.* A. Cærula, *Nobis MS.*

Icon. *Daniell's Scenery of Southern Africa. Nobis* male and female.

Habitat. Caffraria.

854. 47. * *A. Perpusilla*, Nob. (the Kleenebock.) Male about twenty-six inches long, twelve inches high; head shorter; forehead more elevated than the preceding; a suborbital sack as before; no lachrymary sinus; ears short and round; horns black, conical, slender, reclined, slightly turned inwards, nearly two inches long; incisor

teeth broader ; pasterns longer; hoofs smaller ; general colour dull brownish-buff; beneath white; perhaps only a variety of the former.

A. Cærula, *Nob. MS.* A Pigmæa, *Desmarets.* Kleenebock *of the Dutch Colonists*, and Noumetje *of the Hottentots.*
Icon. *Nobis.*
Habitat. Caffraria.

855. 48. * *A. Philantomba*, Nob. (the Philantomba.) Young specimen eighteen inches long ; horns very short, half an inch, the points just emerging from the long hair of the forehead ; ears rounded at tip; long slit on the side of the nose ; general colour dark-brown gray ; legs dark ; pasterns short.

May be the Guevei Kaior *of the Negroes.*
Icon. *Nobis.*
Habitat. Sierra Leone.

Sub-genus XI.—NEOTRAGUS. *Horns in the males only, horizontal, very small, with a few annuli or semi-annuli, black, pointed; no suborbital slit ; head round ; nose pointed, with a small muzzle ; tail short ; females two mammæ ; size very diminutive.*

856. 49. *A. Pygmea* (the Guevei.) Adult male about eleven inches high at the shoulders ; nearly twenty inches in length ; horns one inch and a quarter long, high on the head, rather close, bulky at base, with one or two prominent annuli, points sharp and black; a small lachrymary opening, but no slit; ears short, round; general colour bright bay, beneath whitish ; female duller in colours; smaller.

Royal Antelope, *Pent.* King of the Harts, *Bosman.* A. Pygmea, *Shaw.* Cervula Parvula Africana, *Seba.* Chevrotain de Guineé, *Buff.*

Icon. *Shaw. Nobis* from specimens in Leverian and Bullock's Museums.

Habitat. Guinea, Central Africa.

857. 50. * *A. Madoka*, Nob. (Salt's Antelope.) Animal very small; horns one inch and a quarter long, very slender, recumbent, points slightly turned forward, six or seven semi-annuli at base; ears broad, oval; hair of the forehead very close, short, and fine; no lachrymary sinus; colour of the head pale fulvous; pasterns long; hoofs very long, pointed, horn colour.

A. Madoka, *Nobis MS.* A. Saltiana, *Blainv. Desmar.* Madoka, *in Abyssinia.*

Icon. *Nobis* and *Blainville* of the fragments in the Royal College of Surgeons.

Habitat. Abyssinia.

Sub-genus XII.—Tragelaphus. *Horns in the males only? with ridges forming angles, which turn somewhat spirally, seated high on the frontals, reclining; small or naked spot for a muzzle; no lachrymary opening; colours remarkably diversified with white spots and streaks; form elegant, though receding from the typical structure of true Antelopes, and assuming that of goats; females with four mammæ.*

858. 51. *A. Sylvatica.* (The Boschbock.) The adult male about two feet eight inches high, and five feet three inches in length; head seven inches; horns ten inches long, marked with an obsolete ridge in front, and one in rear, horizontal, spiral and sub-lyrate, black, and closely annulated at base; general colour brilliant chestnut brown above, and marked with a narrow streak along the spine; several round spots on the cheek; shoulder, loins, and

thigh, of a pure white, as also the whole of the lower parts ; tail six inches long.

Boschbock, *Sparr.* and *the Dutch Colonists.* A. Sylvatica, *Auctor.* Bosbock, *Allaman in Buff.*

Icon. *Buff. Daniell Sket. Scen. of S. Africa. Nobis.*

Habitat. The forests of Caffraria.

859. 52. *A. Scripta.* (Harnessed Antelope.) Adult male two feet eight inches high, four feet eight inches long; horns seven inches long, reclining, straight, wavy, with two ridges twisting spirally round the axis ; general colour bright fulvous bay, two narrow lines passing from the withers obliquely downwards, one to the flank, the other to the groin, intersected at right angles across the back by three others, and four or five similar across the croup ; several round spots about the face and thighs all pure white.

Le Guib, *Buff. Adanson?* A. Scripta, *Pall.* and *Auctor*, Harnessed A., *Pent. Shaw.*

Icon. *Buffon. Nobis* male and female.

Habitat. Central and Western Africa about Senegal ; doubtful in Caffraria.

860. 53. * *A. Phalerata,* Nob. (Ribbed Antelope.) Male about two feet four inches high, four feet long ; horns three or four inches long, reclining, conical, not compressed, without ridges or transverse protuberances ; forehead broad ; a small black muzzle ; general colour rufous ; a black line edged on each side by one of white, along the spine to the tail ; a second white line from the middle of the shoulder to the groin, between them nine perpendicular lines forming ribs, but not intersecting the inferior ; on the thigh many, and on the cheeks and face several, round spots all of white.

A. Phalerata, *Nob. MS.* Le Guib, var *Desmar. in note.*

Icon. *Nobis* male, female, and young.

Habitat. Western Africa, about the river Congo; is found in the bushy plains.

Sub-genus XIII.—NÆMORHEDUS. *Structure assuming a caprine form; skull solid, heavy ; horns in the males only ? short, round, bent back, annulated at base ; a small muzzle ; a pouch upon the intermaxillary bone of some ; hair coarse, loose, dark ; legs robust. Reside in mountainous and woody regions of Asia and Indian Archipelago.*

861. 54. *A. Sumatrensis* (the Cambing Ootan.) Adult male two feet four inches high, four feet six inches long; muzzle broad and black; suborbital sinus opening at a naked space, with a round puncture; horns six inches long, round, reclining, bent back, with ten or twelve wrinkles at base; tail short; general colour black; neck covered with long white hairs ; under jaw and gullet white ; hair coarse ; forms robust.

Cambing Ootan, *Marsden Sum.* A. Interscapularis, *Licht.* A. Sumatrensis, *Auctor.* Cambtan, *F. Cuv. Desmar.*

Icon. *Marsden Sumatra. Fred. Cuv. Nobis.*

Habitat. Mountain forests of Sumatra. Malayan Peninsula ?

Var ? *A. Duvaucelii* (Duvaucel's Antelope) described from a drawing sent by Mr. Duvaucel from India. Muzzle smaller, horns more reclined, with fewer annuli; colour ashy gray ; mane on neck, standing up, and shorter ; lips, chin, and throat, white.

862. 55. *A. Goral* (the Goral.) Male two feet high, three feet one inch long ; horns four inches and a half long, black, subulate, bent back, smooth, with five or six annuli at base ; eyes large, dark ; body round ; general

colour gray-brownish beneath, and whitish under the throat; tail short; female with tubercles covered with a tuft of dark hair instead of horns.

A. Goral, *Hardwick, Lin. Trans.* Bouquetain de Nepal, *Duvaucel, MS.*

Icon. *Lin. Transactions. Nobis.*

Habitat. Mountains of Nepaul.

Sub-genus XIV.—RUPICAPRA. *Structure caprine; horns in both sexes, vertical, round, striated, with few wrinkles at base, taper, suddenly uncinated backwards; limbs strong; inguinal pores; two mammæ; two glandular apertures behind the horns; dark streak through the eyes; hair longer, with a small quantity of wool beneath; stature middle sized. Reside in the mountains of Europe and Asia.*

863. 56. *A. Rupicapra* (the Chamois.) Adult male about two feet three inches high, four feet six inches long; horns seven or eight inches long, uncinated backwards and pointed; in old males wrinkled at base, longitudinally striated; cheeks and throat, fawn colour; a black streak through the eyes; general colour brownish gray; wool beneath grayish; tail short.

Rupicapra, *Pliny.* A. Rupicapra *Pall.* and *Auctor.* Capra Rupicapra, *Linn.* Chamois, Yzard, *of the French.* Gemsebock *of the Germans.*

Icon. *Buff. Schreber. Nobis* in all its states.

Habitat. The secondary ridges of the Alpine Mountains of Europe and Asia.

Var. α. The Yzard. Smaller, gray-brown, cheeks and buttocks fawn colour. Inhabits the Pyrenees.

Var. β. The Persian Chamois, smaller; horns bent back into a regular hook from their root; streak through the eyes, nearly obliterated; hair close and fine; colour rufous yellow.

Sub-genus XV.—APLOCERUS. *Structure approaching ovine forms; horns resembling Næmorhædine group, simple, sub-recumbent, conical, obscurely annulated, the points smooth, bent back ; no lachrymary opening ; no muzzle ; tail short. Reside in the mountains of America.*

864. 57. *A. Lanigera,* Nob. (Woolbearing Antilope.) Adult male equal in size to a large sheep ; nose ovine ; chaffron nearly straight ; horns about five inches long, sub-reclined, conical, with two or three obscure annuli ; eye-lashes white; tail short; structure exceedingly robust; fur, long, fine, abundant, concealing beneath a very fine wool ; colours entirely white; hoofs black.

Rupicapra Americana, *Blainv.* Ovis Montana, *Ord.* A. Americana, *Desmar.* A. Lanigera, *Ham. Smith, Lin. Trans.* Mazama Sericea, *Raffinesque.* Mazama Dorsata, *Raffinesque ?*

Icon. *Nobis. Lin. Trans.*

Habitat. Mountains in the north-west of America.

865. 58. * *A. Mazama,* Nob. (Ovine Antelope.) A doubtful species which may be a variety of climate of the above. Structure resembling the former, smaller, less robust ; horns similar ; tail thick and short ; fur close and fine, pale rufous-brown.

Mazame, *Seba?* A. Mazama, *Hamilton Smith, Lin. Trans.*

Icon. *Seba ? Nobis.*

Habitat. Rocky forests and mountains of tropical America.

866. 59. * *A. Temmamazama* (the Chichiltic.) Like-wise a doubtful species ; size of a kid ; horns five inches and a half long, black, wrinkled, slender, bent back at a slight angle ; general colour, pale chestnut-brown, with

some white beneath ; tail carried erect, five or six inches long.

Temmamaçame seu Cervus Maçatl Chichiltic, *Seba ?* Ovis Pudu. *Molina ?*

Icon. *Nobis.*

Habitat. Mountains of New Mexico.

Sub-genus XVI.—ANOA. *Horns placed on the edge of the frontal crest, on the same plane with the face, exceedingly robust, slightly depressed, sub-triangular, short, straight, wrinkled and pointed; facial line straight; no suborbital opening ?*

867. 60. * *A. Depressicornis* (the Anoa.) Head nine inches long, straight ; horns, ten inches long, straight, very robust, slightly depressed at base, flat on the anterior side, sub-triangular two-thirds of their length, tapering suddenly to a sharp point, the rest nearly of equal thickness, rudely and irregularly wrinkled, and of a dark gray colour. Those of the female ? more slender, rounded at back ; face covered with close gray hair ; a broad muzzle.

A. Compressicornis. *Leach, MS.* The Anoa, *Loten, MS.* Idem, Var. B. of Buffalo, *Pen.*

Icon. The head and horns, *Nobis.*

Habitat. The Island of Celebes.

Genus II.—CAPRA.

Incisors $\frac{0}{8}$; canines $\frac{0\cdot0}{0\cdot0}$; molars $\frac{6\cdot6}{6\cdot6}$ = 32. Horns common to both sexes, or rarely wanting in the females, in domesticated races, occasionally absent in both; they are directed upwards, or depressed backwards, more or less angular, nodose ; no muzzle, no lachrymary sinus, nor inguinal pores ; eyes light coloured, pupil elongated ; tail short,

355

flat, and naked at base; below the chin, bearded. Reside in the primitive and highest mountains of the ancient continent. The domestic varieties are more or less subject to modifications in their general characters.

868. 1. *Capra Ibex* (the Ibex.) Adult male, two feet eight inches high at the shoulder; five feet long from nose to tail; horns flat, with two longitudinal ridges at the sides, crossed by numerous transverse knots. They are subvertical, curved backwards, about thirty inches long, dark coloured, and very robust; ears short, pointed; legs strong; general colour, red-brown in summer, and gray-brown in winter; beard, short and dark; inside of ears, and under part of tail, white. Female, horns short, more erect, with three or four knots in front; general colour, earthy brown and ashy: the young gray.

Ibex, *Pliny, Gesner.* Capra Ibex, *Linn. et Auctor.* Bouquetin *of the French.* Steinbock *of the Germans.*

Icon. *Buff. Meisner. Nobis.*

Habitat. The snowy regions of the Alps, Pyrenees, Asturias, Apennines, Tyrol, *&c.*

869. 2. *C. Jaela,* Nob. (Abyssinian Ibex.) Adult male somewhat higher at the shoulder than the former; horns three feet long, subvertical, forming a semi-circle backwards, sub-triangular, round in front, with twenty-three irregular prominent knots, extending along the external surface, with several smaller at base, and interposed among the upper, of a dirty horn colour; beard short; general colour, dirty brownish-fawn, with a dark streak along the back; long hair under the throat.

Var? the Siberian Ibex, Ibex Alpium Sibiricarum *of Pallas,* pale gray and brown, black line on the back and down the front of the legs, black space on the upper arm, and under parts white.

Jaela, *Chaldaic.* Jaal, *Arabic.* Akko *of Deuteronomy?*
Icon. *Nobis.*
Habitat. The Mountains of Abyssinia, Upper Egypt,
Mount Sinai, and probably Persia.

870. 3. *C. Caucasica* (Caucasian Ibex.) Adult male
equal in stature to the Alpine ; horns triangular, the anterior edge obtuse, irregularly marked with transverse
knots, and uniform wrinkles, but fewer and more distant
than in the former, the horns twenty-eight inches, dark-
brown, and less curved ; general colour dark brown
above, white beneath the breast, and line on the back
dark. Female, horns nearly erect, slender, short, and
wrinkled.

C. Caucasica, *Guldenstædt. Gmel.*
Icon. *Guldenstædt, Act. Petrop.* 1779.
Habitat. The summits of the Caucasian Mountains.

871. 4. *C. Ægagrus* (the Ægagrus.) Adult male nearly
equal to the Alpine Ibex, in proportion longer, but lower ;
horns forming an acute angle to the front, rounded at the
back, transversely ribbed, forming an undulating anterior
edge, three feet long ; head black in front ; beard brown ;
general colour brown and gray, varying with the seasons ;
the female, with short or no horns.

Capricerva Paseng, *D. Garcia, ab Horto.* Monardes
Paseng, *Kæmpfer.* C. Ægagrus, *Pallas* and *Auctor.* Paseng, *G.* and *F. Cuvier.*

Icon. *Kæmpfer. Pallas.*
Habitat. Mountains of Persia, Caucasus, the Chora-
zan, Candia ? The Alps ?

Var ? *C. Hircus.* Domestication of the Ægagrus, is
supposed to have produced the greater number of breeds,
spread over every part of the globe ; we refer for the most
remarkable to the text.

872. 5. *C. Jemlahica* (the Jemlah Goat.) The male
nearly equal in size to the Ibex; horns placed obliquely on
the frontals, high above the orbits, nearly in contact, depressed, nearly flat, nine inches long, inclined outwards,
then suddenly tapering, and turned inwards; anterior edge
marked with seven small protuberances, from whence pass
as many wrinkles, transversely to the rear; the colour
ashy buff; facial line nearly straight; ears small; no
beard, but the sides of the head and whole body covered
with very abundant long hair of a dirty buff colour; dark
streak down the face and along the spine.

C. Jemlahica, *Nobis MS.*

Icon. *Nobis.*

Habitat. The Jemlah chain of the Hymalaya Mountains,
east of the Burrampootra.

Var. The Cossus and beardless goats of Blainville;
perhaps the Capricorn of Buffon, and *C. Depressa.* See
the text.

Genus III.—Ovis.

Incisors $\frac{0}{8}$; canines $\frac{0.0}{0.0}$; molars $\frac{6.6}{6.6}=32$. Horns common
to both sexes, sometimes wanting in the females; they are
voluminous, more or less angular, transversely wrinkled,
pale coloured, turned laterally in spiral directions, first
towards the rear, vaginating upon a porous bony axis;
the forehead and chaffron arched; they have no lachrymary
sinus, no muzzle, nor inguinal pores; no beard properly so
called. The females have two mammæ; the tail rather
short; ears small; legs slender; hair of two kinds, one
harder and close, the other woolly. In a domestic state
the wool predominates, the horns vary or disappear, the
ears and tail lengthen, and several other characters undergo
modifications. The genus is gregarious in the mountains
of the four quarters of the globe.

873. 1. *Ovis Ammon* (Asiatic Argali.) Adult male about three feet high at the shoulder, and five feet in length; horns sometimes near four feet in length, and fourteen inches in circumference at base, placed on the summit of the head, touching in front, and covering the occiput, bending out backwards and laterally, then forwards and outwards; at base triangular ; surface wrinkled ; general colour fulvous-gray above, white beneath, with a whitish disk on the buttocks; hair close, concealing the wool beneath; female smaller, with slender wrinkled horns, nearly straight.

O. Ammon, *Erxl. Gmel.* Capra Ammon, *Lin.* Ovis Argali, *Bodd.* Stepnie Barani, *Gmel.* Ophion *of the Ancients.* Artak, *Rubruquis.* Dishon *of the Pentateuch?* Pygargon *of the Septuagint?* Weisfarsh *of the old German writers.*

Icon. *Pallas. Nobis.*

Habitat. The mountains and *Steppes* of Northern Asia, Tartary, Siberia, the Kurile Islands.

Var? *O. Pygargus*, Nob. (American Argali.) Adult male three feet high at the shoulder ; four feet six inches in length; horns more spiral, fifteen inches in circumference at base, bent more forward, the tips generally broken off, more round at the base ; no long hairs under the throat; colour dun rufous-gray ; a large white disk on the rump ; tail short ; eyes pale bluish-gray.

Wild Sheep of *California.* Venegas, *Clavigero.* Culblanc *of the Canadians* ; O. Montana, *Geoff.* Big-horned Sheep *of the Americans.* O. Pygargus, Nob. *MS.*

Icon. *Shaw. Natural Miscel. Geoff. Nobis* male and female.

Habitat. The Rocky Mountains, and North-west coast of North America.

874. 2. *O. Tragelaphus* (Bearded Argali.) Adult male

three feet six inches at the shoulder; five feet nine inches from nose to tail; head one foot three inches; horns two feet long, wrinkled, angular, black, thirteen inches and a half in circumference at base, and turned spirally back and downwards; a large beard from the cheeks and under jaw, divided into two lobes; neck short, lined with a standing mane; knees covered by long dense hairs bent back; general colour rufous-brown; external hoofs of the fore-feet longer than the internal; six incisor teeth.

Tragelaphus, *Caius in Gesner*. Fishtall and Lerwee *of Shaw*.

Icon——

Habitat. The mountains of Mauritania (Morocco.)

Var. Size of the Common Ram; horns eleven inches in circumference, bending outwards and backwards; no tuft or mane on the shoulders; long tufts of hair round the fore-knees; tail six or seven inches long; general colour pale rufous.

Mouflon D'Afrique, *Geoff*. Bearded Sheep, *Pent*. Ophion, *Plin*.

Icon. *Mem. de l'Institut. d'Egypt*.

Habitat. The mountains of Upper Egypt.

875. 3. *O. Musmon* (the Musmon.) Adult male in size about the Common Ram, somewhat higher on the legs; horns curved back, forming little more than a half circle, not so voluminous as in the Argali, points turned inwards; general colour brown or liver-coloured gray, with some white upon the face and legs; a darker streak along the back and on the flanks, and often black about the neck; a tuft of hair beneath the throat; females usually hornless and smaller.

Musmon, *Plin. Gesner*. Mouflon, *Buff. F. Cuv*. Ovis Aries, *Desmarets*. (Viewed as the parent of the domestic races, the following is a variety.)

Icon. *Buff. F. Cuv. Mam. Lithog. Nobis* male and female.

Habitat. The mountains of Corsica, Sardinia, and Candia? It was formerly common in those of Asturias, and probably in most of the high chains of Europe.

Var. *O. Aries* (the Domestic Sheep.) Both sexes in general furnished with more wool than hair; the horns frequently wanting, when present, less robust, more angular, wrinkled, spirally contorted in various directions; colour most usually white. For the distinctive marks of the principal breeds or races, we refer to the text.

Genus IV.—DAMALIS.

Incisors $\frac{0}{8}$; canines $\frac{0\cdot0}{0\cdot0}$; molars $\frac{6\cdot6}{6\cdot6}$ = 32. Horns common to both sexes, or in the males only, situate upon the frontal crest, variously bent, and the osseous core provided with a basal cavity communicating externally by a sinus passing beneath the horny sheath; the head heavy, long; the neck short; the spinous processes of first vertebræ of the back mostly elevated, and the croup often depressed; the body bulky; the legs stout; the tail pendulous, more or less lengthened; a mane and beard or tuft usual, and the dewlap wholly or partially developed; the stature of the species in general large.

Sub-genus I.—ACRONOTUS. *Horns common to both sexes, with double flexures more or less pronounced, approximated at base, annulated below, smooth and turned back at the tips; head narrow, long; muzzle small or none; small lachrymary opening; no tufts on knees; inguinal pores; the shoulders in general much elevated; the croup depressed; tail terminated by a tuft reaching to the houghs; two or four mammæ; not remarkable for speed: confined to Africa.*

361

876. 1. *D. Bubalis* (the Bubalis.) Adult male larger than the Stag; horns about thirteen inches long, robust, black, nearly in contact at base, oblique, grooved, then diverging, bent forwards, and the tips turned back; the eyes high in the head; a distinct lachrymary sinus; the shoulders very high; croup much depressed; hair short, smooth, wholly yellowish-dun.

Βουξαλος, *Arist. Opp.* Bubalis, *Plin.* Le Bubale, *G. Cuv.* Icon. *Buffon. Nobis.*

Habitat. Northern Africa.

877. 2. *D. Caama* (the Caama.) Adult male five feet high at the withers; shoulder not so elevated as in the former; seven feet six inches from nose to tail; female considerably less; the head longer; horns placed upon a ridge above the frontals, very close at base, robust, black, diverging, turned forwards and the points backward, five or six prominent knots on the anterior surface, black spot at their base; from the forehead a black streak to the nostrils; the chin a narrow line on the ridge of the neck; streak down the fore-legs, and one on the middle of the thigh, black; general colour of the fur pale fulvous or lively ochre; large triangular spot of white on the buttocks, as also the inferior parts of the body; mammæ two.

Hartebeest, *Sparrman.* Le Caama, *G. Cuv.*

Icon. *Sparrman. Nobis* male, female, and calves.

Habitat. Caffraria.

878. 3. *D. Suturosa* (the Collared Damalis.) An adult female about four feet long; tail one foot; body long, bulky; stature low, heavy; head large; neck short; eyes small; no suborbital sinus; horns large, annulated, round, with double flexures, nearly vertical at base, then abruptly bent backwards and outwards, and the tips again upwards and to the rear; tail flat at base, stiff, tapering,

with a tuft at the end reaching to the heels ; mammæ four; general colour gray-brown and yellowish above, white beneath about the feet, lips, croup, and tail; forehead marked with a dark space, with a white spot above, and two smaller of the same colour behind the eye and ear ; three bars of longer hair forming a kind of collars descend from behind the ears to the throat, from the nape to the sides of the neck, and the third forming a crest on the ridge of the neck descends to the throat. On the back and several other parts there are tufts of long hair, which is in part directed forwards and in part towards the tail.

Antilope Suturosa, *Otto.*

Icon. *Berlin Transactions.*

Habitat. Probably Africa. The species seems to approach Gazella Mytilopes.

879. 4. *D. Senegalensis* (the Koba.) Male head fourteen inches and a half long, facial line convex, dark streak down the nose ; muzzle broad and black ; cheeks paler brown ; lachrymary sinus not evident; horns on the summit of frontals above the plane of occiput, nineteen inches and a half long, five inches and a half from tip to tip, curved backwards and inwards, seven inches in circumference at base, and marked with five or six semi-annuli, and then with sixteen annuli ; size equal to a stag; general colour dark rufous, dirty white beneath, and tail with long hair.

A. Senegalensis, *Auctor.* A. Koba.

Icon. Skull in *Buffon.* Head *Nobis.*

Habitat. Central Africa.

880. 5. *D. Lunata,* Nobis (the Sassayby ?) An adult female four feet six inches long, about three feet high at the shoulder, two feet eight inches at the croup ; horns robust, on the summit of the frontals turning outwards, and forming two semi-circles with the points inwards, with

twelve indistinct annuli; neck short; body bulky; head broad; dark streak down the face; general colour a deep blackish purple-brown above, more fulvous beneath; ears six inches and a half long; small lachrymary sinus; facial line straight; tail middle-sized, covered with long black hair; mammæ?

The Sassayby, *Daniell.* D. Lunata, *Nobis MS.*

Icon. Sassayby, *Daniell.* D. Lunata, *Nobis.*

Habitat. The Booshwana country, South Africa.

Sub-genus II.—BOSELAPHUS. *Horns common to both sexes, heavy, very robust, placed on the summit of the frontals, transversely wrinkled, straight or slightly bent with tips forward, brown or gray in colour, twisted on their own axis, which is in a prolonged direction with the plane of the face, a ridge more or less prominent forming one spiral turn round them; a large sinus in the base of the nucleus, the rest partially porous; a muzzle; no suborbital sinus; mane on the neck; broad and deep dewlap edged with long hair; females an udder of four mammæ; stature very large; species confined to Africa.*

881. 6. *D. Oreas* (the Impoofo.) Adult male above five feet high at the shoulder, nine feet long, and weighing eight hundred pounds; forehead square; muzzle broad; facial line straight; horns about two feet long, straight, with a ponderous ridge ascending in a spiral form to near the tips; proportions of the body like a bull, above seven feet in girth behind the arms; neck thick; shoulders very high; larynx very prominent; dewlap fringed with long hair; a crest of bristles from the forehead passing upwards and recurrent along the ridge of the neck; croup depressed; tail two feet long, with a large tuft of coarse hair; hide black; general colour rufous-dun and ashy; females smaller; horns more slender and longer.

Ant. Oreas, *Pallas, &c.* Coudou, *Buff.* Canna, *Gordon.* Eland Gazelle, *Sparrm.* Impoofo, Poffo *of the Caffres.*
Icon. *Sparrman's Travels,* &c. *Nobis,* male and female.
Habitat. Gregarious in South Africa.

882. 7. * *D. Canna* (the Canna.) Adult male somewhat smaller than the Impoof, more slender; head shorter; horns without prominent spiral ridge, but obtusely angular in front and feint, twisting this angle into a spiral curve: they are more parallel, very closely wrinkled, and bent back beneath the facial line with the point forward, seventeen inches long in a male, twenty-two inches in a female; narrow dark streak down the forehead; small lachrymary or rather prolonged inner canthus of the eyes, with a dark angular spot beneath; shoulders not much elevated; mane on neck not recurrent; general colour a mixed tone of dark gray brown; sternum white; limbs nearly black.

Bastard Eland *of the Dutch Colonists.* D. Canna, *Nobis MS.* Y'Gann *of the Hottentots.*
Icon. In the Banksian collection by Mr. Foster. *Nobis. Daniell's Sketches, &c. of Southern Africa.*
Habitat. South Africa, principally beyond the Gareep.

Sub-genus III.—Strepsiceros. *Horns in the male only, smooth, without wrinkles, pale coloured, with dark tips, forming regular spiral curves, and issuing from the summit of the frontal crest; the nucleus with a cavity at base, and porous above; a broad moist muzzle; real dewlap; long mane on the neck; a beard on the chin; white streak over the eyes; ears broad; shoulder elevated; tail covered with long hairs; females having an udder of four mammæ; stature large. Group confined to Africa.*

883. 8. *D. Strepsiceros* (the Koodoo.) Adult male four feet high at the shoulder; above eight feet long; horns

bulky, compressed, with an anterior ridge, forming with the horn two complete spiral circles, the tips turned outwards and forward; colour pale, tips dark with a white point, and three feet long; chaffron straight; muzzle very broad; ears oblique, very broad tips, pointed; neck thick; withers elevated; dewlap anteriorly square; forehead black, a white line passing over the orbits, unites on the chaffron; chin white-bearded; long fringe of hair on the dewlap, and on the neck a standing mane; general colour of the fur a buff-gray, marked with a white line along the spine, and intersected by four or five others running downwards towards the belly, and four more across the croup; buttocks white; colour beneath rufous; tail white above, edged with rufous and black at the end; female hornless, and with fewer and fainter white markings.

Strepsiceros, *Caius apud Gesn.* Condoma, Coesdoes, *Buff.* Coudou, *Vosmaer.* A. Strepsiceros, *Auctor.* Striped Ant. *Penn.*

Icon. Mr. Daniell.

Habitat. The Cape Colony in the rocky plains of the Karoo Mountains.

NOTE. To this group belong the *A. Torticornis* of Herman, and probably the horn figured by Afzelius, for which we refer to the text in the work.

Subgenus IV.—PORTAX. *Horns in the males only, placed on the sides of the frontal crest, short, robust, sub-angular, without rings; cavity in the nucleus? a complete muzzle; deep suborbital sinus; elevated shoulders; depressed croup; bulky short body; mane on neck; tuft of hair on throat; small dewlap, and vaccine feet and tail; stature large; confined to Asia.*

884. 9. *D. Risia* (the Neelghau.) Adult male about four feet four inches at the shoulder; four feet at the

croup; shoulders high; neck arched; head long, and pointed; forehead arched; horns rising at the sides of frontals, subtriangular, thick at base, with a ridge towards the front, bending forward and upwards, black, smooth, seven inches long; lachrymary sinus considerable; muzzle broad; ears broad, marked with two black streaks on the inner surface; mane erect, black, reaching upon the withers; long dark tuft on the throat to dewlap; general colour slaty-gray, browner on the legs; pasterns often marked with one or more white rings; tail with a black tuft; female smaller, rufous ashy-gray.

Ant. Picta, *Auctor*. White-footed Antelope, *Pent*. Ris'ya or Rishya *of the Sanscrit*. Neelghau *of Northern Indostan*. Gaw-zan *of the Penjab*.

Icon. *Buffon, Pent. Nobis* male and female.

Habitat. Northern India.

Tribe V.—*Bovidæ*.

Horns persistent, common to both sexes, vaginating upon a bony nucleus, not solid but more or less porous, and cellular; the horny sheath increasing by ringlets at the base; the horns round, without annuli, striæ, or ridges; invariably placed upon or at the sides of the frontals, never straight, but at first always bending outwards or forwards; a broad muzzle, almost always naked; no lachrymary sinus; neck short; breast and shoulder deep, more or less dewlapped; structure powerful; vertebræ of the tail often prolonged below the hough; no inguinal pores; females always bearing an udder; stature large; manners gregarious.

Genus I.—Catoblepas.

Incisors $\frac{0}{8}$; canines $\frac{0\cdot0}{0\cdot0}$; molars $\frac{6\cdot6}{6\cdot6} = 32$. Head square; horns flat and broad at base, nearly joining on the crest of the frontals; lying outwards, turning down with the points uncinating upwards; muzzle broad; nostrils as in the Ox, but provided internally with a moveable valve; glandulous excrescence on the cheeks ; a mane on the neck; considerable beard beneath the throat; a small dewlap; bristles round the orbits and on the lips; ridge of hair on the chaffron; carcass round; tail hairy, as in the horse; legs clean and firm; gregarious. Reside in Africa.

885. 1. *C. Gnu* (the Gnoo.) Adult male three feet ten inches high at the shoulder, five feet six inches long; head square; shoulder deep; body round; a pillow of fat on the haunches; legs long and clean ; horns dark, broad upon the summit of the head, tapering out sideways over the eyes, and turning up into a pointed hook ; black bristly hair upon the face ; a tuft of similar hair beneath each eye, concealing a gland; the ears are short ; white bristles surround the eye, and spread on the legs; a vertical mane on the neck, black in the centre, white at the sides ; a bushy beard on the under jaw, and dark-brown fringe along the throat, down to between the fore-legs; tail lined with long white hair ; general colour of the fur deep brown; hoofs pointed, blue-black; females smaller; base of horns less approximated, covered with coarse hair; calves pure white.

Antelope Gnu, *Auctor.* The Gnu *of English.* A. Niou *of French Authors.* Gnoo *of the Hottentots.* Wilde Beest *of the Dutch Colonists.*

Icon. *W. Daniell. F. Cuiver. Nobis* in all its states.

Habitat. The Karoo Plains of South Africa.

886. 2. *C. Taurina* (the Kokoon.) Adult male nearly four feet six inches high at the shoulder ; much lower at the croup, and five feet from the breast to the rump ; head, neck, and shoulder, excessively thick and strong; head

shorter and broader in proportion; eyes very high in the head; horns less broad at base, marked with irregular rugosities at the roots, more distant at base, black, bent down sideways behind the ears, and then suddenly turned upwards, but not to the front; forehead high between their bases; head one foot ten inches long; ears ten inches; tail three feet three inches, covered with long black hair; neck with a long flowing dark mane, reaching beyond the withers; a cartilaginous protuberance covering the chaffron, and furnished with long black hair; large circular glandulous naked spot, distilling a viscous humour beneath each eye; chin covered with dark bristly beard, descending down the dewlap to the breast; general colour dark ashy-gray.

Cocong, *Lichtenstein's Travels.* A. Taurina, *Burchell.*
Kokoon, *Somerville* and *Daniell.*

Icon. *Daniell's Sketches of Southern Africa*, 1820. *Nobis.*
Habitat. Country of the Caffres, South Africa.

887. 3. * *C. Gorgon* (the Brindled Gnoo.) A. male? larger than the Gnoo; horns placed close together, white, round, standing up, bent outwards, and the points turned towards each other, and black; forehead and chaffron covered with irregular depressions; nose and mouth flat and square; ears short; long flowing mane on neck, extending beyond the withers; no beard on the under jaw; long black hairs on the throat and dewlap; tail black, but shorter than in the former; general colour dirty-dun and sepia-gray, variegated with obscure streaks or brindles; four or five cross streaks on the upper arm.

Bastard Wilde Beest *of the Dutch Colonists.*
Icon. *Nobis. Howitt.*
Habitat. The interior of South Africa.

NOTE. In Mr. Brook's collection a horn thirteen inches and a quarter long, base nearly flat, very open, forming a triangular figure, and terminating in a rounded point, bending back and then forward, the base extremely rugous and pearly, the point smooth, all shining black. In our MS. collection noted as *C. Brooksii.*

Genus II.—Ovibos.

Body low and compact; legs short, clean; feet hairy under the frog or heel; forehead broad, flat; no suborbital sinus; a muzzle, but not naked though square; horns common to both sexes, in contact on the summit of the head, flat, broad, then tapering and bent down against the cheeks with the points turned up; the ears short, placed far back; eyes small; tail short; mammæ two? hair very abundant, long, and woolly; stature large. Reside in northern latitudes.

888. 1. *O. Moschatus* (the Musk-Ox.) Adult male size of a small cow; horns and characters as noticed above; colour of the hair brownish-black, hanging low down to the ground; feet often white; in the female the horns do not form a complete scalp; the frog in the hoof soft, transversely ribbed, and partially covered with hair; the external hoof larger and round, the internal pointed and crooked; swell of musk very powerfully.

O. Moschatus, *Blainv. Desm.* Bos Moschatus, *Gmel.* Musk-Ox, *Pent* .Bæuf Musqué, *Cuvier.* Mistus, *Northern and Chippeway Indian.*

Icon. *Pennant, Howitt, Nobis, Parry.*

Habitat the latitudes of North America, adjoining the polar region, and south to the province of Quivira.

NOTE. The Fossil Musk Ox, *O. Pallantis*, with the horns pressed against the temples behind the orbits, found on the coasts of Siberia, is not definitively ascertained to be a separate species.

Genus III.—Bos.

Skull very strong, dense about the frontals, which are convex, nearly flat or concave; horns invariably occupying the crest, projecting at first laterally; osseous nucleus throughout porous, even cellular; muzzle invariably broad, naked, moist, black; ears in general middle-sized; body long; legs solid; stature large.

Sub-genus I.—BUBALUS. *Animals low in proportion to their bulk; limbs very solid; head large; forehead narrow, very strong, convex; chaffron straight; muzzle square; horns lying flat or bending laterally with a certain direction to the rear; eyes large; ears mostly funnel-shaped; no hunch; a small dewlap; female udder with four mammœ; tail long, slender.*

889. 1. *B. Caffer* (the Cape Buffalo.) Adult male about five feet six inches at the shoulder, nine feet from nose to root of tail; horns spreading horizontally on the head, in contact at base, eight or ten inches broad, very ponderous, dark coloured, and above five feet from tip to tip, the internal nucleus very cellular, the points turned up; the incisor teeth loose; ears wide, rather hanging; under-jaw bearded; back straight; hide black, almost naked, and the end of tail furnished with a few distichous bristles; in the young much black longish hair, particularly about the ridge of the back.

B. Caffer, *Sparrm. et Auctor.* Cape Ox, *Pent.* Qu'araho *of the Hottentots.* Zamouse? *in Bornou.*

Icon. *Sparrman's Travels. Buff. Daniell. Nobis.*

Habitat. The interior of Africa.

890. 2. * *B. Pegasus?* (the Pagasse.) A young male, the horns lying across the summit of the head, the tips turned up; colour darkish, with obscure transverse ridges; head very short, thick, abrupt at the nose; forehead wide; eyes large and full; the neck with a dense mane; ears long, flaccid, pendulous; tail to below the houghs covered with long woolly black hair; general colour deep brown; feet white.

Pacasse, *Gallini* and *Carli*. Empaguessa, *Merolla*. Empacasse, *Lopez, Marmol*. Pegasus, *Pliny*. Wadan ? *Captain Lyon's Travels*. B. Pegasus, *Nobis MS*.

Icon. Drawing in the Collection of Prince John Maurice of Nassau in the Berlin Library. *Nobis* a young specimen.

Habitat. Congo, Angola, Central Africa.

891. 3. * *B. Arnee.* Adult male said to be near seven feet high at the shoulders, three feet broad at the breast, and the horns from five and a half to six and a half feet long (each); face nearly straight; breadth of head descending from the summit of the frontals to the foremost molar; horns triangular, rising obliquely, wrinkled, brownish, slightly hanging forwards, with the points turned inward and backward; hide white; colour black; very hairy; tail with the tuft reaching little below the houghs.

B. Arnee, *Shaw*. Arnee, Arnaa, *in Indoostan*. Pfang ? *of the Burmans* ? Taurelephantus, *Ludolph*.

Icon. *Shaw's Zoology ? Oriental Field Sports.* Horns, *Nobis*.

Habitat. The woody valleys at the southern foot of the Hymalaya Mountains and in the Birman Empire.

The domesticated race, China, the Peninsula of Malaya, and Indian Archipelago.

892. 4. *B. Bubalus* (Domestic Buffalo.) Adult male five feet six inches high at the shoulder, eight feet six inches long; horns directed sideways, compressed, with a ridge in front, reclining towards the neck, and the tips turned up, placed at the side of the frontal ridge, and very solid; the forehead convex; mammæ of the male placed on a transverse line; hide dark or black; tail long, slender, tufted at end; hair coarse, scattered, black.

B. Bubalus, *Auctor*. Bhain *in Indoostan*. Buflus *of the Middle Ages*. Buffle, *Buffon*. Yamus, *Arabic*. Buwol, *Polish*. Busan, *Tartaric*.

Icon. *Buffon. Fred. Cuv. Nobis.*

Habitat. In a wild state, India, &c. Domesticated, Persia, the Levant, Turkey, Hungary, Italy, and North Africa.

For the varieties and breeds we refer to the text.

Sub-genus II.—BISON. *Forehead slightly arched, much broader than high; horns placed before the salient line of the frontal crest, the plane of the occiput forming an obtuse angle with the forehead, and semi-circular in shape; fourteen or fifteen pair of ribs; the shoulders rather elevated; the tail shorter; the legs more slender; the tongue blue; and the hair soft and woolly.*

893. 5. *B. Bison* (the Bison.) Adult male six feet high at the shoulder, and ten feet three inches from nose to tail; head broad; horns distant, short, robust, slightly turned forwards, dark coloured; forehead arched; eye large, full dark; body with fourteen pair of ribs; mammæ disposed in a square; anterior half of the animal, excepting the chaffron, covered with a heavy coat of mixed woolly and long harder hair, a foot long in winter, shorter in summer; the woolly gray, the long browner; throat and breast bearded.

B. Bison, Le Bison, *G. Cuvier, Gilibert.* Bison, *Pliny.* Bison, *Bisam.* Wizend *of the Germans.* Subr, *Polish.* Aurochs *of the Modern Germans.*

Icon. *Ridinger. Gilibert. Nobis.*

Habitat. At present the forests of Southern Russia in Asia, Carpathian and Caucasian Mountains, and the Desert of Kobi.

894. 6. * B. *Gaurus* (the Gaur). Adult male six feet high at the shoulder; twelve feet long to the end of tail: above seven feet six inches in girth; head resem-

bling the Common Ox; forehead more arched; horns robust, not bent back, spinous processes of the withers, much elevated, *externally projecting?* forehead covered with whitish wool; eyes small, pale-blue; hair smooth, close, shining, brown; tail short, tufted.

Le Gaour, *Mem. du Mus. d'Hist. Nat. vol.* ix. Gaur, *Dr. Johnson's Sketches of Indian Sports.* B. Gaurus, *Nob.* Gor *of Firdousi,* mistaken for the Wild Ass.

Icon. ——

Habitat. Rhamghur district, and other high mountain forests of India, Æthiopia? *Pliny,* 4. viii. c. 21.

895. 7. *B. Americanus* (American Bison.) Adult male above five feet high at the shoulder, four feet at the croup, eight feet long from nose to tail; form heavy in front, weak behind; body with fifteen pair of ribs, and only four Coccigian vertebræ; eye, round, dark, and full; chaffron short; forehead broad; muzzle wide; horns small, round, lateral, black, very distant, turned sideways, and upwards, hair woolly, very abundant on the head and shoulders; short and close on the hind quarters; in winter brownish black, in summer lighter; tail eighteen inches, with long tuft of dark hair.

B. Americanus, *Auctor.* B. Bison, *Linn. Erxleb.* Bison, *Fred. Cuv., Warden.* Buffalo *of the Anglo Americans.*

Icon. *Buff. Pennant. Nobis.*

Habitat. Interior of North America.

896. 8. *B. Poephagus* (the Yak.) Adult domestic variety, three feet ten inches high at the shoulder; seven feet long from nose to tail; forehead flat; lips tumid; muzzle small; occiput convex covered with frizzled hair; horns round, smooth, pointed, lateral, bending forward and upwards; withers very high, but not hunched; mammæ four, placed transversely; ribs fourteen pair; hair

on the neck and back, very woolly, whitish and black; tail with very long hair; sometimes hornless.

Poëphagus, *Ælian.* B. Grunniens, *Pallas* and *Auctor.* Sarlyk Ukur and Yak, *Tartar.* Ghau-nouk and Gawdashti, *Persian.* Soora Goy, *Indœ.* Si-nyn, *Chinese.*

Icon. *Shaw's Zool. Trans. Soc. of Calcutta. Nobis. Pallas:*

Habitat. Mountains of Central Asia.

897. 9. *B. Gavæus* (the Gayal.) Adult male, four feet nine inches high, nine feet six inches long; horns strong, short, distant, lateral, compressed, turned upwards and forward; head broad and flat, narrowing suddenly on the chaffron; ridge of the frontals covered with frizzled white hair; eyes not large; ears long, broad, turned sideways; neck slender; a middle-sized dewlap, fringed with long hair; ridge of the withers much extending half way down the back; tail to the houghs: tuft at the end; general colour brown, with some white about the feet.

B. Gayal, *Linn. Trans.* Gauvera? *in Ceylon.* Shial (the wild) and Seloc *of the Cucis.* Catin? *in Siam.* J'hongnuaht, *of the Mugs.* Nunel *of the Birmas.* Gabay, *in the Shastras.* Bos Silhetanus, *Cuv.*

Icon. *Transactions of Asiat. Soc. Calcutta. Nobis.*

Habitat. The mountain forests east of the Burrampootra, Silhet, Chatgoon.—*Ceylon?*

Obs. For the fossil species, referrible to the Bisontine group, we refer to the text.

Sub-genus III.—TAURUS. *Forehead square, from the orbits to the occipital crest, somewhat concave, not convex or arched, as in the former; the horns rising from the sides of the salient edge or crest of the frontals; the plain of the occiput forming an acute angle with the frontal and of quadrangular form; the curve of the horns outwards, upwards and forwards ; no mane ; a deep dewlap ; thirteen pair of ribs ; tail long ; udder four teats in a square.*

898. 10. *B. Urus* (the Urus.) In a fossil state of collossal size; the later Uri possessed of the above characters, but the horns turning downward and forward, excepting in two figures quoted in the text, in which they are forward with the tips turned up, white, the ends black : the form of a domestic bull, entirely black, excepting the chin, which is white.

Var. Bos Scoticus, smaller than the usual domestic bull, entirely white ; horns dark, pointing downwards ; a large breed of the same in Hamilton Park.

Bos Urus, *Herberstein, Cuvier.* The true Aurox *of the Germans,* Thur, *Polish.* Wild Bull *of Scotland, Pen. Shaw.*

Icon. *The Black Species. Herberstein. Nobis. The Scottish. Pent. Bewick. Nobis.*

Habitat. Formerly the forests of Middle Europe, Lithuania, Massovia, &c., probably the temperate parts of Tartary, the white species ; England.

Var. Bos Taurus, the Domestic Ox, has the same characters, varied by circumstances, for which we refer to the text, where the principal varieties are enumerated.

Order VIII.—CETACEA.

Body pisciform, terminated by a caudal appendage, cartilaginous and horizontal; two anterior extremities formed like fins, the bones of which are very much flatted and short; head joined to the body by a very short and thick neck; cervical vertebræ very slender, and partly soldered together; two pectoral or abdominal teats ; ears with very small external openings; skin more or less thick, without hair ; brain large, hemisphere well developed ; bone of the

376

internal ear separate from the head, or adhering by liga-
ments; two rudimentary bones lost in the flesh, represent
the pelvis and posterior extremities.

Animals altogether *aquatic,* comprising the largest species
in the world, *carnassial* for the most part; swim by the
assistance of the tail, which moves up and down, and not
from right to left like that of the Fish; viviparous; mammiferous.

Inhabit almost all seas; the very large species the more
northern; the herbivorous nearer the equator.

Family I.—Sirenia (Herbivorous Cetacea.)

Molars with flat coronal; sometimes tusks in the upper
jaw; teats two, pectoral; mustachios; nostrils, properly
so called, at the end of the muzzle; nasal apertures in the
osseous head situated above; body very massive.

Genus I.—Lamantin. Manatus.

*Rondelet, Lin., Scopoli, Storr, Lacep., Cuv., Geoff.,
Illig.* Trichecus, *Linn., Erxleb., Schreb., Shaw, Gmel.*
Manati, *Bodd.*

Incisors $\frac{2}{0}$; canines $\frac{0\cdot0}{0\cdot0}$; molars $\frac{9\cdot9}{9\cdot9}=38$. Incisors very
small, exist only in the fœtus; adults have but thirty-two
teeth, four molars being lost when young; two transverse
hillocks on the coronal of the molars; head not distinct
from the body; eyes very small, placed above, between the
auditory foramina and the end of the muzzle; auditory
foramina hardly visible; tongue oval; hinder part of the
body very thick, depressed, and rounded at the end; no
caudal fin properly speaking; some vestige of claws on the
edges of the pectoral; mustachios, composed of a bundle
of enormous hairs directed downwards, and forming a
kind of corneous tusk on each side; naked skin, thick and
rugous; six cervical vertebræ; six pair of ribs, thick and

377

clumsy, the two first only united to the sternum; stomach divided into several pouches; bifurcated cæcum; inflated colon.

These animals live in troops, and feed only on vegetable substances.

Inhabit the shores of the Atlantic towards the mouth of the great rivers of the western coast of Africa and the eastern coast of South America.

899. 1. *Manatus Americanus* (American Lamantin.) Osseous head, rather elongated in proportion to breadth in the region of the muzzle and nostrils; nasal foramina thrice as long as broad; zygomatic apophysis of the temporal very high; lower edge of the lower jaw strait; sometimes twenty feet long.

Manati Phocæ genus, *Clus. Exot. p.* 132. Manate ou Vache Marine, *Dampier's Voyage, tom.* I. *p.* 46. Sea Cow, *Sloane's Jamaica, vol.* II. *p.* 529. *La Condamine's Voyage, p.* 154. Grand Lamantin des Antilles, *Buff. Hist. Nat. tom.* XIII. *p.* 377 and 425. *Ejusd. Suppl. tom.* VI. *p.* 396. *Cuv. Ann. Mus., tom.* XIII. *p.* 282. *Ejusd. Recherch. sur les Oss. Foss. tom.* IV. *Mem. sur les Phoques Desm. Nouv. Dict. d'Hist. Nat. tom.* XVIII. *p.* 213. *F. Cuv. Nouv. Dict. des Sci. Nat.*

Icon. *French. Encyc. pl.* 112 *fig.* 2 and 3. *Clus. ut supra. Buff. ut supra, pl.* 57. *Cuv. Ann. Mus. ut Sup. pl.* 19. *Mem. sur les Phoque. Nouv. Dict. d'Hist. Nat tom.* XVIII. *pl. G.* 9.

Inhabits the river of the Amazons, the Orinoco, Cayenne, the Antilles. Grown rare in frequented places.

Obs. The Baron cannot affirm if the Lamantin, placed by some authors on the coasts of Peru, be of this species or not.

900. 2. *M. Senegalensis* (Lamantin of Senegal.) Head

short in proportion to breadth; breadth of nasal foramina equal to three-fourths of length; zygomatic apophysis of the temporal, slightly elevated; lower edge of lower jaw curved. About eight feet long.

Lamantin du Senegal, *Adanson's Voy. Dapper, Afric.* Trichecus Australis, *Shaw, Gen. Zool. Buff. tom.* xiii. *Suppl. tom* vi. *Cuv. Ann. Mus. Recher. Sur less. Oss. Foss.* Icon. *Shaw. Oss. Foss.*

Manners unknown.

Habitat. Mouth of the Senegal River, and other great rivers of the western coast of Africa.

Genus II.—Dugong. Halicore. Illig.

Incisives *(adult)* $\frac{2}{8}$; canines $\frac{0\cdot0}{0\cdot0}$; molars $\frac{3\cdot3}{3\cdot3} = 14$, *(young)* Incisives $\frac{4}{8}$; canines $\frac{0\cdot0}{0\cdot0}$; molars $\frac{3\cdot5}{3\cdot5} = 32$. Four upper incisors, two of which cylindrical, strait and strong, form real tusks; two very small behind these, only found in the young. Anterior face of the lower jaw truncated obliquely; eight alveoli on two lines, containing teeth never developed. First molar in the adult cylindrical and worn obliquely, and in a hollow at point; second cylindrical, with flat coronal; third formed of two cylindars united, and truncated at top. Body pisciform, terminated by a horizontal fin, with two lobes; head not distinct from body; muzzle very thick, truncated, and mobile, furnished with very thick spiny hairs on the edge of the lips; nostrils very small, separated in front of the eyes; eyes small; tongue soft, partly fixed; fins short, without distinct fingers or claws; seven cervical vertebræ; eighteen pair of ribs; stomach two pouches, and two cæcal appendages; heart bifurcated, each ventricle forming a particular lobe; penis with voluminous and bifid gland.

901. 3. *H. Indicus* (Dugong of Indian Seas.) General

colour bluish-gray; seven or eight feet long; greatest circumference three or four; length of tusks an inch and a half.

Dugong. *Renard poiss. des Indes.* Dugong *Buff. tom.* xiii. Indian Walrus, *Pent.* Trichecus Dugong, *Erxleb.* Trichecus Dugong, *Gm.* Rosmarus Indicus, *Bodd., Camper, Raffles. Home, F, Cuvier.*

Icon. *Renard, Buff.* Cranium, *Camper, Trans. Philos.*

Habitat. Indian Seas.

Obs. The Malays distinguish two varieties, the second shorter and thicker than the first.

Genus III.—STELLÈRE. STELLERUS, CUV.

Incisives $\frac{0}{0}$; canines $\frac{0\cdot0}{0\cdot0}$; molars $\frac{1\cdot1}{1\cdot1} = 4$. No teeth implanted, but a molar plate on each side of the jaws, not attached by roots, but by a number of small vessels and nerves; body inflated towards the centre, diminishing gradually to the caudal fin; head obtuse, without distinct neck; no external ears; lips double; a cartilaginous membrane to cover the eyes; palmated fins like the Sea-Tortoise; caudal fins very broad, crescented, and pointed at each end; skin covered by a very thick epidermis, composed of serrated tubes, perpendicular to the dermis; stomach simple. Eats fucus.

Habitat. The most northern part of the South Sea, western coasts of North America, &c.

902. 4. *Stellerus Borealis* (Northern Stellère.) Round head; no tusks; twenty-three feet long.

Manatus, *Steller.* Trichecus Manatus, *Var.* Boreal. *Gmel.* Trichecus Borealis, *Shaw.* Whale-tailed Manati, *Penn.* Grand Lamantin du Kamtschatka, *Sonnini.*

Icon. —— ?

Habitat. See Genus.

380

Family II.—Cete (Common Cetacea.)

Teeth sometimes pointed, sometimes obtuse; all of one sort on the edges of the jaws; sometimes no teeth, but transverse corneous laminæ in the vault of the palate; two anal mammæ; nostrils opening on the top of the head to eject water, called *spiracles;* olfactory nerve small; larynx pyramidal, penetrating the back; nostrils flatted; eyes with thick sclerotica; smooth tongue; no hairs, eyelash, or mustachios; skin smooth and shining, covering a thick coat of fat; stomach with five and sometimes seven pouches; spleen divided into many separate lobes.

Tribe I—*Small-headed Cetacea.*

Head in the usual proportion with body.

Genus I.—Delphinus, Linn. (Dolphin.)

Teeth of one sort, canine form, compressed and indented on their trenchant edge; number very variable, two hundred at most, two at least, or none; jaws more or less advanced, beak-formed; no tusks; spiracles with a common and crescented aperture above; sometimes an adipore dorsal fin; sometimes a longitudinal dorsal fold of skin; tail flatted horizontally, and bifurcated; no cæcum; carnassial.

Sub-genus I.—Delphinus (Dolphins proper) Cuvier. *Muzzle elongated into a moderate beak; large at base, rounded at point; jaws widened posteriorly, with edges furnished with numerous teeth: a single dorsal fin.*

903. 1. *D. Delphis* (Common Dolphin.) Jaws moderately elongated, of equal length, forty-two to forty-five teeth each side, fine, round, pointed, arched, equidistant, from forty-two to forty-five in each jaw; dorsal fin placed be-

yond one half the middle of the back ; upper parts black; under white.

Delphinus Delphis, *Linn.*, *Lacepède*, *Bonnaterre*, &c.
Icon. *Encyc. Cetologie, pl.* 9, 10. *fig.* 2.
Habitat. The Seas of Europe.

904. 2. *D. Tursio* (Great Dolphin.) Jaws moderate, the lower a little longer; teeth strait, obtuse, three and twenty in each side above, and one and twenty below; dorsal fin as in the last ; back blackish ; belly white.

Delphinus Tursio. Nesarnak, *Bonnat. Cetol.* Delphinus Delphis, *Hunter.* Coudin or Caudrieu, *Duhamel. Lacep.*
Icon. *Encyc. Cetol. Hunter.*
Habitat. European Seas.

905. 3. *D. Nesarnak.* Compressed muzzle ; teeth twenty or twenty-three on each side ; thick, strong, obtuse, and couched obliquely from front to back, below, and the reverse above ; body thick.

Nesarnak, *Oth. Fabricius.*
Icon. ——
Habitat. Sea of Groënland.

906. 4. *D. Niger* (Black Dolphin.) Muzzle flat and elongated; above twelve teeth on each side ; very small dorsal fin near the caudal ; general colour black ; commissure of the lips, edge of the pectoral and dorsal fins, white.

Delphinus Niger, *Lacep.*
Icon. ——
Habitat. Japanese Seas.

N. B. Known only from a figure seen by M. de Lacepède.

907. 5. *D. Rostratus* (Slender-beaked Dolphin.) Long slender muzzle ; teeth twenty-two to twenty-six on each side ; conical, a little curved, with rugous surface.

Dauphin à bec Mince.   Delphinus Rostratus, *Cuv. Rapp. sur les Cetac.   Ann. du Mus.   Desm. Nouv. Dict. d'Hist. Nat.*

Icon. ——

Habitat. ——.   The great freshness of a head in possession of Mr. Sowerby induces M. de Blainville to think the species European.

908. 6. *D. Frontatus* (Fronted Dolphin.)   Slope of the frontal convexity more abrupt ; beak more defined ; teeth twenty-one, twenty-two, or twenty-three on each side ; seven feet long.

Frontatus, *Cuv. Oss. Foss.*   Dauphin, *de Geoffroy, Desm.*

Icon. —— ?

Habitat. Doubtful.

NOTE. Of this Sub-genus *Delphinus Boryi, D. Linensis, D. Dalius, D. Orca, D. Feres, D. Canaaensis, D. Bertini,* being all marked as doubtful by M. Desmarest, and so considered by the Baron, we forbear to insert them. The same may be said of M. Rafinesque Smaltz's Sub-genus OXYPTERUS, characterized by two dorsal fins, and of which the only species is *D. Mongitori,* observed but once, and having no description or figure. Should the above character be correct, this animal must be distinguished not only from the Dolphins, but from all the other Cetacea.

Sub-genus II.—PHOCÆNA (MARSOUIN.) *No beak; muzzle short and convex; numerous teeth in each jaw; dorsal fin.*

909. 1. *D. Phocæna* (Dolphin Marsouin.)   Body and tail elongated; muzzle rounded; teeth compressed, trenchant; twenty-two to twenty-five in each side of two jaws; dorsal fin about the middle of the back, almost triangular and rectilinear; colour blackish above and white below; total length four or five feet.

Phocæna, *Rondelet.*   Delphinus Phocæna, *Briss. Linn. Gmel.* Bonnaterre, *G. Cuvier.*   Dauphin Marsouin, *Lacep.* Merschswein *of the Germans.*   Porpus *with us,* &c.

383

Icon. *Encyc. Cetol. Cuvier Menag. Nat. Lacep. Cet.*
Habitat. All our seas.

910. 2. *D. Gladiator* (Sword-fish.) Body and tail elongated;
head convex; muzzle short and rounded; jaws equal; teeth
sharp and curved; dorsal fin near the neck, and elevated
one-fifth of the length of the body; twenty-three to twenty-
five feet long.

Swerd-fisch, *Anderson.* Poisson à Sabre, *Pagès.* Del-
phinus Maximus, *Olafsen.* Dauphin Epée, De Mer. *Bon-
naterre.* D. Gladiateur, *Lacep. Cuv. Règne. An.*
Icon. *Lacep.*
Habitat. Sea of Spitzbergen, Davis's Straits, coast of
New England.
Obs. This is joined to the following by the Baron.

911. 3. *D. Grampus* (Grampus.) Colour blackish above;
white belly; twenty-five feet long.

Epaulard, *des Saintongeois.* Orca, *Oth. Fred. Muller.*
Butkopf *of the Dutch.* Our Grampus. D. Grampus,
Hunter. D. Orca, *Linn., Gmel., Shaw.* Dauphin Epaulard,
Bonnaterre. Cachalot d' Anderson, *Duhamel.*
Icon. *Shaw, Lacep., Duhamel.*
Habitat. North Atlantic.

912. 4. *D. Peronii* (Peron's Dolphin.) Forms, proportions, and size of the Porpus; back bluish-black; belly,
sides, end of muzzle, fins, and tail, bright white.

Dauphin du Peron, *Lacep.* Dauphin Leucoramphe,
Peron.
Icon. ——
Habitat. Great Austral Ocean.

913. 5. *D. Commersonii* (Commerson's Dolphin.) Silvery

white; extremities of muzzle, fins, and tail, blackish; a little larger than *D. Phocæna*.

Dauphin de Commerson; *Lacep.*

Icon. ——

Habitat. From Cape Horn to the further point of America.

914. 6. *D. Griseus* (Gray Dolphin.) Head like the Porpus; dorsal fin much elevated and pointed; gray on the upper parts, gradually blending with the white underneath; two-thirds smaller than the Grampus.

Dauphin Gris. Delphinus Griseus, *Cuv. Rapport sur les Cet. and Ann. du Mus. Schreb. Goldfuss.*

Icon. *Ann. du Mus. tom.* xix.

Habitat. Taken in the neighbourhood of Brest.

915. 7. *D. Globiceps* (Round-headed Dolphin.) Summit of the head very convex; muzzle rounded; dorsal fin not high, sloped behind; pectorals long and pointed; teeth nine to thirteen each side : blackish gray or shining black.

D. Globiceps, *Cuv. Rapp. sur les Cet. Ann. Mus. Schreb. Goldfuss.*

Icon. *Ann. du Mus. Schreb. Goldfuss.*

Habitat. The Atlantic Ocean.

Obs. D. Ventricosus and *D. Rissoanus* are doubtful. The first the Baron joins to *Epaulard*. The second he considers as approximating to the first, and of course to Epaulard.

Sub-genus III.—DELPHINAPTERUS (LACEP.) *Head obtuse; muzzle not prolonged, beak-formed; number of teeth middling; no dorsal fin.*

916. 1. *D. Leucas* (the Beluga.) Head like the Porpus; teeth short, blunt, nine on each side in both jaws, the lower oblique from front to back, the upper the reverse;

385

very small dorsal eminence instead of fin; yellowish white; twelve to eighteen feet long.

Witfisch oder Weissfisch, *Anderson, Crantz, Muller*. Delphinus Albicans, *Oth. Fabr.* Delphinus Pinna in dorso nulla, *Briss.* Dauphin Beluga, *Bonnat. Ency. Cet.* D. Leucas, *Gm.* Delphinopterus Beluga, *Lacep.* Beluga, *Shaw.* Huitfisch *of the Danes.*

Icon. *Shaw.*

Habitat. Northern Seas.

Sub-genus IV.—HYPEROODON (LACEP.) *Teeth one in each jaw, sometimes none; lower jaw usually more voluminous than the upper.*

917. 1. *D. Hyperoodon* (Dolphin of Honfleur.) Head convex, beak round and flatted; no teeth in either jaw; palate furnished with small points or false teeth; lower jaw very thick in proportion to upper; orifice of spiracles crescented, points turning back; dorsal fin near the middle of the body; leaden gray above, whitish below; twenty-three feet long.

Dauphin Butskopf. Delphinus Butskopf, *Bonnat. Baussard desc. des deux cet. Journal des Phys.* Hyperoodon Butskopf, *Lacep.*

Icon. *Journ. de Phys. March,* 1789, *pl.* 1 *et* 2.

To this Sub-genus are referred *D. Anarnak, D. Chemnitzianus, D. Hunteri, D. Edentulus, D. Epiodon.* None sufficiently authenticated to be considered as specifically different from *D. Hyperoodon.*

Genus II.—MONODON, L.

Incisors $\frac{1\cdot1}{0\cdot0}$; canines $\frac{0\cdot0}{0\cdot0}$; molars $\frac{0\cdot0}{0\cdot0}$ = 2. One or two large tusks implanted in the incisive bone, straight, long, and pointed, in the direction of the axis of the body; general form like the Dolphin's; orifice of spiracles united on the highest part of the head behind; longitudinal dorsal projection, not; fin pectorals, oval; manners like the Dolphin's; carnassial. Inhabits the Northern Seas.

918. 1. *Monodon Monoceros* (Common Narwhal.) General form ovoïd; length of head one-fourth that of the animal; left tusk unique (the right not being developed), of spiral form, one half as long as the body; back uniformly grayish in the young, blackish or marbled in the old; twenty or twenty-two feet long.

Monodon, *Artedi.* Narwhal, Oder Einhorn *Anderson, Muller.* Monodon Narwhal, *Fabri.* Monodon Monoceros, *Lin., Erx., Gm.* Narwhal, *Bonnat.* Narwhal Vulgaire, *Lacep., Shaw, Vulgo.* Sea-Unicorn.

Icon. *Encyc. Cet. Lacep. Shaw.*

Habitat. Eightieth degree of North latitude.

Obs. M. *Microcephalus* and M. *Andersonianus,* not authenticated. The last is represented as having smooth tusks.

Tribe II.—*Large-headed Cetacea.*

Genus I.—Physeter, Linn.

Lower teeth eighteen to twenty-five each side of the jaw; upper jaw wide, elevated, without corneous laminæ or teeth, or with short and undeveloped ones; lower jaw elongated, narrow, corresponding to a furrow of the upper; thick and conic teeth entering corresponding cavities in the upper; orifice of spiracles united at the end or near the upper end of the muzzle; dorsal fin in some species, simple eminence in others. Large cavities with cartilaginous walls in the upper region of the head, communicating with diverse parts of the body by particular canals, filled with an oil which fixes and crystallizes when cool; carnassial. Inhabit the Polar Seas.

Sub-genus I.—CATODON (LACEP.) *Orifice of spiracles placed at the very end of the upper part of muzzle ; no dorsal fin.*

919. 1. *P. Macrocephalus* (Great-headed Cachalot.) Lower teeth twenty to twenty-three on each side, curved, and a little pointed at the extremity; small conical teeth concealed in the upper gums ; tail straight and conical ; longitudinal eminence on the back, above the anus; upper part of the body blackish or slate-blue, a little spotted with white ; belly whitish ; forty-six to sixty feet long.

Shaw, Gen. Zool. Cachalot Macrocephale, *Lacep.* Grand Cachalot, *Bonnat. Cetol.*

Icon. *Encyc. Cetol. Shaw, Lacep.*

Habitat. North Seas. Have been found even in the Adriatic.

Obs. Between *Ph. Trumpo* and the above species, the Baron finds no distinction. Between *Ph. Catodon* (Svinewal) and *Macrocephalus* he considers any difference in teeth to be the result of age. *Ph. Macrocephalus* of Gmelin, which forms the Sub-genus *Physalus* of Lacepède, rests only on a bad figure of Anderson's, and is considered doubtful both by the Baron and M. Desmarest.

Sub-genus II.—PHYSETER (LACEP.) *Orifice of the spiracle situated at the end or near the end of the upper part of muzzle ; a dorsal fin.*

920. 1. *Ph. Microps* (Small-eyed Cachalot.) Lower teeth twenty-one on each side, arched, the points directed backwards and a little inwards ; dorsal fin large, straight, and pointed ; pectoral fins broad ; eyes very small ; sixty-six to eighty feet long.

Physeter dorso pinnâ longâ, &c. *Artedi.* Cachalot Microps, *Bonnat.* Physetère Microps, *Lacep.*

Icon. *Bonnaterre.*

Habitat. Northern Seas nearest the pole

921. 2. *Ph. Sulcatus* (Furrowed Cachalot.) Teeth of lower jaw pointed and straight ; inclined furrows on each side of this jaw ; dorsal fin conical, situated above the pectorals, which it equals in length ; dimensions unknown.

Physeter Sulcatus, *Lacep. Mem. du Mus.*

Icon. ——

Habitat. Seas of Japan, and perhaps the North Pacific.

Obs. Taken from a Chinese figure, communicated to M. Lacepède by M .Abel Remusat.

Orthodon and *Mular* are not separated by the Baron from *Microps.*

Genus II.—BALÆNA, Linn.

Teeth none ; upper jaw keel-formed, furnished on each side with whalebones or transverse corneous laminæ, slender, serrated, and attenuated at the edges ; orifices of the spiracles separated, and situate towards the middle of the upper portion of the head ; a dorsal fin in some species, nodosities on the back in others ; short cæcum ; feeds on small fish and mollusca, *&c. ;* inhabits the Northern Seas, but some species frequent the temperate zones.

Sub-genus I.—BALÆNA, Lacep. *No dorsal fin.*

922. 1. *B. Mysticetus* (Common Whale.) Body thick and short ; tail short ; no boss on the back ; upper jaw furnished with about seven hundred transverse laminæ or whalebones ; eighty to one hundred feet long.

Balæna Major, *Sibbald.* Balæna Vulg. Gröenlandica, *Briss., Oth. Fabricus.* Balæna Mysticetus, *Linn., Erx., Gm.* Baleine Franche, *Bonnat. Lacep.*

Icon. *Ency. Cet. Lacep.*

Habitat. Atlantic Ocean and Polar Seas in the neighbourhood of Groënland.

923. 2. *B. Glacialis* (Nord-Caper.) Lower jaw rounded; high and broad body, and tail elongated; no boss on the back; general colour gray, more or less clear; under the head a vast oval surface of a shining white, with a few blackish spots ; dimensions unknown.

Balæna Islandica, *Briss.* B. Glacialis, *Klein.* Nord-Caper, *Anderson.* Baleine Nord-Caper, *Bonnat.* B. Mysticetus, Var. B. *Gm.* Baleine Nord-Caper, *Lacep.*

Icon. *Lacep.*

Habitat. North Atlantic, between Spitzbergen, Norway, and Iceland.

924. 3. *B. Nodosa* (Knotted Whale.) A boss on the back, situate near the tail ; pectoral fins white, long, and remote from the end of the muzzle.

Pflokfisch, *Anderson, Crantz, Dudley.* B. Gibbosa, Var. B. *Gm.* Baleine Tampon, *Bonnat.* Baleine Noueuse, *Lacep.*

Icon. ——

Habitat. The coasts of New England.

925. 4. *B. Gibbosa* (Bossed Whale.) Five or six bosses on the back, near the tail ; whalebones white.

Knoten fisch, *Anderson*, Balæna Mæra, *Klein*? Baleine, à six bosses, *Briss., Crantz., Muller.* Baleine à bosses, *Bonnat.* Baleine Bossue, *Lacep.*

Icon.——

Habitat. Sea of New England.

Obs. B. Japonica and *B. Lunulata*, are considered doubtful by M. Desmarest, resting only on Chinese sketches, communicated to M. Lacepède by M. Abel Remusat.

Sub-genus II.—BALENOPTERA, (Lacep.) *Whalebones ; a dorsal fin.*

926. 1. *B. Gibbar* (The Gibbar.) Jaws pointed and

equally advanced ; whalebones short ; no folds under the throat or belly ; whalebones bluish ; body brown above and white underneath ; as long as the Common Whale.

Fin-Fisch, *Martens.* Baleine Gibbar, *Rondelet.* Balæna Tripinni Ventre lævi, *Briss.* Balæna Physalus, *Linn.*, *Erx.*, *Gm.* Gibbar, *Bonnat.* Baleinoptère Gibbar, *Lacep.*

Icon. *Encyc. Cet. Martens, Lacep.*

Habitat. The Arctic Icy Sea, also the North Atlantic.

927. 2. *Bal. Boops* (Jubarta.) Nape elevated and round ; muzzle advanced and a little rounded ; longitudinal folds under the throat and belly ; tuberosities almost demi-spherical in front of the spiracles ; dorsal fin curved behind ; fifty-four feet long.

Jubartes, *Klein.* Jupiter-Fisch, *Anderson*, Baleine à Museau Pointu, *Briss.* Bal. Boops, *Lin.*, *Erx.*, *Gm.* Baleinoptère Jubarte, *Lacep.*

Icon. *Encyc. Cet. Lacepède.*

Habitat. Seas of Groënland, but occasionally found in many seas in both hemispheres.

Obs. The *Balæna Musculus*, is not sufficiently distin-guished. *B. Rostrata* of Hunter, Fabricius, and Bonnaterre, the Baron thinks differs only in dimensions from the Jubarta. *B. Rostrata* of Pennant is the Hyperoodon. *B. Punctata*, *B. Nigra*, *B. Cærulescens*, and *B. Maculata*, depend only on the Chinese sketches before mentioned, of M. A. Remusat, and are marked doubtful by M. Desmarest.

END OF THE SYNOPSIS OF THE MAMMALIA.

Principal Errata in the Synopsis of Ruminantia.

Page	Line		for			read	
298	21	*for*	Ruguere	.	.	*read*	Kirguise
	25	,,	slightly turned		.	,,	tumid.
299	16	,,	less turned	.	.	,,	tumid
	22	,,	Pennich-cat	.	.	,,	Pennich-Catl.
310	3	,,	probably these		.	,,	probably this
315	8	,,	coloured ; triangle		.	,,	coloured triangle
318	15	,,	Hondurus	.	.	,,	Honduras.
321	9	,,	prolonged with		.	,,	with prolonged

London : Printed by WILLIAM CLOWES, 14, Charing Cross.

A

SYNOPSIS OF THE SPECIES

OF THE

CLASS REPTILIA.

By J. E. Gray, Esq. F.G.S., &c. &c.

———

N.B.—In the following List those species alone are included, which have either been seen* by the author, or which have been well figured or described.

REPTILES.

Vertebrated animals, respiring by lungs, having warm red blood, heart with one ventricle and two auricles and the skin covered with scales.

* In forming this list, the collection of reptiles of the British Museum, the College of Surgeons, and of Mr Bell, in London, of the Gardens of Plants and Ecole de Medicine at Paris, of the Royal Museum of Leyden and Berlin, and the Free Town of Francfort, have been studied with attention through the kindness of their several keepers.—J. E. Gray. Oct. 1830.

Section I.   CATAPHRACTA.

Body covered with two shields.   The vent longitudinal or circular : the ossa quadrata, and the pterygoid processes forming part of the skull; the organs of generation simple and single in both sexes.

Order I. TESTUDINATA, or TORTOISES.

Ribs, vertebræ and sternum united together in a bony case, enclosing the body, and protecting the head and limbs ; the jaws toothless.

The case is usually covered with horny shields, which are granular in the newly-hatched animal, and which increase in size by the addition of layers on the under side of the edge, and having the first formed granular part marked on their surface ; this part is called the areola.   The shields of the upper shell are called, from their position, discal, consisting of the vertebræ or central series, and costal, composed of the two lateral series and marginal plates, which have often an additional small narrow plate placed on the back of the neck, called the nuchal plate.

The under shell is usually covered with six pair of shields, one called the gular (which have sometimes an additional intergular plate between them), the second the humeral, the third the pectoral, the fourth the abdominal, the fifth the femoral, and the sixth the anal plates ; the suture between the upper and lower shell is generally covered by the end of the pectoral and abdominal plates, and there is a small plate placed at each end of it, the front called the axillary, and the hinder the inguinal plate.

Genus I. TESTUDO.

Feet club-shaped, claws five before, four behind blunt ; jaws horny.   Shells hemispherical, solid, covered with horny plates ; the hinder marginal plate broad, incurved.

2

I. *Shell and sternum immoveable, sternal plates in the gular pair being separate.* Testudo.

Indian Tortoise. Test. Indica, Lin. *Test. Retusa,* Merrem. *Test. Elephantopus,* Harlan. *Test. Nigra,* and *Test. Californica,* Gaimard. *Test. Gigantea,* Schw. and *Test. Dussumieri,* Schegel. Shaw, t. 3. Frey. Voy. t.

Shell black, nuchal plate, often deficient; pectoral plates short. India, naturalized in California, Isle of France and Galapago Isles. Length three feet.

Hercules Tortoise. Testudo Hercules, Spix. Braz. t. 14. *Test. Carbonaria,* Spix. t. 16.

Shell subquadrate, depressed, contracted in the sides, black, areola yellow, rarely exceeds three quarters of an inch in width. Nuchal plate none. Sternum behind roundly lobed, leg red spotted, length twenty-four inches. South America, Var. *Truncata.* Shield and bones beneath elevated, conical, truncated.

Tabular Tortoise, Shaw. *Test. Tabulata,* Walb. Schoepf. t. 12. f. 2. t. 13. *T. Sculpta,* Spix. t. 15. *T. Cagado,* Spix. t. 17. Young, *T. denticulata,* Lin.

Shell oblong, depressed, deeply grooved; black brown areola large; the sternum behind rather acutely lobed. Brazils.

Schweiger Tortoise. Test. Schweigeri, n.

Shell oblong, depressed, pale brown, obscurely brown dotted; beneath brown rayed; areola small central; sternum before acutely cut. Hab. —— ?

Gopher Tortoise. Test. Polyphemus, Bartram. *T. depressa,* Lesueur, Mss. Guerin. Icon. t. 1. f. i. *T. Carolina,* Leconte Mus. Paris.

Shell oblong, depressed pale brown grooved; nuchal plate

broad, square ; front of sternum projecting ; tail obsolete. North America.

Bell-shaped Tortoise. Test. Marginata. Schoepf. t. 11, 12, f. 1. *T. Graji.* Herm. *T. Gracœ,* Lacep. *T. Campanulata,* Walb.

Shell oblong, ventricose ; hinder margin flattened, expanded toothed ; shields grooved ; blackish yellow varied ; areola yellow, contracted ; nuchal plate, very slender, long. Egypt ?

Radiated Tortoise, Shaw. *Test. Radiata,* Shaw. Zool. t. 2. *J. Coui,* Daud. *T. Calcarata,* Merrem.

Shell hemispherical ; black, yellow rayed, areola sunk ; nuchal plate none. Madagascar.

Sulcated Tortoise. Test. Sulcata, Shaw, Miller, Cym. Phys. t. *Test. Calcarata,* Merr. *Test. radiata Var.* Gray, Syn. Rept.

Shell sub-globular, yellow ; shield rather convex, deeply grooved ; areola small, superior ; nuchal plate none. Senegal. Dongola. Dr. Ruppell.

Leopard Tortoise. Test. Pardalis, Bell. Zool. Jour. t. 15. *T. armata,* Boie Mss. *Test. Bipunctata,* Cuv. R.A.

Shell hemispherical ; yellowish, black spotted ; nuchal plate none. Cape of Good Hope.

Starred Tortoise. Test. Stellata, Test. Actinodes, Bell. Z. Jour. t. 23. *Test. Elegans,* Shaw, t. 6. f. 1. and Schoepf. t. 26. *T. Stellata,* Schw.

Shell globular ; shields convex, grooved ; black, yellow-rayed ; areola yellow, large ; nuchal plate none. India. Ceylon.

4

Geometrical Tortoise, Shaw. *Test. Geometrica,* Lin. Shaw. t. 1. f. 1. Jun. *T. Luteola,* Daud.

Shell globular oblong, black ; shields yellow-rayed ; nuchal plate long ; β nuchal plate short. *Test. Tentoria,* Bell. Z. Jour. t. 26. Cape of Good Hope, Madagascar.

Greek Tortoise. Test. Græca, Lin. *T. Carolina,* Herm. *T. Geometrica,* Brunn. *Test. Hermanni,* Schn. Edw. t. 204. Shaw. t. 1.

Shell oblong, globose ; black, yellow-rayed ; nuchal plate, slender ; caudal plate incurved. Var. β hinder margin expanded, Schoepf. t. 9. Inhab. South of Europe, Var. no nuchal plate. *Test. Zolhafa,* Forsk.

Areolated Tortoise, Test. Areolata, Thumb. Schoepf. t. 23.

Shell oblong, rather depressed ; shields yellowish, deeply grooved ; areola brown ; sutures deep ; nuchal plate, narrow. Var. β greenish ; hinder lobe of sternum mobile. Cape of Good Hope, variable in number of dorsal and marginal plates.

Marked Tortoise. Test. Signata, Walb. Schoeff. t. 28. *T. Cafra ?* and *T. Juvencella ?* Daud.

Shell rather depressed ; dorsal shields flattish ; yellowish, black-rayed ; areola blackish ; nuchal shields narrow. Africa ?

II. *Dorsal shell and sternum both solid ; sternal plates eleven, the gular pair produced united into one.* Chersina. Gray.

Angular Tortoise, T. (*Chersina*) *Angulata,* Dumeril. *Test. Bellii,* Gray Spec. Zool. t. 3, f. 1. *Test. Pusilla,* Lin. *T. miniata,* Lacep.

Shell oblong, ventricose ; black, shields grooved ; areola yellow ; nuchal plate very narrow. Cape of Good Hope.

5

III. *Hinder part of dorsal shell mobile, united to the anterior by a carious ligamentous suture; sternum solid, shields twelve.* Kinyxis. *Bell.*

Home's Tortoise. Test. (Kinyxis) Homeana. Bell. Lin. Trans. xiv. t. f. 2.

Shell brown; hinder edge reflexed; upper edge of fifth vertebral plates prominent; nuchal shield long, narrow; front of sternum expanded. Demerara. Guadaloupe.

Bells Tortoise. Test. (Kinyxis) Belliana, Gray.

Shell brown, edge hardly expanded; centre of hinder vertebral plates convex; nuchal shield long, narrow; front of sternum, narrow. Inhab. ?—Mus. Gray.

Worn Tortoise. Test. (Kinyxis) Erosa.

Shell brown; hinder edge reflexed, denticulated; centre of fifth vertebral plates prominent; nuchal shield none. Young, back rounded. *K. Castanea.* Bell. l. c. t. f. 1. *Test. Denticulata,* Shaw. Zool. t. 13. *Test. Erosa,* Schw. Inhab.

IV. *Dorsal shell solid; front lobe of sternum mobile.* Pyxis. *Bell.*

Radiated Box Tortoise. Test. (Pyxis) Aranoides, Bell. Lin. Trans. xv. t. 16.

Shell hemispherical, varied black and yellow; length 6, breadth 4 inches. Inhab.

Genus II. Emys.

Feet palmated; claws five—four, sharp; nostrils pervious; jaws horny; shell solid, covered with horny plates, marginal plates twenty-three or twenty-five, hinder pair free; sternal shields eleven or twelve; neck retractile; pelvis only attached to the vertebra.

6

I. *Shell hemispherical ; sternum rounded before, divided by a transverse suture ; both lobes mobile, and united to the back shell by a cartilaginous suture ; sternal plates twelve.* Cistuda, Gray.

American Box Terrapin, E. *(Cistuda) Carolinæ, Test. Carolinæ,* Lin. *Test. Brevicaudata,* Lacep. *Test. Clausa,* and *T. Irregulata,* Daud. *Terrapene Carolinæ, Ter. Nebulosa,* and *Ter. Guttata,* Bell. Edw. t. 205.

Shell hemispherical, brown yellow varied, convex, lowly keeled ; marginal plates twenty-six or twenty-seven. North America ; Var. *β. fusca,* brown, not spotted.

Eastern Box Terrapin. E. *(Cistuda) Amboinensis,* Illust. Ind. Zool, t. f. 1. *Emys Couro,* Schw. *Terrapene Bicolor,* Bell. Zool. Jour. ij. t. 14. *Test. Amboinensis,* Daud.

Shell hemispherical, slightly tri-keeled, blackish, margin expanded, broad ; nuchal plate linear ; sternum yellow and black varied ; head with two yellow lines on each side. Java and Penang, Var. *Leveriana.* Shell oblong, with back flattened, not keeled.

Three-banded Box Terrapin. E. *(Cistuda) Trifasciata,* Gray. *Sternothærus Trifasciatus,* Bell. Zool. Jour. iij. t. 13.

Shell oval, keeled, dull yellow, varied with black and reddish, with three dorsal streaks ; sternum behind nicked ; head yellow, with two brown bands on each side. Inhab.—

European Box Terrapin. E. *(Cistuda) Europea. Testudo Europea,* Schn. *Test. Flava,* Lacep. *T. Punctata,* Gotw. *T. Orbicularis,* Linn. ? *T. Lutaria,* Merrem. *T. Meleagris,* Shaw, Zool. t. 5. Jun. *T. Pulchella,* Schoeff. t. 26.

Shell depressed, oval, yellow spotted ; sternum behind nicked

7

before rounded ; body and head black and yellow spotted ; tail long. South of Europe. These are the only Water Tortoises known which have concave sternums.

II. *Sternum united to the back shell by a horny suture ; solid, before and behind lobed ; sternal plates twelve.* Emys.

A. Margin of shell acutely toothed ; sides of sternum rounded. Found in the Old World.

Spinous Terrapin. Emys Spinosa, Gray, Illust. Ind. Zool. t.

Shell suborbicular, pale brown ; edge sharply toothed ; areola with a central spine. Penang.

Dhor Terrapin. Emys Dhor, Gray, Illust. Ind. Zool. t.

Shell suborbicular, hinder edged, acutely toothed ; areola punctate ; brown, black spotted. India.

Spengler Terrapin. Emys Spengleri. Test. Spengleri, Walb. t. 3. *Test. Serrata,* Shaw, t. 9. *T. Tricarinata,* Bory, Atlas. t.

Shell oblong, pale brown ; tri-keeled ; hinder margin deeply toothed. Africa.

Thick-necked Terrapin. Emys Crassicollis, Bell, Gray, Illust. Ind. Zool. t.

Shell ovate, black, tri-keeled ; hinder edge toothed. Sumatra.

Thuryi Terrapin. Emys Thuryi, Gray, Illust. Ind. Zool. t.

Shell oblong, black, slightly tri-keeled ; hinder edge slightly toothed ; head blackish, eyebrow and chin yellow lined· Bengal.

8

Hamilton's Terrapin. Emys Hamiltonii, Gray, Illust.
Ind. Zool. t.
Shell oblong, black, yellow-rayed; hinder edge toothed;
head, and body, and limbs yellow spotted. India.

Eyebearing Terrapin. Emys Occilifera, Kuhl. Beytr.
Shell hemispherical; behind toothed; above with blackish
eyed rings, beneath yellow; a doubtful species.

*B. Shell margin entire; sternum with the sides keeled.
Living in the Old World.*

Tented Terrapin. Emys Tecta, Bell, Gray, Illust. Ind.
Zool. t.
Shell ovate, oblong, solid; olive, beneath varied brown and
red; animal red and yellow lined.

Lined Terrapin. Emys Lineata, Gray, Illust. Ind. Zool. t.
Shell oblong, olive; head blueish ash; chin and cheeks yellow varied; nape and eyebrow scarlet lined. India.

Batagur Terrapin. Emys Batagur, Gray, Illust. Ind. Zool.
Shell suborbicular, depressed, pale olive; margin dilated;
body ash; chin and lips yellowish. Var. β *Test. Baska.*
Shell oblong, ovate.

*Common Terrapin, or Mud Tortoise. Emys vulgaris. Test.
Lutaria,* Lin.?
Shell ovate, olive, black dotted, clouded and ringed with
blackish and orange, head and limbs orange lined. Adult
smooth, young rugose. South of Europe. Spain.

Caspian Terrapin. Emys Caspia, Ruppell. *Test. Caspia,*
Gmel. Trav. t. bad.
Shell oblong, slightly contracted on the sides, olive yellow
netted; beneath black, irregular yellow spotted, head olive
brown, with narrow sub-concentric yellow lines. Caspian
Sea. Dr. Ruppell.

9

C. Shell margin sub-entire; sternum sides rounded. Living in the New World.

Rough Terrapin. Emys Scabra, Lacep. t. 8. f. 2. *Emys Dorsata,* Schoepft.

Shell ovate oblong, acutely keeled; head with a fork band over each eye. South America.

Dotted Terrapin. Emys Punctularia, Schweiger, (not figured.)

Shell oblong, convex, brown; shield blackish edged; head with red lines and spots. South America.

Marbled Terrapin. Emys Marmorea, Spix, t. 10.

Shell oval, greenish, varied with black and yellow; shields yellow edged; head with yellow lines. Brazils.

Muhlenberg's Terrapin. Emys Muhlenbergii, Schw. *Emys Bipunctata,* Say. Schoeff, t. 31.

Shell oval, low, side contracted; shields grooved, varied with yellow; head blackish, with two yellow spots. North America.

Spotted Terrapin. Emys Guttata, Schw. *Test. Punctata.* Bosc. Schoeff, t. 5.

Shell low, oval, smooth, black, yellow dotted; head yellow spotted. North America.

Painted Terrapin. Emys Picta, Schw. Schoeff, t. 4.

Shell oblong, flattish, smooth, olive; shields pale edged; head and neck yellow lined. Young. *Cinereous Tortoise,* Brown, Illust. 48. North America.

Specious Terrapin. Emys Speciosa, Gray, Syn. Rept.

Shell oval, flattish, keeled; shields grooved, black, yellow spotted; beneath yellow with black areola. North America?

10

Concentric Terrapin. Emys Concentrica, Schw. Schoeff.
t. 15. *Test. Palustris,* Gmel. *T. Centrata,* Bosc.

Shell oval, side edges recurved ; shield, brown ringed ; head and limbs ash, black speckled. North America. *Var. polita* shell, black polished.

Reticulated Terrapin. Emys Reticulata, Gray. *Test. Reticulata,* Latr. Daud. t. 22, f. 3.

Shell oblong, olive, pale netted ; beneath yellow, with four round black spots on side of the margin, and one at each end of the sternocostal suture. North America.

Banded Terrapin. Emys Vittata, Gray, Syn. Rept.

Shell oblong, olive ; behind doubly toothed, varied with unequal yellow lines ; beneath yellowish, with squarish eyed spots on the sutures of the marginal plates, and roundish ones on the end of the symphysis; head and feet yellow lined. North America ?

Furrowed Terrapin. Emys Decussata, Gray.

Shell oblong, pale brown, shields rugulose ; beneath yellowish, with subocellated spots, on the sutures of the marginal plates ; animal greenish. North America.

Lettered Terrapin. Emys Scripta, Gray. *Test. Serrata,* Daud. t. 21. f. 1. Schoeff. t. 3. f. 5. Jun.

Shell oblong, longitudinally rugose, brown, sides yellow banded ; beneath yellow, with a series of roundish spots on the middle of each marginal plate, or two on the gular plates, and four on each sternocostal symphysis. North America.

Serrated Terrapin. Emys Serrata, Gray.

Shell oblong, longitudinally rugulose, irregularly cross-banded ; beneath yellowish, with a series of subocellated spots on the sutures of the marginal plates, and dark-edged pale lines on the symphysis. North America.

11

Ornamented Terrapin. Emys Ornata, Gray.

Shell oblong, longitudinally rugose olive ; vertebral shields, irregularly ringed ; beneath, pale yellow, with an irregular black band down the centre, and on each side, and squarish-eyed spots on the suture of the marginal shields. South America. Young, beautifully orange and green-ringed.

Rugose Terrapin. Emys Rugosa. Test. Rugosa, Shaw, Zool. t.

Shell oblong, rugulose; black, with yellow lines and dots ; beneath yellow with black dots and lines ; Var. β shell livid, black dotted. North America.

Lake Erie Terrapin. Emys Lesueurii, Gray, Syn. Rept.

Shell oval, smooth, tuberculately keeled ; behind, toothed ; olive, with anastamosing black-edged pale lines ; beneath, yellow, shields, black edged, margin, black ringed ; head, yellow lined. β *Emys Geographica,* Lesueur, Jour. Acad. N. Philad. t. North America.

Bell's Terrapin. Emys Bellii, Gray, Syn. Rept.

Shell oblong ; dorsal line, depressed ; olive, netted with green and black dots ; beneath blackish, yellow dotted, surrounded by an irregular yellow margin.

White Spotted Terrapin. Emys Kinosternoides, Gray, Syn. Rept.

Shell oblong, pale brown, with black-edged white keel, and side-spots, and with a dentated white edge ; beneath yellow spotted ; sternum rounded at both ends. (——— ?)

Ring-bearing Terrapin. Emys Annulifera, Gray, Syn. Rept.

Shell oblong, depressed ; pale brown, yellow lined with

12

yellow ring placed across the sutures of the vertebral plates ; beneath pale brown, varied with darker brown. (——— ?)

III. *Sternum united to the back shell by a horny suture ; before rounded, behind rounded, or slightly nicked, divided by one or two cross sutures ; sternal plates, eleven gular pair united.* Kinosternon. *Spix.*

Three-keeled Kinosternon. E. (*Kinosternon*) *Scorpoides.* *Test. Scorpoides,* Lacep. *Test. Tricarinata,* Daud. *Kinosternon Shavianum,* Bell. *Kin. Longicaudatum,* Spix, t. 12 ? Shaw Zool. t. 15. Jun. *Test. Retzii,* Daud. Shaw, t. 11.

Shell oblong, olive polished, rather compressed three-keeled, keels continued, dorsal shields long, hexangular, sub-imbricate ; sternum as broad as the opening of the shell. South America.

Three Ridged Kinosternon. E. (*Kinosternon*) *Triporcata,* Gray. *Terrapena Triporcata,* Wiegmann, Mus. Berl.

Shell, oblong, olive, with three very high compressed ridges ; sternum cross shaped, very narrow ; the largest species of the genus.

Pennsylvanian Kinosternon. E. (*Kinosternon*) *Pennsylvanica.* *Test. Pennsylvanica,* Bosc. *Test. Subrufa,* Lacep. Edw. t. 287. Var. *Test. Glutinata,* Daud. *Test. Boscii,* Merrem.

Shell oval, brown ; dorsal shields, flattened ; sternum, rather narrower than the opening of the back shell ; behind nicked. North America.

Musky Kinosternon, E. (*Kinosternon*) *Odoratum.* *Test. Odorata,* Bosc. *Cistuda Odorata,* Say. *Sternotheros Odoratus,* Bell.

Shell ovate, keeled ; sides declivate ; sternum very narrow ;

13

before acute, behind acutely nicked ; head brown, yellow lined on each side. North America.

IV. *Sternum very narrow, cross-shaped, rounded at both ends, united to the back shell by very long bony processes. Sternal plates five pair, and a pair over the symphysis.* Chelydra Schw.

Serrated Alligator Tortoise. Emys (Chelydra) Serpentina. Test. Serrata, Penn. *Chelydra Lacertina,* Schw. Young. *Test. Serpentina.* Lin. Shaw, t. 29.

Shell oblong ; vertebral shields flat ; upper edge of costal plates, prominent ; behind bluntly six toothed. Young sharply three-keeled and toothed. North America.

III.—CHELYS, Gray.

Feet palmated ; claws five—four, sharp ; nostrils sub-tubular ; shell solid, covered with horny plates ; hinder pair of marginal plates separate ; sternal plates thirteen, with an intergular one ; neck contractile on the side ; pelvis attached to the sternum and vertebræ.

I. *Lips horny ; sternum front, and sometimes hinder lobes mobile ; intergular plate marginal.* Sternotherus, Bell (part.) Gray.*

Chestnut Chelys. Ch. (Sternotherus) Castaneus, Gray. *Emys Castaneus,* Schw. *Testudo Subniger,* β Daud. *Sternotherus Castaneus,* Bell. Zool. Jour. t. 14.

Shell oval, sharply keeled, convex, chestnut ; shields radiately rugose ; areola black. Var. β hinder lobe, mobile.

---

* I have used as the name of the sub-genera those which have been given by authors to what they considered genera in this family; but I have generally restricted their signification, so as to render them more definite.

14

Test. Subniger, Lacep. 2. O. t. 7. f. r. *Test. Nigricans,*
Merrem.

II. *Jaws horny; sternum solid; intergular plates six
sided between the gular and pectoral plates.* Chelodina.

Long-necked Chelys. Ch. (Chelodina) Longicollis, Gray.
Testudo Longicollis, Shaw. New Holland. t. 7.

Shell oblong flat ; sternum yellow, shields brown edged.

III. *Jaws horny; sternum solid, intergular plate small,
marginal; nostrils short, tubular.* Hydraspis. *Bell.*

Species of the Old World or its Islands.

Cape Chelys. Ch. (Hydraspis) Subrufa. n. *Test. Sub-
rufa,* Lacep. Q. O. t. 12. *Test. Indica,* Daud.—Jun. *Test.
Galeata,* Schoeff. t. 3. f. 1. *Test. Scabra,* Retz. *T.
Senegalensis,* Daud.; and *Emys Olivacea,* Schw.

Shell oblong, low, pale brown, smooth ; nuchal plate,
none ; vertebral plate depressed. Cape of Good Hope.
Senegal.

*Adamson's Chelys. Ch. (Hydraspis) Adamsonii, Emys
Adamsonii.* Schw.

Shell low, oval, very broad behind, yellow, black spotted ;
nuchal plate none ; vertebral plates keeled, first urn-shaped,
hinder triangular. Africa. Nigritia.

*Macquary's Chelys. Ch. (Hydraspis) Macquarii. Emys
Macquarii.* Cuv. R. A.

Shell oval, depressed, contracted in front, behind subdentate, dark olive ; shields rugulose ; dorsal line sunk ; nuchal
plate narrow. New Holland.

15

Species of America.

*Flat Headed Chelys.   Ch. (Hydraspis) Planiceps.   Test.
Planiceps,* or *T. Platycephala.* Schn. Berb. Naturf. t.
16. *T. Martinella,* Cuv. R. A. *Emys Discolor,* Thumb.
E. Caniculata? Spix, t. 8. *E. Carunculata,* Cuv. R. A.
Emys Geoffroyana, Schw.

Shell oblong, flattened, black brown, centre concave; nuchal plate narrow; sternum yellow edged. *E. Aspera,*
Cuv. R.A. is a variety with larger warts in the neck.

Depressed Chelys. Ch. (Hydraspis) Depressa, Gray. *Test.
Depressus.* Pr. Max. t. not Spix.

Shell elliptical, pale brown, black rayed; nuchal plates
narrow; head and neck black spotted.   Brazils.

Radiated Chelys. Ch. (Hydraspis) Radiolata, Gray. *Emys.
Radiolata,* Mikan. Pr. Max. *T. Depressa,* Spix, t. 3. f. 12.

Shell oval, slightly narrowed before, yellowish, black rayed;
nuchal plate distinct; neck beneath yellow, eyelids streaked.
Brazils.

Red Footed Chelys.   Ch. (Hydraspis) Rufipes, Gray,
Spix, t. 6, Jun. *Emys Stenops,* Spix, t. 9, f. 3, 4.

Shell elliptical, keeled in front, brown, beneath yellowish;
nuchal shield linear; head and neck thick, olive brown,
beneath yellowish.   South America.

Green Chelys. Ch. (Hydraspis) Viridis, Gray. *Emys Viridis,* Spix, t. 2, f. 4.; t. 3, f. 1.

Shell elliptical, olive green, brown dotted; nuchal shield
linear; intergular shield large, cordate.   Brazils.

16

Expanded Chelys. Ch. (Hydrapsis) Expansa, Gray. *Emys Expansa*, Schw. *E. Amazonica*, Spix, t. 1. *E. Tracaxa*. Spix, t. 5, f. 1. 2.

Shell ovate, depressed, brown, black spotted; margin very much expanded; nuchal shield none; nose longitudinally grooved, Var. β *Emys Erythrocephala*, Spix, t. 7, is perhaps the same. It is from a stuffed specimen. Brazils.

Dumeril's Chelys. Ch. (Hydraspis) Dumerilliana, Gray. *Emys Icterocephala*, Spix, t. 4.

Shell ovate, black, discal shields flat, hinder margin expanded; head globose; nose convex, smooth. Brazils.

Cayenne Chelys. Ch. (Hydraspis) Cayennensis, Gray, *Emys Cayennensis*, Schw.

Shell ovate, convex, tuberculately keeled; smooth, yellow green, black varied; nuchal shield none; head brown, with two yellow spots. β *Emys Gibba*, Schw. Shell black. Guiana.

Demerara Chelys. Ch. (Hydraspis) Lata, Bell MSS.

Shell suborbicular, depressed, black; head and neck black orange spotted. Demerara. *Bell.*

Consult also, 1, *Emys Bitentaculata*, Cuv. MSS.—2, *Emys Barbatula*, Gravenhorst, Delic. t. 5. f. 3. 4.—3. *Emys Constricta*, Cuv. MSS.—4. *E. Maximiliana*, Michan, and—5, *Hydraspis Pachyura*, Boie MSS.

IV. *The beak very broad, depressed, covered with soft lips; the nostrils long, tubular; intergular plate marginal;* Chelys, *Dumeril.*

Brazilian Matamata. Chelys Matamata, Test. Matamata, Brong. Jour. H. N. 7, t. 13. *Ch. Fimbriata*, Schn.

Shell oblong; shields elevated acute, forming three keels. South America.

IV.—TRIONYX. Geoff.

Feet palmated, claws three, sharp ; the shell covered with a soft skin expanded on the edge into a flexible margin. *Living in fresh water in tropical climates.*

I. *The margin of the shields cartilaginous and the sternum narrow.* Trionyx, Gray.

Fierce Trionyx. Trionyx Ferox, Merrem. Pen. Phil. Trans. lxi. t. 10. *Trionyx Georgicus*, Geoff.

Shell rather convex, obscurely keeled ; front and hinder margin wanting ; sternum callosities four, two hinder, large, united into one. *Trionyx Spiniferus*, Lesueur, Mem. Mus. xi. appears to be the young, as may also be *Test. Brongniartii*, Sch. and *T. Carinatus*, Geoffroy.

Armless Trionyx. Trionyx Muticus, Lesueur, Mem. Mus. xv. t.

Shell elliptical, confounded with the neck, and armless in the front ; sternal callosities four, two hinder large, united together. North America.

Egyptian Trionyx. Trionyx Ægyptiacus, Geoff. Rept. Egypt. t. 1. *Test. Triunguis*, Forsk.

Green, white spotted ; shell convex, slightly keeled ; sternum callosities four, the hinder ones triangular, separate. Egypt.

Indian Trionyx. Trionyx Indicus, Gray, Illust. Ind. Zool. t.

Olive green, with black-edged irregular pale tortuous and forked streaks ; sternal callosities four, the hinder ones rounded triangular. India.

Hurum Trionyx. Trionyx Hurum, Gray, Illust. Ind. Zool. *Trionyx Gangeticus*, Cuv. Oss. Fos.

Dull brown ; head green, with two yellow spots on each side

18

over the eyes ; sternum with four callosities, the lateral truncate on the inner hinder angles, and the hinder ovate-triangular. India.

Javan Trionyx. Trionyx Javanicus, Geoff. Ann. Mus. xiv. t. Illust. Ind. Zool. t.

Dull green, with numerous narrow lines of minute white granular specks. Head green, with five or six diverging black lines, and a streak between the eyes. Sternal callosities two, linear, lateral. When young, back black ocellated. Java and India.

Flat Trionyx. Trionyx Subplanus, Geoff. Ann. Mus. xiv. Illust. Ind. Zool. t.

Shield nearly flat ; front edge smooth ; sternum without any callosities. India.

Euphratic Trionyx. Trionyx Euphraticus, Geoff. *Test. Rascht*, Oliv. Voy. t. 41. This needs a better description. Euphrates.

Margin of the shield with a series of small bones in front and behind, limbs covered, when withdrawn, by the valves on the side of the sternum. Emyda. *Gray.*

Punctuated Trionyx. Tri. (Emyda) Punctatus, Lacep. Q. O. t. 7. f. 1. Schoepf. t. 30. A. B. *Trionyx Coromandelicus*, Geoff. Gray, Illust. Ind. Zool. t.

Deep green, occiput white spotted ; sternum with seven large callosities, the hinder of them united. India.

N. B.—M. Cuvier informed me on his late visit to this country that there has been lately received at the Paris Museum a species of this genus, with 4 *claws* on *each foot ;* it will form a section (*Tetraonyx*), and it may be indicated by the name of *Trionyx (Tetraonyx) Cuvieri.*

c 2   19

5. CHELONIA, Brongn.

Feet fin-shaped, compressed; shell with a bony margin all round, and a ring-like sternum.

Living in the sea ; eating algæ and mollusca.

I. *Shell covered with a leathery skin. Sphargis,* Merrem. *Coriaceous Turtle, Ch. (Sphargis) Coriacea. Testudo Coriacea.* Lin. Pen. Brit. Zool. 4. t. 1.

Shell with three longitudinal dorsal ridges; when young the skin granular, and fins very long. See Schoepft. t. 29. Mediterranean Ocean.

II. *Shell covered with horny scales.* Chelonia.

A. shields depressed. Chelonia.

Green Turtle. Chelonia Mydas. Testudo, Lin. Lacep. 2. O. t. 1. f. 1

Shell cordate, greenish brown, slightly keeled, shelving on both sides; lower jaw deeply denticulated. Varies first in the number of the dorsal shields, and thus *Test. Atra,* Lin., and *Test. Multiscutata,* Kuhl. Second in the number of sternal shields, thus *Test. Cepediana,* Daud. Sometimes the shields are variously coloured, and thus *Chelonia Virgata,* Dumeril, and *Chelonia Radiata, Ch. Maculosa* and *Ch. Lachrymata,* Cuv., R. A. *Testudo Japonica,* Thumberg, appears to be only a variety with the shields so thin, that he mistook the sutures of the bones for the edge of the scales.

Loggerhead Turtle, Chelonia Caretta, Testudo Caretta, Lin. Schoepf. t. 16.

Shell convex, vertebral shields gibbous; head very large; when young the nose produced, *Testudo Nasicornis,* Lacep. Edw. t. 206, Var. Vertebral shields sometimes seven. *Chelonia Olivacea,* Eschscholtz Atlas, t. 3. Mediterranean and other seas.

20

B. *Shields Imbricate.*

4. *Imbricated Turtle.* *Chelonia Imbricata*, Schw. *Testudo*, Schoepf. t. 18, *a*.

Shell elliptically keeled ; shields yellow, spotted and rayed. In the young, the ends of the shields are obliquely truncated, and scarcely imbricate, *Caretta Nasicornis*, Merrem. Schoepf. t. 17, f. 1. Indian and Atlantic Ocean.

Ord. 3. Emydosauri.

Vertebræ of the back and the ribs free, mobile ; the mouth toothed ; the feet digitate, webbed ; above and beneath covered with imbeded squarish plates, with small scales on the sides.

Gen. 1. Crocodilus.

Toes 4, 5, claws 3, 3 ; tail compressed, above keeled, serrated ; teeth conical in a single series ; tongue fleshy.

1. *The head depressed; the canine teeth of the lower jaw received in a pit in the upper jaw.* Alligator.

* *Feet digitate.* South America.

Spectacled Alligator. Croc. (Alligator) Sclerops, Schw. Seba, 1, t. 104, f. 10, Spix. Braz.

With a cross rib between the orbits ; nape with a band of shields. South America and Brazils. The frontal cross rib varies in shape.

Eye-browed Alligator. Croc. (Alligator) Palpebrosus, Cuv.

Eyelids bony ; nape with a band of shields, each bearing a crest. Var. the nuchal shields are sometimes irregular. *Croc. Trigonatus*, Seba, 1, t. 108, f. 3. Guiana.

+ *Feet, subpalmated.* North America.

Pike-muzzled Alligator. Cayman Croc. (Alligator) Lucius. Cuv. *Croc. Cuvieri*, Leach Zool. Misc. t.

Head depressed, parabolic, nuchal shields 4. North America.

21

II. *The head oblong ; canine teeth of the lower jaw received in a notch in the edge of the upper ; feet palmated.* Crocodilus.

Common Crocodile. Crocodilus Vulgaris, Ann. Mus. x. t. 3.

The head equal, broad ; nuchal plates six ; cervical six, sub-equal ; the dorsal shields quadrate, six rowed. Africa and India.

Ruppell's Crocodile. Croc. Octophractus, Ruppell, MSS.

The head equal, long ; nuchal plates four, cervical eight, unequal ; the dorsal shields slightly keeled. Africa, Dongola, and Senegal.

Double-crested or Indian Crocodile. Crocodilus Biporcatus, Cuv.

The forehead with two nearly parallel longitudinal ridges ; nuchal shields six ; dorsal shields ovate, eight rowed. India and its islands.

Rhombic Crocodile. Crocodilus Rhombifer, Cuv.

The forehead convex, with two diverging ridges ; nuchal shields six ; dorsal scales quadrate, six rowed ; scales of the limbs thick, keeled.

Siam. Crocodile. Crocodilus Galeatus, Cuv. Hist. Anim. Par. 6, a.

Crown of the head with a two-toothed elevated crest ; nuchal shields six. India. Siam.

Two-shielded Crocodile. Crocodilus Biscutatus, Cuv.

Nuchal shields two ; the middle dorsal scales quadrate, the outer irregular, scattered.

West Indian Crocodile. Crocodilus Acutus, Cuv. Geoff. Ann. Mus. Par. II. t. 37.

Head produced, convex at the base ; middle dorsal scales
22

quadrate, outer irregularly scattered; nuchal scales six. West Indies.

Armed Crocodile. Crocodilus Cataphractus, Cuv. Oss. Foss. f.

Head produced; nape with four bands of bony shields like the Alligators. Mus. Col. Surgs.

Flat-headed Crocodile. Croc. Planirostris, Graves.

Head, base, flat; shields tubercular; nuchal plates six, distant. Africa?

Intermediate Crocodile. Crocodilus Intermedius, Graves.

Head produced, sub-cylindrical; dorsal shields, six-rowed, nuchal, six. America?

III. *Head long, produced, muzzle cylindrical; teeth and feet like Crocodiles.* Gavialis.

Common Gavial. Croc. (Gavialis) Gangeticus.

The nuchal shield united to the dorsal plates; the crown and orbits appear to dilate with age. See Faugas. Hist. Mont. St. Petri, t. 46, 48; from whence *Croc. Tenuirostris,* Cuv. Ganges.

Order III.? ENALIOSAURI. *Conybeare.*

Vertebræ of the back and the ribs free, mobile; the mouth toothed; the feet fin-shaped. Only found in the fossil state; perhaps not of this section.

Gen. 1. ICHTHYSAURUS, Koenig.

Head large; teeth conical, placed in a groove; neck short.

For the anatomy and species of this genus, consult Home Phil. Trans. Conybear. Geolog. Trans. and Cuv. Os. Fos. vol 5.

Ichthyosaurus Communis, Conybeare. Cuv. Os. Fos. t. 29. f. 1, 9.

Ichthyosaurus Platyodon, Conybeare, teeth compressed. Cuv. Os. Fos. t. 28. f. 4, 5.

Ichthyosaurus Tenuirostris, Conybeare, Cuv. Os. Fos. t. 29. f. 8, 9.

Ichhyosaurus Intermedius, Cuv. Os. Fos. t. 29. f. 2, 5.

Ichthyosaurus Latifrons, Koenig. Icon. Sect. t. f. ined. Mus. Brit.

Ichthyosaurus Grandipes, Sharp. Geol. Trans. 1830.

Gen. 2. PLESIOSAURUS, Conybeare.

The head small; neck very long, of many vertebræ; the teeth placed in pits.

For the anatomy of this genus, consult Conybeare, Geolog. Trans. and Cuv. Os. Fos. v.

Plesiosarus Dolichodeirus, Conybeare, Geolog. Trans. Cuv. Os. Fos. v. t. 31 and 32. f. 1, 2, &c. Lyme Regis.

Plesiosaurus Recentior, Conybeare, Kimmeridge.
Plesiosaurus Carinatus, Cuv. Os. Fos. v. 486.
Plesiosaurus Auxois, Cuv. L. c. v. 486.
Plesiosaurus Pentagonus, Cuv.
Plesiosaurus Trigonus, Cuv.

Gen. 3. SAUROCEPHALUS, Harlan. *Saurodon*, Hay.

Teeth like incisors, placed in pits with a regular hole on the inner edge of the Alveola.

Saurocephalus lanciformis, Hay. Amer. Phil. Trans. y. t. 16. *Saurocephalus, Leanus*, l. c. t. 16.

Section II. SQUAMATA.

Tongue free, organs of generation double, ossa quadrata separated from the skull.

Order I. SAURI.

Mouth not dilatatile, skin covered with various unequal scales.

24

Division I.

*Tongue long, deeply bifurcated, teeth placed on the inner edge of the jaws.**

Gen. I. MONITOR.

Tongue retractile, head covered with small polygonal shields. *Inhabiting the old world, living near water.*

1. *Tail compressed, nostrils subapical, or medial, open;* Monitor, nob.

Two Banded Monitor, Mon. bivittatus, nob. *Tupinambis bivittatus.* Kuhl.

Black, with five cross rows of white spots or rings, back of head spotted, above banded ; cheeks, with two black streaks ; nostrils subapical. India. Jun., more spotted. *Tup. exilis.* Reinw. Mss. Mus. Brit.

*Laced Monitor, Mon. Varia,*n. *Lacerta Varia,* Shaw. White. Voy. t. p. 253. Phil. Voy. t. at p. 279.

Nostrils one-third from the tip of the nose ; toes short ; sub-equal. New Holland. *Tup. Marmoratus,* Oppel. of Manilla, appears to be a variety. Mus. Brit.

Yellowish Monitor, Mon. Flavescens, nob. *Uran. Russelii,* Schegel.

Yellowish, brown netted, scales very large ; toes very short, nearly equal, India. Mus. Ruppel.

Eyed Monitor, Mon. Ocellatus, Von Heyden Ruppel Atlas, t.

Yellowish brown, with pale eyed spots, dorsal scales oblong,

---

* All the species of the genera of this section require a complete re-examination, the following is only an attempt at noting the species which I have seen named in collections.

25

surrounded by small granular scales, ventral scales square. Dongola, Senegal. Mus. Ruppel.

Green Spotted Monitor, Mon. Chlorostigma, nob.

Black; head, body, and tail, white spotted; scales, rather large; toes very unequal, nostrils one-third from the tip. Young, belly whitish, black spotted. Rawack, Mus. Paris.

Timor Monitor, Mon. Timorensis, nob. *Tup. virido-maculatus,* Daud.

Like T. Bengalensis, but the scales are larger, especially those of the tail, toes longer, nostrils more than one-fourth from the tip. Timor. Mus. Paris.

Green Monitor. Mon. Viridis.

Nostrils medial, scales large, yellow, with large dark spots, Mus. Leyd.

Elegant Monitor. Mon. Exanthematica, n. *Lac. Exanthematica,* Bosc. t. 5, f. 3. *Tup. Cepedianus,* Daud. t. 29. *Lac. Argus,* Daud. from Seba J, t. 85. f. 3. *Uran Elegans,* Mer.

Nostrils medial, with a black streak from the eye; head with black edged white spots, throat black spotted, belly black banded. Africa.

Brown Spotted Monitor. Mon. Bengalensis, n. *Tup. Bengalensis,* Daud. *Uran Punctatus,* Mer.

Nostrils medial, head black, dotted; back, black dotted and slightly eyed; throat black dotted; scales rather large; toes short. India. *Lac. Dracœna,* Lin. fron. Seba. 1, t. 101, f. 1, now in Mus. Par. The *Tup. Caudivertus,* Daud. has the nostrils medial, and is very like this species.

Clouded Monitor. Mon. Nebulosus.

Nostrils medial, above green brown, white dotted, loins and base of tail, with bands of eyed spots ; beneath white, black netted, toes unequal, rather long, head long, angular. Java. Mus. Par.

Ornamented Monitor. Mon. Ornatus, Tup. Ornatus, Daud. Ann. Mus. II. t. 48. Lac. Capensis, Sparrm.

Nostrils medial, above black, beneath white ; throat, with nine black bands, nape, with four or five curved lines, and with seven bands of round white spots. Africa. Mus. Leyd.

Nilotic Monitor. Mon. Niloticus, Lac. Niloticus, Hasselq. Geoff. Rept. Egypt, t. 3, f. 1.

Pale, varied and eyed with brown ; toes very unequal. North Africa and India. Mus. Brit.

Heraldic Monitor. Mon. Heraldicus.

Nostril submedial, rather nearer the eyes ; head with regular and variously placed white lines ; nape and back with a ring of eyed white spots Toes short, subequal. Bengal. Mus. Paris.

Varied Monitor. Mon. Pulcher, Leach. Uran Elegans, Merrem. Lac. Monitor, Linn. Tup. Stellatus, Daud. n. t. 31 ? Seba, 1. t. 94, f. 1, 2. y. t. 30. f. 2. y. 68. f. 2.

Nostrils submedial, brown ; head and neck with concentric white rings of dots ; back, white dotted, with nine bands of round white dots ; beneath white, brown dotted. Central and South Africa. Mus. Brit.

II. *Tail round, nostrils large, valvular, placed near the orbits.* Psammosaurus.
Grey Ouran. Mon. (P.) Scincus, Licht. *Tup. Grisseus,* Daud. Geoff. Rept. Egypt. t. 3.f. 2. t. 4. f. 14, 15.

27

Grey, dotted with brownish scales, nearly square, edged with small grains. North Africa. Mus. Brit.

White throated Ouran. Mon. P. Albogularis. Tup. Albogularis, Daud. iij. t. 32.

Head, and neck, beneath, and sides, whitish, brown, dotted with two white lines, from the eye to the nape. Inhab.

N.B. The *Dragon* figured by Klein Tent Herpet, t. is an animal of this genus which has had the head united to the skin of the body, just behind the fore legs, so that it appears only to have two legs.

Gen. II. HOLODERMA. Wiedeman.

Tongue contractile? Head broad, depressed, covered with crowded, irregular, many sided, convex, tubercular shields. Palate not toothed; body squamulose with parallel transverse ridges of larger distant tuberiform long scales; abdominal scales, four sides, smooth; femoral pores none; tail round.

Horrid Monoxillo. Holoderma Horridum. Wied. *Tachydermun Horridum,* Wagler, II. t. 18.

Brown, yellow spotted, tail with close yellow rings. New Spain.

Gen. III. TEIUS.

Tongue contractile, head shielded, palate mostly toothed. Femoral pores distinct. Found only in America.

I. *Abdominal scales square, as long as broad, smooth ; dorsal scales keeled. Tail compressed, keeled.* Crocodilurus, Spix. Ada Gray.

Great Dragon, Lacépède. *Common Ada, Teius (A.) Crocodilus,* Merrem. Lacep. t. 9. Daud. t. 28.

Back with scattered keeled scales. Guiane.

28

Double Crested Ada. Teius (A.) Bicarinata, Lac. Bicarinata, Gm. *Tup. Lacertinus*, Daud. *Croc. Amazonicus*, Spix, t. 21.

Back with equal keeled scales, belly with eight series of scales. South America. Young, side of body spotted. *Croc. Ocellatus*, Spix, t. 22. f. 1.

II. *Abdominal scales smooth, longer than broad; dorsal scales smooth; tail round; collar small.* Teguixin.
Variegated Safeguard. Teius. (T.) Monitor, Merrem. *Lacerta Teguixin*, Lin. Seba, i. t. 96. f. 23. t. 97. f. 1, 5. t. 99. f. 1. *Tup. Nigropunctatus*, Spix, t. 20. Pr. Max. Beytr. t.

Black, with bands of yellow spots; beneath yellow; tail, black and yellow-banded. *Tup. Maculatus*, Daud. *Tup. Monitor*, Spix, t. 19., appears to be the young.

According to Mr. Caup's account of his genus *Exypuestes*, which he says, has the scales of the Monitor, and the tongue, teeth, head, plates, and country of *Teius*, it must belong to this subgenus. He described the back as finely and equally scaled, and the palate toothless.

III. *Abdominal scales smooth, broader than long; tail round; dorsal scales smooth; collar none. The young streaked with black on the sides.* Ameiva.

The Ameiva. Teius Ameiva Vulgaris, Licht. Spix, t. 23. *Lac Ameiva*, Lin. *Ameiva Argus*, Fitz.

Green, black spotted, sides with four bands of black-edged white spots. Young *Teius Ocellifer*, Spix.

Lettered Ameiva. Teius (A.) Litterata, Daud. Seba, i.

29

t. 83. and *Lac. Gutturosa,* Daud. from Seba. ij. t. 103, f. 3, 4.

Blue green, varied with oblong black dots; sides with white-eyed black cross-bands; neck beneath plaited. South America, not Germany.

Blue-headed Ameiva. Teius (A.) Cæruleocephalus, Seba. i. t. 91, f. 3.

Head blue; dorsal line white, with two yellow lines along each side; thighs white spotted.

Blue Ameiva. Teius (A.) Cyaneus, Lacep. i. t. 31, Seba. ij. t. 105, f. 2.

Bluish, sides with roundish white spots.

Side-streaked Ameiva. Teius (A.) Lateristriga, Cuv. Seba, i. t. 90, f. 7. (Spix, t. 24, f. 1 ?)

With a dark white-edged line on each side.

Striped Ameiva. Teius (A.) Lemniscatus, Lac. Lemniscatus, Gmel. Seba, i. t. 92, f. 54. Daud. ij. t. 36, f. 1.

Dusky blue, back with eight white lines, tail round, limbs white spotted, with three white lines.

Three-streaked Ameiva. Teius (A.) Tritæneatus, Spix. t. 24, f. 2.

Checkered Ameiva. Teius (A.) Tesselata, Say, Long. Exp. ij. 50.

Black, with nine or ten longitudinal, and eighteen or twenty cross lines; dorsal scales small, beneath whitish. Tail long, brownish olive, black spotted. North America, Say.

Blue and black Ameiva. Teius (A.) Cyanomelas, Pr. Max. Beytr. v. t.

Head short; tail and back black, with a central broad band, and the sides with two narrow whitish blue streaks. Brazils.

30

Green Ameiva. Teius (*A.*) *Viridis,* Merrem. *Lacerta. Teyou,* Daud. *Ameiva Teyu,* Lecht.

Top of head and dorsal line green ; body olive violet, with six white longitudinal lines ; throat and belly silvery white.

Collared Ameiva, Teius (*A.*) *Collaris. Agama Collaris,* Say, Long. Exp.

Olive, with five or six alternate broad dusky and narrow yellow or grey spotted fibrous bands ; sides greenish yellow ; sides of neck fibrous, varied with red and black bands, beneath pale. North America, Arkansa Territory.

IV. *Abdominal scales lanceolate, keeled ; tail round ; collar distinct.* Kentropyx, Spix.

Intermediate Centropyx. Teius (*C.*) *Intermedius,* Schlegel. Back with three pale lines ; dorsal scales minute. Surinam, Mus. Leyd.

Spurred Centropyx. Teius (*C.*) *Calearatus,* Spix. t. 22, f. 2.

Scales of the back and throat granular. Males with two small spine-like scales on each side of the vent ; scales of the belly, legs, and tail, keeled, green ; sides of the back blackey. Brazils.

Striated Centropyx. Teius (*K.*) *Striata. Lacerta Striata,* Daud. *Pseudo-Ameiva Striata,* Fitz. Merrem. Wetter. Ann. i. t. 1, Pr. Max. Beytr. xiii.

Grey, sides blue, with two longitudinal brown lines ; abdominal scales twelve rows ; dorsal scales keeled.

Genus IV. LACERTA.

Tongue contractile ; head shielded ; palate toothed ; lateral line, none ; the bones of the skull advanced over the temples

31

and orbits; femoral pores distinct. *Inhab. temperate parts of Old World.*

1. *Back with granular scales; belly with a collar of large plates; femoral pores numerous.* Lacerta.

A. collar separate the whole length from the chest-plates by small granular scales; the frontal plate nearly as broad before as behind.

Eyed Lizard. Lac. Ocellata, Edw. t. 202.

Belly with eight or ten rows of plates; occipital plate large, Young. *Lac. Lepida,* Daud. iij. t. 37. f. 1. South of Europe. *Lac. Rhombica,* Merrem, *Lac. Jamaicensis,* Daud., are from Edwards' figure.

Green Lizard. Lac. Viridis, Lin. Daud. t. 34., and *Lac. Bilineata,* Daud. t. 55. f. 1.

Scales of back slender, keeled; of tail, sharply keeled; collar free, serrated; occipital plate rudimentary; abdominal plates six-rowed; hinder legs not reaching the armpits. Europe. Length, twelve inches.

Wall Lizard. Lac. Muralis, Merr. Daud. iij. t. 8. f. 1. *Lac. Vivipara,* Jacq. Act. Helv. j. t. 1.

Scales of back and sides smooth; of the tail, octant above, slightly keeled; collar entire, adnate; belly, scales six-rowed; hinder legs reaching the armpits; length five inches.

Schreber's Lizard. Lac. Schrebersiana, Edw. An. Sci. t. 5. f. 5. *Lac. Fusca,* Daud.

Occipital plate rudimentary; abdominal plates six-rowed; temples covered with granular scales and a cheek-plate; hind legs long. Mus. Par.

Cape Lizard. Lac. Lalandii, Edw. An. Sci. t. 5. f. 6, and t. 8. f. 5.

Occipital plate rudimentary; abdominal plates six rowed;

temples covered with granular scales only ; hind legs short, ith medial scales before the vent. Cape. Length thirteen inches.

Duges Lizard. Lac. Dugesii, Edw. t. 6. f. 2.

Occipital plate rudimentary ; belly plates, six-rowed ; temples with granular plates only ; hind legs long ; one medial plate before the vent. Madeira.

Edwards' Lizard. Lac. Edwardsiana, Duges.

Abdominal plates eight-rowed ; scales of back imbricate, pointed ; four sub-maxillary plates on each side ; limbs slender ; thighs cylindrical. Shore of the Mediterranean.

B. *Collar free the whole length ; the frontal plate small, and narrowed behind.*

Olivier's Lizard. Lac. Olivieri, Ardouin. Sav. Rept. Egypt. t. 2. f. 3.

Abdominal plate six or eight-rowed ; one large medial scale before the vent. North Africa.

Savigny's Lizard. Lac. Savignii, Ardouin. Sav. Rept. Egypt. t. 1. f. 8.

Abdominal plates six or eight-rowed ; three larger medial plates before the vent. Egypt.

Shielded Lizard. Lac. Scutellata, Ardouin. Sav. Rept. Egypt. t. 1. f. 2.

Abdominal plates twelve or fourteen-rowed ; interparretal plate rudimentary ; three large medial scales before the vent ; the hinder larger than the side ones. Egypt.

Dumerill's Lizard. Lac. Dumerillii, Edw.

Abdominal plates twelve or fourteen-rowed ; interparretal plates rudimentary ; three larger medial scales before the vent, the hinder smaller than the side ones. Senegal. Mus. Par.

Knox's Lizard. Lac. Knoxii, Edw. t. 6. f. 8.

Abdominal plates twelve or fourteen-rowed ; interparietal plates large. South Africa.

Variable Lizard. Lac. Variabilis, Licht. *L. Variabilis, L. Arguta, L. Velox,* and *L. Cruenta,* Pallas.

Abdominal plates fourteen or sixteen-rowed; head rather acute ; scales small, rather triangular or roundish, smooth ; caudal scales closer keeled. Perhaps *L. Deserti,* Lepech, and *L. Corunea,* Merrem. Tartary.

C. Collar united in the middle to the abdominal plates, free on the sides.

Panther Lizard. Lac. Pardalis, Licht.

Abdominal scales ten rowed ; head rather acute ; scales of the back smooth ; of the tail rhombic above, keeled ; tail little longer than the body. Egypt.

Bosc's Lizard. Lac. Boscii, Daud. iij. t. 36. f. 2. *Lac. Velox,* Edw. Savig. Rept. t. 1. f. 9.

Abdominal plates ten-rowed ; scales keeled ; cervical ones small ; dorsal large, rhombic, blunt ; tail twice as long as the body. Egypt.

Red-spotted Lizard. Lac. Rubropunctata, Licht. Savign. Rept. Egypt. t. 1. f. 1 ?

Abdominal plates ten-rowed ; head acute ; scales of the back small, smooth ; of the tail large, sub-quadrate, olive, keeled ; tail twice as long as the body. Egypt.

Spotted Lizard. Lac Guttulata, Licht. Sav. t. 1. f. 8 ?

Abdominal plates eight-rowed ; head rather acute ; scales of

34

back very small; of tail sub-rhombic, above keeled; tail long. Egypt.

Grammic Lizard. Lac. Grammica, Licht. Sav. t. 1. f. 7.

Abdominal plates fourteen or twenty-rowed; head acute; scales of the back small, scarcely keeled; of tail rhombic, keeled; collar sometimes obsolete; tail half as long again as the body. Egypt.

Indian Lizard. Lac. Leschenaultii, Edw. t. 6. f. 9.

Abdominal plates six-rowed. Inhab. Coromandel.

11. *Back with granular scales; belly with small plates; collar indistinct.* Psammodromus. Fitz.

Spanish Lizard. L. (Psammodromus) Hispanicus, Fitz MSS. See also Var. of *Lac. grammica*, Licht.

Blackish blue, lighter underneath. Spain.

III. *Back and tail with keeled scales, belly with smooth imbricate scales, collar none, femoral pores numerous.* Algyra Cuv.

Common Algyra. Lac. (Algyra) Cuvieri, Lac. Algyra, Lin.

Tail long, round, brown, beneath yellowish, with two yellow streaks on each side; length four inches. Spain.

IV. *Back and tail with lanceolate imbricate keeled scales, abdomen with smooth imbricate scales, collar none, femoral pores none.* Tropidosaurus, *Boie*.

Mountain Lizard, Lac. *(Tropidosaurus) Montanus*, Boie. Olive dark metallic green, blackish on the sides, and with a narrow black dorsal, and two narrow white lines on each side. Java. I have seen this genus both at Paris and Leyden, it has much more the appearance of a scinct than a lizard, but I could not see its tongue, which I suppose caused Cuvier to place it in this group.

D 2   35

Gen. V. TACHYDROMUS. Oppel.

Tongue contractile; head shielded; back, belly, and tail, with square keeled scales; lateral line distinct, with small scales; femoral pores one on each side the vent; body very long; feet very far apart and small.

Chinese Tachydrome. *Tachyodromus Sexlineatus*, Daud. iij. t. 39.

Silvery, with six brown bands. Tail three or four times as long as the body. Java and China. Common in insect boxes from China. The other species, *Quadrilineatus*, Daud. in the Paris Museum, is in too bad a state to distinguish it from the former. Is this *Lac. Sept.* of Linne?

Section II.

Tongue short, contractile, end slightly lobed.

Gen. I. IGUANA.

Teeth three lobed or toothed, placed in the inner edge of the jaw. Body and head compressed, palate mostly toothed.

A. Ribs simple; throat dilatile; head short; back crested; palate toothed; femoral pores numerous. Iguana, nob.

1. *Tail equally scaly; toes unequal; head shields flatish; dewlap denticulated.* Iguana.

Common Guana. *Iguana Tuberculata*, Laur. *Squamosa*, t. 5, I. *Viridis*, t. 6, and I. *Emarginata*, t. 8, Spix.

Sides of neck with convex scales; front edge of dewlap toothed; sides of lower jaw with orbicular plates. Var. Nose plates prominent, hornlike. *Iguana Cornuta*, Em. t. 4. f. 4. Young, dorsal spines lower. *Iguana Cærulea*, Daud. and *I. Lophyroides*, Spix, t. 9. South America, Mus. Brit.

36

Smooth-necked Guana. Iguana Delesatissima, Laur. *1 Nudicollis,* Merrem. Mus. Bresl. t. 13. f. 1.

Sides of neck smooth; front of dewlap slightly toothed; sides of lower jaw with small plates. Brazils and Gaudaloupe.

2. *Tail equally scaly; toes unequal; head shield, flatish; dewlap entire.* Brachylophus. (*Les Brachylophes.*) Cuv.

Banded Guana. Iguana (Brachylophus.) Fasciatus, Brongn. Bul. Soc. Philom. t. 61. f. 1.

Green, brown banded. South America, not Java, as said by Brongniart. Mus. Brit.

4. *Tail compressed, with rings of spinous scales; toes nearly equal; head shields convex.* Amblyrhynchus, Bell.

Rough headed Guana. Iguana (A.) Cristatus, n. Bell. Zool. Jour. i. t. 12.

Brown, sides and belly yellow? Galapagos.

Black Guana. Iguana (A.) Ater, nob.

Nearly black; beneath dusky. Galapagos.

B. Ribs simple, throat with a cross fold.

I. *Head long, scutellated; back crested; tail with rings of spinous scales; back with square scales; femoral pores distinct.* Cyclura nob.

Palate toothed. Ctenosaura. *Weigmann.*

Cycluroid Guana. Iguana. (Ctens.) Cycluroides. n. *Ctenosaura Cycluroides.* Wiegmann, Iris xxi.

Dorsal crest continued; scales of back and sides, obsoletely

37

keeled ; of sacrum mucronate. Palatine teeth many, small ; tail very long and round, crested with a series of spines. Mus. Berl.

Spiny Tailed Guana. Iguana. (Ctenosaura) Acanthura. Lac. Acanthura. Shaw. Zool. iij. 326. *Cyclura Shawii,* Gray Ann. Phil.

Pale brown ; dorsal crest interrupted over the sacrum ; tail cylindrical, with a ring of depressed scales between each ring of spiny ones ; palatine teeth few, large. Mus. Brit.

Armed Iguana. Iguana (Ctenosaura) Armata.

Head long, pyramidical ; tail with rings of spines, having two rings of depressed scales between them at the base ; teeth three or five lobed ; palatine teeth small, in one row. Mus. Bell.

Allied Iguana. Iguana (Ctenosaura) Similis.

Grey, black dotted, body with four oblique dark bands ; occiput forming a concave band behind ; dorsal crest low but continued over the sacrum. Teeth blunt, three lobed ; palatine on two raised lines on each side. Head two, body nine inches. Mus. Bell.

Lancet Toothed Iguana. Iguana (Ctenosaura) Lanceolata.

Grey, black dotted. Tail black banded, base black spotted ; head scales rather large, convex ; dorsal crest, completely interrupted over the sacrum. Palatine teeth, in two bunches ; the teeth of the jaw, long lanceolate sharp-edged, two or three lobed ; femoral pores six. Length of body and head fourteen, of tail twenty-four inches. Mus. Bell.

Bell's Iguana. Iguana (Ctenosaura) Bellii. Gray MS.

Grey, black netted ; dorsal and caudal crest of short, broad scales, interrupted over the sacrum. Scales of the occiput and
38

sides of the head, convex ; scales of the mastoid ridge, small, conical. Palatine teeth in two bunches; tail, base smooth, middle and end with distant rings of slightly elevated scales, having four rings of small scales between each ; toes short, thick. Mus. Bell. Length of body eighteen inches.

Iguana Cyclura of Cuvier may be probably one of these species, but his account of it is too general to distinguish it.

Palate toothless. Cyclura. *Harlan.*

Keeled Guana. Iguana (Cyclura) Carinata. Cyclura Carinata. Harlan. Acad. Nat. Sci. Phil. iv. t. 15.

Deep dirty brown; dorsal crest interrupted over the shoulders and loins ; scales of body uniform ; legs and feet minutely square. Tail above keeled, with three rings of depressed scales between each ring of spiny scales. Inhab. Turk's Islands, Bahama.

Round Guana. Iguana (Cyclura) Teres, n. *Cyclura Teres.* Harlan. l. c. t. 16.

Dark green, teeth small, uniform pointed ; dorsal crest interrupted over the loins ; scales of the sides square, of the thighs and legs bristly. Tail cylindrical, with two rings of depressed scales between each rings of spines. Tampico.

Clouded Guana. Iguana (Cyclura) Nubila. Gray.

Blue with oblique bands of roundish spots ; tail compressed, with four rows of small depressed scales between each ring of spines. South America ? Brit. Mus. Figured here under the name of *Clouded Lizard. L. Nebulosa.*

II. *Head covered with convex scales, and often some super-ciliary plates ; back, covered with small scales slightly crested, femoral pores none.* Ophyessa, Boie.

1. *Tail, simple, compressed, dorsal scales, minute,* (Xiphura, nob.)

39

Sword Tailed Ophyessa. Oph. Superciliosa, Lacerta, Lin. from Seba, 1. t. 199. f. 4. *Lophyrus Xiphurus*, Spix, t. 10, good.

Head with sub-equal, keeled scales, occiput with cross rows of convex tubercles. Back with small, belly and limbs and tail with larger keeled scales. Brazils. Mus. Brit.

Pearly Ophyessa. Oph. Margaritaceus, Lophyrus, Spix, t. 12, f. 2.

Back and tail slightly crested, supercilliary scales, small. Brazils. *Loph. Auronitens*, Spix, t. 13, scarcely appears to differ from this except in age.

Lozenge Ophyessa. Oph. Rhombifer, Lophyrus, Spix, Braz. t. 11, adult. *Loph. Albomaxillaris*, Spix, t. 13, f. 2. Young, *Agama Catenata*, Pr. Max. t.

Brown, beneath paler, limbs pale banded, back with two sinuous, brown bands, forming a series of rhombic spots, super-orbital scales small. Brazils.

9. *Tail round, slender, with imbricate scales.* Plica, nob.

Brazilian Ophyessa. Oph. Braziliensis.

Head with rather convex, keeled scales, superciliary ones rather smaller, bluish; chest, inside of legs, belly, and vent, whitish. Scales of back and throat granular, of belly rather larger, of limbs and tail largest, keeled; tail round; back and loins with a small toothed crest. Brazils.

Painted Ophyessa. Ophyessa Picta, Agama, Pr. Max. t. *Lophyrus Ochrocollaris*, Spix, t. 12, f. 2. *L. Panthera.* Spix, Braz. t. 13, f. 1. Young.

Brown; body, face and limbs darker banded; beneath pale, Superciliary scales oblong; occipital plate, distinct; back slightly crested. Brazils.

40

CLASS REPTILIA.

Umber Ophyessa. Ophyessa Plica. Lac. Plica, Lin. *Agama Plica,* and *A. Umbra,* Daud. *Loph. Agamoides,* Gray. Ann. Phil.

Back with minute scales, sides of neck behind the ear with four bundles of spines.

3. *Tail round, tapering, with spinose, whorled scale; ears spinose in front, back and belly scaly.* Tropidurus, *Pr. Max.*

Palate toothed, Oplurus, Cuv.

Collared Tropidurus. Trop. Torquatus, Pr. Max. t. *Agama Taraguira,* Licht.

Green; black stripe on each side the neck; eyelid black rayed, back with lines of spot when young. Mus. Brit. See *Agama Nigricollis,* Spix, t. 16, f. 2, but Spix says it has no palatine teeth. This may be, *Ephimotes,* Cuvier 47, but certainly it has no pores.

Tubercular Tropidurus. Trop. Tuberculata, Gray. *Agama Tub.* Spix, t. 15, f. 1, ☿ 2, ♂

Green, black spotted; streak on side of neck black; scales of back and feet keeled; the dorsal ones the largest. Brazils.

Cuvier's Tropidurus. Trop. Cuvieri, Gray. *Oplurus Torquatus,* Cuv. ij. 48. Seba. j. t. 94. f. 4. Cop. Shaw Zool t. 69.

Dorsal scales minute, keeled; caudal scales large, keeled; with a black half collar on each side the neck. Braz.

*** "*Palate toothless,*" Spix.

Half-banded Tropidurus. Trop. Semitæniatus. Agama sem., Spix, t. 15. f. 1.

Green-brown, smooth; dorsal line blue; dorsal scales minute; abdominal rather larger; caudal largest, keeled.

Cycluroid Tropidurus. Trop. Cyclurus. Agama Cyclurus. Spix, t. 17. f. 1.

Olive brown, longitudinally keeled; back and tail black-banded; beneath brown. See also Seba, t. 97. f. 4. Pr. Max. Bon. Trans. xiv. t. 15.

4. *Tail tapering, with large verticillate spinose scales; head and body with minute granular scales.* Uranocentron, *Caup.* Doryphorus, *Cuv.*

Long-tailed Doryphorus. Oph. (Uran) Azureus. Agama Azureus, Daud. iv. t. 46.

Bright blue, spotless; tail long. Inhab. Surinam.

Short-tailed Doryphorus. Oph. Uran. Brevicaudatus. Agama Brevi, Daud. iv. t. 47. *Lac. Azurea,* A. Gmel.

Blue, black-banded; tail short, depressed. Mus. Col. Surg.

III. *Head with numerous small regular shields; back and tail with large keeled spines, converging towards the back, and forming a crest; femoral pores none.* Leiocephalus, Gray.

Keeled Leiocephalus. Leiocephalus Carinatus, Gray. Ann. Phil.

Green, blackish-banded; dorsal scales large, broad, keeled; keels forming oblique ridges; the vertebral keel the highest. ———— ? Mus. Brit.

IV. *Head covered with small shields; back and tail with large keeled scales; belly with smooth scales; tail long, with imbricate keeled scales; femoral pores large.* America. Tropidolepis, *Cuv.* Scelophorus, *Wiegmann.*

42

Northern Tropidolepis. Tropidolepis Undulatus, Gray. *Agama Und.*, Daud. *Lacerta Hyacenthina*, and *Lac. Faciata*, Green.

Ash, with irregular brown cross-bands; head with regular flat plates. North America. Mus. Brit.

Thorny Tropidolepis. Tropidolepis Aculeatus, Gray.

Head with convex irregular scales: eyebrow plates band-like, back with a crest of short compressed scales; scales of body and limbs dagger pointed. Martinique.

Cuvier places *Agama Nigricollis*, and *A. Cyclurus*, Spix, in this genus; but Spix says they have no femoral pores, they are more like *Ophyessa*.

Collared Tropidolepis. Trop. Torquatus, Gray. *Scelophorus*, Weigmann. Iris xxi. *Tecoixin* Hernandez.

Olive brown, with a black collar, in an orange band; dorsal scales large, keeled; rhombic, smooth, tip denticulated. Mexico. Mus. Brit.

Spinous Tropidolepis. Trop. Spinosus, Gray. *Scelophorus*, Weigmann.

Grey brown, back with a series of brown spots; scales rhombic, tooth-edged; apex longly pointed, acute; abdominal ones, smooth truncate. Mexico.

Side Streaked Tropidolepis. Trop. Pleurostictus, Gray. *Scelophorus*, Wiegmann.

Olive grey; sides black and yellow spotted; scales rhombic, keeled, tooth-edged, mucronate; those in base of tail largest. Mexico.

Green Tropidolepis. Trop. Grammicus. Gray. *Scelerophorus.* Weigmann.

Shining coppery olive, with black flexuous cross lines;

43

scales rhombic keeled, mucronate. Variety, scales smaller Mexico. Mus. Brit.

Æneous. Tropidolepis. Trop. Æneus, Gray. *Sclerophorus*, Wiegmann.

Shining coppery ; scales ovate, lanceolate, keeled, toothed. Mexico.

Banded Tropidolepis. Trop Scalaris, Gray. *Sclerophorus*, Wiegmann.

Grey brown ; side band white, with white edged brown semi-lunar spots ; scales ovate, keeled entire. Mexico. Mus. Brit.

Bell's Tropidolepis. Trop. Bellii, Gray.

Metallic green, scales of the back and upper part of the body, and tail long, strongly keeled, dagger pointed, the keels forming 14-16 ridges ; of the belly broad, blunt, those of the limbs and tail smaller, becoming larger near the end ; length 10 inches. Mus. Bell.

Banded Tropidolepis. Trop. Fasciatus, Gray.

Pale brown; back with wavy cross bands ; head varied, pale and dark brown ; scales of back tail and limbs moderate keeled, keels 18-20 ; continued, oblique, entire; of the limbs and tail smaller. Mus. Bell.

V. *Head, short ; round, dilated behind, covered with convex scales, and crowned with spines. Body and tail short, depressed, covered with irregular keeled spines ; belly with larger scales, fringed on each side ; pores distinct, palate toothless.* Phrynosoma. Weigmann. America.

Douglass' Toad Lizard. Phrynosoma Douglassi, Gray. *Agama* Bell. Lin. Trans. good.

Yellow ; brown spotted. Mus. Brit.

44

Orbicular Lizard, or Horned Toad Lizard. Phrynosoma Cornuta. Gray. *Agama cornuta.* Harlan. Journ. Acad. N. S. Phil. good.

Brown ; with two large hornlike spines behind the head. Mus. Brit.

Common Toad Lizard. Phrynosoma Buffonium, Weigm. Isis, xxi. Seba, 1. t. 83. f. 1, 2.

Scales of the belly keeled ; spines on side of body two rowed, short, straight, three sided. Surinam. Mus. Berl.

Orbicular Toad Lizard. Phrynosoma Orbiculare, Weig-mann.

Abdominal scales smooth ; spines on side of body one rowed, large, recurved, rather compressed. Mexico. Mus. Berl.

C. Ribs forming complete circles ; head scutellated ; throat extensile.

I. *Head hooded; back crested* ; *pores o. ; toes margined on the side.* Basilicus, nob. (part Cuvier.)

Mitred Basilisc. Bas. Americanus, Laur. *Lac. Basili-cus,* Lim. Seb. i. t. 100. f. 1. Guiana. Mus. Brit.

Banded Basilisc. Basiliscus Vittatus, Boie.

Blue, with two series of spots down the sides of the back, and white streak along each side of the dorsal ridge, another edging the two lips and extended to the shoulders ; beneath white ; throat brown marbled. Dorsal ridge scarcely ele-vated ; hood small. Length about nine inches. Mus. Berl.

II. *Head squarish, occiput compressed, produced into a crest ; back lowly crested ; pores none.* Chamœleopsis, Wiedemann, MSS. Mus. Berl.

Hernandy's Chameleopsis. Chamœleopsis Hernandesii, Wiedemann, MSS. Gray. Mus. Brit.

Black, white speckled ; scales minute, thin   Mexico. Mus. Brit.

III. *Head elongate, simple ; toes, last joint but one dilated, pear-shaped beneath.* Anolis, Merrem. Anolius, Cuv.

Equestrian Anolis. Anolis Equestris, Merrem. Cuv. R.A. t. 5. f. 2.

Purple ash, scapular band white ; dorsal line toothed.

Edwards Anolis. Anolis Edwardsii, Merrem. Edw. Glean. i. t. 245. f. 21.

Uniform slate colour, and nearly allied to Equestris.

Two-spotted Anolis. Anolis Bimaculata, Daud. *Lac. Bimaculata*, Sparmann, Swed. Acad. v. t. 4. f. 1. bad.

Cuvier's Anolis. Anolis Cuvieri, Merrem. *Anol. Velifer*, Cuvier, R.A. t. 5. f. 1.

Lined Anolis. Anolis Lineatus, Daud. Rept. t. 48. f. 1. *Lac. Strumosa*, Lin. Seba, ij. t. 24. f. 4 ?

Dotted Anolis. Anolis Punctatus, Daud. t. 40. f. 2.

Gouty Anolis. Anolis Podargicus, Daud. Catesby Car. t. 3.

Violet Anolis. Anolis Violaceus, Spix, Braz. t. 17. f. 2. Brazils.

Green Anolis. Anolis Viridis, Pr. Max. Good. Brazils.

All these species are so very much alike that it is impossible to distinguish them without long descriptions ; indeed the genus requires a complete revision, which I hope shortly to be able to publish : the following is very distinct.

Slender Anolis. Anolis Gracilis, Pr. Max.

Head elongate, slender, two keeled in front ; back slightly keeled. Dull brown, with seven cross-bands of white spots ; throat orange. Brazils.

IV. *Head angular; body with smooth scales, larger beneath; back simple; tail round; toes simple; femoral pores distinct.* Polychrus, Cuv.

Common Marble. Polychrus Marmoratus, Lac. *Lac. Marmorata,* Lin. Spix, Wagler, t. 21.

Eyelids black rayed; scales above, small, blunt, beneath lanceolate, keeled, of the tail largest. *P. Virescens* and *P. Strigiventer,* Wagler, and *P. Acutirotris,* Spix, t. 14. appear to be scarcely distinct, though they are said to have no femoral pores. *Polychrus Fasciatus,* La Porte, if from Molucca, must be an *Agama.* See also *Agama Molinaii,* Lesson. Bul. Sci. from Chili.

Genus III.—GECKO.

Head and body depressed; scales small, tubercular, usually larger beneath; toes five, sub-equal, generally furnished with transverse scales beneath; claws retractile; throat simple; palate toothless; eyes large, nocturnal.

Toes with a single row of broad cross scales beneath; last joint compressed, attached. Platydactylus nob.

A. Toes free, clawless, dilated; thumb small. Phelsuma.

Clawless Platydactile. Gecko Inunguis, Cuv. R. A. t. f. 1.

Femoral pores none. Violet; beneath white, with a black streak on each side. Isle of France.

Eyed Platydactyle. Plat. Ocellatus. Gecko, Cuv. R. A. t. f. 4.

Femoral pores none. Grey, with white-eyed brown spots. Isle of France.

Cepedian Platydactyle. Plat. Cepedianus. Gecko, Merr. Cuv. R. A. t. f. 5.

47

Orange, marked with blue, and a white streak on the side; femoral pore distinct.   Isle of France.

Ornamented Platydactyle. Plat. Ornatus. Phelsuma, Gray.
Ann. Phil.

Brown; back with six rows of red oval spots. New Holland.

B. Toes, first and fourth clawed, free, dilated; pores none.
Tarentola, *Gray.*

Fasciculated Gecko. Plat. Fascicularis. Gecko, Daud.
Grey; head rough; back with twelve series of groups of three or four small acute spines.   South of Europe.

Annulated Gecko. Plat. Ægyptiacus. Gecko, Cuv. *Gecko Annulaire*, Geoff. Egypt. t. 5. f. 7.
Grey; scales granular, with rows of large, flat, simple, round tubercules.   Egypt.

American Gecko. Platydactylus Americanus, Gray.
Pale, lined and varied with brown; scales granular, with twelve or fourteen close rows of blunt, simple, equal tu-bercles.   New York. Mus. Par.

C. Toes clawed, thumb clawless, subanal pores distinct.
Gecko.

Toes free, dilated the whole length. Gecko, Gray.

Common Gecko. Gecko Guttatus, Daud. t. 49. *Lac. Gecko*, Lin.
Body black, with white spots; scales minute, with low rounded simple tubercles.   Java. Mus. Brit.

Chinese Gecko. Gecko Reevesii.
Black, with cross band of white spots, and some obscure rather larger tubercular scales.   China.

48

Madagascar Gecko. Gecko Madagascariensis. Gray.

Scales smooth, with large tubercles on the sides; femoral and subanal pores in two straight diverging lines; young, with a black and white lateral line. Madagascar.

Banded Gecko. Gecko vittatus, Houttyn. Act Ulys. ix. t. 11. f. 2. Daud. iv. t. 50.

Brown, with a white dorsal band, forked over the head and the base of the tail. Java.

D. Toes free, scarcely dilated. Eublepharis, Gray. Stenodactylus, Part. Licht.

Hardwicke's Gecko. Eublepharis, Hardwickii, Gray.

Grey, brown banded; back tubercular. Java.

Ascolobates Stenodactylus. Licht. Savigny, Egypt. *Phyllodactylus Marmoratus.* Mus. Franc. *St. guttatus,* Cuv.

Scales above small; white spotted; tail, black, with irregular white rings. Egypt.

E. Toes dilated, webbed, sides of body and tail margined. Pteroplura, Gray. Ptychozoon, *Kuhl.*

Horsfield's Gecko. Pte. Horsfeldii, Gray. *Lac. Homalocephala,* Creicht.

Brown, with darker cross bands; beneath pale; tail, when reproduced, roundish, not pinnated.

E. Toes and thumb clawed, subwebbed, sides sub-finned. Leache's Gecko. Gecko Leachianus, Cuv.

Grey, with large white spots; scales uniform, smooth.

II. *Toes, with a series of cross scales, divided by a groove, beneath, last joint compressed, sheathed.* Thecodactylus.

E   49

*a. Toes dilated to base, scales under toes transverse, thumb
clawless, femoral pores, none.*   Thecodactylus, Cuv.

Smooth Sheath claw.   Thecadactylus lævis, Daud.

Tail round, conical ; grey, brown marbled ; tail when reproduced, turnip shaped.   Gecko rapicauda, Daud. iv. t. **31.**

b. Toes, base, slender ; scales 2 or 3 at the tips. Phyllodactylus, Gray.

Beautiful sheath claw.   Phyllodactylus pulchellus, Gray,
Spic. Zool.

c. Tail round, femoral pores, none.

Porphyry sheath claw.   Thec. porphyreus, Gecko, Daud.

Pale reddish, marbled with many roundish pale spots, Java.

*d. Scales under the toes radiating, toes, slender, tail
round, slender, femoral pores, distinct.* Ptyodactylus, Cuv.

Egyptian sheath claw.   Thecodactylus Lobatus, Gecko,
Geoff. Rept. Eypt. t. 3, f. 5.   *Gecko Ascolobates,* Merrem.

Pale brown, with larger scattered scales. Egypt. *Ptyodactylus Guttatus,* Ruppel. t. 4, appears scarcely distinct.

e. Toes bare, webbed ; tail, body, and limbs, margined.
Uroplates.

Fimbriated sheath claw.   Thecodactylus fimbriatus, Gecko,
Daud. iv. t. 52.

Pale brown, scales minute, with scattered larger ones. Madagascar, Mus. Brit.   *Gecko Sarroube,* appears to differ very
little, if at all, and *Lacerta Caudiverbera,* Seba, ij. t. 103,
f. 2, is only known from La Feuille's account.

III. *Toes, base dilated, with a double row of oblique scales
beneath, last joint compressed.*   Hemidactylus.

Warty Gecko.   Gecko Verrucosus, Cuv.   Lac. Turcica.
Lin. Edw. t. 204 ?

Reddish grey ; back scattered with roundish tubercles ;
subanal pores only.

Mabuia Gecko.   Gecko Mabuia, Cuv.

Grey, varied with brown ; back with about six or seven rows of small scattered tubercles on each side.   South America.

It is very difficult to determine the species of this genus as the individuals vary in the size of the larger tubercular scales, and in the presence or absence of the femoral pores, which have hitherto been considered good specific characters. They are however named as follows. *Thecodactylus Policaris*, Spix, t. 18. f. 2.   *Gecko Aculeatus*, Spix, t. 18. f. 2, 3. *Gecko Spinicauda*, Houttyn, Act. Ulys. ix. t.   *Gecko Trihedrus*, Daud. *Stellio Platyurus*, Schn.   *Stellio Argyropus*, Tiles. Mem. Acad. Petersb. viii. t. ii. a., and *Hemidactylus Gronosus*, Ruppell. Atlas. t. 5. f. 1.   The two last are said to have no pores.

Margined Hemidactyle.   Hemidactylus Marginatus, Cuv. Sides of body and tail, hinder edge of thighs and legs, slightly fringed.   Bengal. Mus. Ind. Comp.

IV. *Toes slender, compressed, free, and fringed on the edge, end compressed, bent and arched.* Cyrtodactylus, Gray.

a. Tail round ; pores none.   Cyrtodactylus, *Gray.* Gonyodactylus, *Kuhl.*   Stenodactylus, *Licht.*   Gymnodactylus, *Spix.*

Beautiful Cyrtodactyle.   Cyrtodactylus Pulchellus, Gray, Zool. Jour.

Pale, with chocolate cross-bands.   India.

Marbled Cyrtodactyle.   Marmoratus Gonyodactylus, Kuhl. MSS.

Ash brown, marbled with darker spots, beneath ash.   Java.

Eyed Cyrtodactyle.   Cyrtodactylus Ocellatus, Gray.

Pale brown, with ocellated yellowish spots. Mus. Brit.

Brasilian Cyrtodactyle. *Cyrtodactylus Spixii*, Gray. *Gymnodactylus Geckoides*, Spix, t. 18. f. 1.

Brown, back with six series of small angular warts. Brazils.

Piping Cyrtodactyle. Cyrtodactylus ? Pipiens. Stenodactylus, Licht. *Lac. Pipiens*, Pallas.

Tail, not annulated.

b. Tail depressed, subanal pores distinct. Phyllurus, Cuv.

Flat-tailed Cyrtodactyle. Cyrtodactylus Platura. Lacerta, Shaw. *Agama*, Merrem. *Phyllurus Cuvierii*, Bory. White Voy. t. 32. f. 2.

Tail cordate ; grey, brown marbled, scales, with scattered pointed tubercles.

Nilius's Cyrtodactyle. Cyrtodactylus Nilli. Phyllurus, Bory. Dict. Class. t.

Tail bluntish, spathuliform, above brick-red with a black half collar and three bands.

c. Toes free, ending in a small simple round disk, without any plates beneath ; claws retractile. Sphœriodactylus, *Gray.*

Banded Spheriodactyle. Spheriodactylus Sputator. Gecko, Lacep. 2. O. t. 28. f. 1.

Reddish, with brown cross-bands.

See also *Lac. Sputator Var.* Lacep. Rept. i. t. 28. f. 2. *Lac. Sputator*, Sparmann. Nov. Act. Stock. 1784. t. 4. f. 1, 3.

Gen. IV.—CHAMELEON.

Teeth placed on the upper edge of the jaws ; toes united into two groups to the claws ; tail prehensile ; body compressed, covered with squarish scales. Africa or India.

a. With a series of spines along the belly and chin.
Common Chameleon. Chameleo Vulgaris. Lac. Africana.

Occiput keeled, arched ; scales equal.   Grey, banded and streaked with yellow.   Africa and India.

Beaded Chameleon. C. Verrucosus, Cuv. *Ch. Monilifer,* Boie.

Occiput keeled, arched ; with larger scales on the sides.

Senegal Chameleon.   Chameleo Senegalensis.   Ch. planiceps.
Mer. Seba, 1. t. f. 2.   *Ch. Gymnocephalus,* Lacep.

Occiput flat ; side parallel edges not margined.   Senegal. *Cham. Galeoratus* appears to be a variety of this species.

Eared Chameleon.   C. Dilepis. Leach. Gray. Spic. Zool. t.
f. 4.   *Ch. Planiceps.* Merrem.   *C. Bilobus.*   Kuhl.

Occiput flat, rather wider behind, under edged with two flaps ; a white band on each side.   Interior of Africa.

Streaked-sided Chameleon.   Chameleo Lateralis, Gray.

Lead colour, sides white streaked ; occiput compressed, slightly keeled ; scales granular, with a series on the back.

b. With a series of processes on throat, none on belly or tail.
Panther Chameleon. C. Tigris. Kuhl. Gray. Spic.

Scales minute, regular ; brown, with minute black specks.

Cape Chameleon.   Chameleo Pumillus.   Daud. iv. t. 53.
Ch. Margaritaceus.   Mer. Seba, i. t. 32. f. 4, 5.

Scales minute, larger on the sides ; olive varied.   Var. With larger processes on the head.   *Ch. Pardalis,* Cuv. *C. Fimbriatus.*   Wied.   Cape of Good Hope.

Brookes Chameleon.   Chameleo Brookesii, Gray. Spic.
Zool. t. 3.   *Cham. Superciliaris,* Kuhl.

Scales minute ; back with a series of processes on each side ; eyebrows produced into horns.   Mus. Brit.

c. Without any spines on the chest, belly, or tail.

Two Horned Chameleon.   Cham. Bifidus, Brongn.   Daud.
t. 54.   *Ch. Bifurcus,* Kuhl.

Scales square ; occiput flat, dilated behind ; the sides with groups of white scales; male with two compressed horns over the nostrils ; female hornless.   Isle of France.

Hooded Chameleon. Cham. Cucullatus, Gray.

Scales oval, unequal ; occiput compressed, flattened, with 2 flat processes on each side of its hinder edge, nose produced.   Madagascar.   Mus. Brit.

Three Horned Chameleon.   Cham. Owenii, Gray.

Scales square, small ; head short, with three long, conical, curved horns over the nostrils.   Fernando Po.

Parson's Chameleon.   Cham. Parsonii, Cuv. Fos. v.
f. 30, 31. Phil. Trans. viii. t. 8.

Scales ovate, equal ; occiput flat, truncated behind ; eyebrow crest, produced and elevated on each side the muzzle into a long irregular lobe.   Mus. Bell., and Col. Surg.

Gen. V.   AGAMA.

Teeth placed on the edge of the jaws, toes free, long, head and body depressed, covered with imbricate scales. Old World.

A. *Femoral pores, none.*

a. Head, lyrate, back and tail crested.   Lyriocephalus.

Scutated Lizard. Lac. Scutata, Lin. Seba, 1, t. 109.

Nose with two rounded tubercles, body with cross bands of larger scales, Mus. Fort. Pitt. Chatham.

54

b. Head, lyrate, ears distinct, back and tail crested, Agama. Gariocephalus, Boie.

Agama Tigrina, Merrem. Shaw, t. 68. *Lophyrus gigas,* Boie.

Pale brown, with transverse spotted bands.

Lophyrus Kuhlii, Boie, MS.

Crest less elevated, back with five or six oblique white bands.

c. Cytophanes.

Cyrtophanus Cristatus, Boie. *Agama Cristata,* Merrem. Seba, 1, t. 94, f. 4.

Scales of body, smooth, of limbs and tail, larger and keeled.

d. Back scales large, shelving upwards. Calotes, Boie.

Common Calotes. Calotes Ophiomachus, Kaup. *Lac. Calotes,* Lin. Seba, 1, t. 93, f. 2.

Blue, sides white banded, with two rows of compressed spines behind the ears, scales large, rounded, vertical on back.

Indian Calotes. Calotes Tiedemani, Kaup. Isis. xx. t. 8. Young. *Agama Versicolor,* Daud. t. 44. *Ag. Flavigularis,* Daud. and *Agama Indica,* Gray.

Brown, varied with spots, with two or three groups of small conical spines over the ears. India.

Smaller Calotes. Agama Minor, Gray, Zool, Jour.

Tail short, back brown, with larger brown spots.

e. Back scales, small, horizontal. Bronchocela, Kaup.

Blue Calotes, Agama Cristatella, Kuhl. Seba, 1, t. 89, f. 1.

Blue, not banded, nuchal crest, very small, low, scales of body, keeled, smaller than on tail and limbs.

Long-legged Calotes. Agama Vultuosa, Harlan. Jour. Acad. Phil. iv. t. 19. *Agama Calotes,* Kuhl. Seba, 1, t. 89, f. 2. *Calotes Gutturosa,* Cuv. Guerin, Icon. t. 7, f. 3.

Blue; crest short, scales, all keeled. India. *Le Geleote,* Lacep. 1, t. 12, is perhaps this species. *Agama Gutturosa,* according to Merrem and Lichterstein, is an American species, and is not this. Seba, 1, t. 89, f. 1, is an *Iguana. Col. Gutturosa,* Mus. Berl. with a blue neck, is white banded.

Kuhl's Calotes. Calotes Tympanistra, Kuhl.

Blue; crest, very small, scales of belly subequal, keeled. Mus. Berl.

f. Acanthosaura.

Armed Calotes. Agama Armata, Gray, Zool. Jour. When young, with very thin scales. *Calotes Lepidogaster,* Cuv. R. A. Tropidogaster. Mus. Paris.

Head, with long cylindrical spines over eyes, a dorsal crest.

g. Agama.

Tubercular Agama. Agama Tuberculata, Gray Zool. Jour.

Head moderate; scales of back rather smaller than on limbs and tail ; of neck and nape with ridges of rather larger triangular scales ; under side of legs with short strong keeled scales. India. Mus. Brit.

Obsoletely Crested Agama. Agama Dorsalis, Gray.

Head large, dilated behind ; scales rather small, smooth ; nape, and back, and tail with very obscure crest of rather larger scales. India. Mus. Brit.

Occipital Agama. Agama Occipitalis, Gray.

Head moderate ; scales of back broad, ovate, lanceolate, keeled, spinose, of head smooth and thin ; nape with a very obscure crest, and bundle of short three-sided spines. Africa. Mus. Brit.

Common Agama. Agama Spinosa. Lac. Agama, Lin. ?

Brown ; head moderate ; scales of back large, smooth ; of limbs larger, keeled ; of belly moderate, nearly smooth ; nape with a slight crest ; back of head and sides of neck with blundle of long subulate scales. Africa. Brit. Mus.

Sitana. *Cuv.* Semiphorus. *Wag.*

Pondichery Sitana. Agama Pondicerana, Nob.

Body and limbs with keeled imbricate scales ; fulvous with two series of large rhombic brown spots. Male with very large pouch extending to between the front legs Hence the genus *Sitana,* Cuvier, called *Tropidosaura* in Mus. Paris.

b. Head rounded, depressed ; ears distinct ; back with band of larger scales ; tail with whorled bands of large spinous scales. Stellio.

Common Stellio. Lac. Stellio, Lin. *Stellio Vulgaris,* Daud.

Olive spotted with black. Rup. Atlas. f. 2.

c. Head ovate lyrate ; ears distant ; back and tail not crested ; tail tapering, with imbricate scales. Trapelus.

Cape Agama. Trapelus hispidus, Kaup. Iris. xx. t. 7. *Agama Orbicularis,* Merr. ? Seba, j. t. 83. f. 1, 2.

Scales unequal, with trihedral spinous scales. Cape.

Yellow-striped Agama. Ag. Ater, Daud. *Ag. Subspinosa,* Gray, Ann. Phil. 1827. *Trapelus Subhispidus,* Kaup.

Brown, striped with a yellow dorsal line and red spot on each side ; scales small ; of occiput large, ovate, smooth.

57

Mutable Agama. *Ag. Ruderata*, Oliv. Voy. t. 29. *Trapelus Mutabilis*, Cuv. *T. Ægyptiacus*, Geof. Rept. Egypt. t. 5. *Ag. Orbicularis*, Daud. t. 45. *Ag. Deserta*, Licht. ?

Tail one-half as long again as the body ; scales of the head convex ; of the body unequal, irregular. Egypt. Mus. Brit. *Trap. Savignii*, Ardouin. Savigny Rept. Egypt. f. 3 and 4, scarcely appears to differ.

Ruppel's Agama. *Agama Sinaita*, Ruppel. t. 3. *Agama Straminea.* Mus. Berl.

Brownish grey, with paler spots ; dorsal scales unequal, slightly keeled, margined ; edge of ear one-spined ; a scaleless pit before the shoulder. Egypt. Perhaps same as former.

Sand Agama. *Agama Arenaria*, Van Heyden.

Yellowish grey, spotless ; dorsal scales equal, slightly keeled, not margined ; edge of ears one-spined ; but pit before the shoulder with scales. Egypt.

Aral Agama. *Ag. Aralensis*, Licht. *Lac. Sanguinolenta*, Pall. ?

Body dirty straw colour ; reddish, spotted ; scales equal, semicircular, keeled, ending in a spine ; tail black ringed.

Slender Agama. *Agama Agilis*, Oliver. Voy. Savigny Rept. Egypt. f. 5. *Agama Isodactyli*. Mus. Berl.

Scales small, equal, nearly smooth ; tail long, slender, black banded ; legs, especially the hinder, long and slender. Egypt

d. *Head round depressed ; ears covered ; claws sharp ; back and tail with irregular scales.* Phrynocephalus.

  * *Margin of lips produced ; toes fringed ; claws long.*

Eared Phrynocephale. *Ph. Auritus*, Kaup. *Lac. Aurita*, and *Lac. Mystacea*, Pallas. *Lac. Lobata.* Shaw.

Brown with a small lobe on each side the neck.

58

** *Margin of lips simple, claws moderate.*

Pallas's Phrynocephale. P. Caudevolvula. Lac. Caude-volvula, Pallas. *Lac. Guttata,* Lepech. Reis, t. 22.

Scales of body very minute, smooth; tail long, smooth.
Eyed Phrynocephale. Ph. Ocellatus. Agama. Licht.

With unequal sized black-eyed white spots.

Ural Toad Lizard. P. Uralensis, Kaup. *Lac. Uralensis,* Lepech. Rees, t. 22. *Lac. Helioscopea,* Pallas?

Scales of back unequal. Buchara.

e. Head roundish; throat three pouched; sides with wing-like expansions supported by the ribs. Draco, Lin.

Short-Pouched Dragon. Draco Abbreviata, Gray, Zool. Jour. *Draco Fimbriatus,* Kuhl?

Scales large; thighs fringed behind; throat pouches short, central reaching to the thorax. Wings spotted beneath.

Banded Dragon. Draco 5 Fasciatus, Gray, Zool. Jour. *Draco Viridis,* and *Draco Fuscus,* Daud. t. 41?

Wings ash, with four continued blackish bands.

Lined Dragon. Draco Lineatus, Daud.

Back ash; wings brown, with numerous white lines; scales of tail large, blunt, and uniform.

Timor Dragon. Draco Timorensis, Kuhl.

Middle dorsal scales larger, keeled. Brownish; wings and body glaucous, and brown marbled; pouch black. Kuhl.

Black Pouched Dragon. Draco Hæmatopogon, Boie. Is peculiar for a black spot on each side of the throat pouch.

B. *Femoral Pores distinct.*

f. Head long; eyes large; tail round; back and tail with irregular imbricate scales. Gemmatophora.

Bearded Gemmatophore. Gem. Barbata. Agama, Cuv.

Back with cross bands of larger scales; scales of throat, sides, and back of ears elongated, bristle pointed. New Holland, Mus. Bell. and Col. Surgeons.

Muricated Gemmatophore. G. Muricata, Lac. *Muricata,* Shaw. Zool. *A. Grandoculis,* Lac. and *A. Jacksoniensis,* Kuhl.

Black brown, with a series of pale spots; back with longitudinal series of elevated scales. New Holland.

g. Head long, having a large pleated frill on each side of the neck behind the ears. Back and tail with irregular scales. Clamydosaurus.

Frilled Lizard. Clamydosaurus Kingii, Gray, Kings. Voy. Pale brown, scales keeled, of back limbs and tail larger, of belly small, and of sides smallest. Frill with large keeled scales on both sides. New Holland. Mus. Brit. Mr. Frazer.

h. Head squarish; back and tail compressed, crested. Lophura.

Amboina Lophura. Lophura Amboinensis. Lac. Javanica, Hornsted. *Lac. Lophura,* Shaw, Zool. t. 62.

Tail above compressed, fin-shaped.

Cuvier's Lophura. Lophura Cuvieri, Gray, Mus. Paris.

Brown green spotted; body with four or five oblique bands of blue-eyed spots; scales of body and tail small subequal; back and end of tail with a series of distant, short, compressed spines. Tail compressed, interrupted by distant rings, upper and lower edge two keeled. Cochin China.

Lesueur's Lophura. L. Lesueurii, Gray, Mus. Paris.

Dark brown, varied with pale netted lines; scales of head conical, with scattered acute conical tubercles on the head and

60

neck; back with a crest of short compressed spines, forming a double crest about two-thirds the length of the back, sides with eight or nine cross bands of oval larger keeled scales, and others similar across the base of the tail and limbs. Paramatta, Lesueur.

Beautiful Lophura. Lophura Concinna, Physignathus Concinnus, Cuv. R. A. t. f. *Ph. Iguanoides.* Mus. Paris.

A crest of conical compressed scales the whole length of the back; head with granular scales. Body and tail with small squarish scales, the scales of the tail becoming larger near the end; tail much compressed; young with scarcely any crest. Dark blue, with some oblique white bands on the side, and whitish beneath, with some rather larger compressed scales on the side of the chin.

i. Head short, arched; back, with minute scales, not crested; tail with whorled scales. Uromastyx.

* Scales of tail large, spinose, Uromastyx. Cuv. Mastigura. Flern.

Common Uromastyx. Uromastyx Spinipes, Merrem. *U. Acanthinurus,* Bell. Zool. Jour. *Stellio Spinipes,* Daud. Geoff. Egypt, t. 2.

Olive dull greenish brown, subcaudal segment, with subdentate scales, placed in two or three series. North Africa.

Two Coloured Uromastyx. Uromastyx Dispar.

Tail like the former, male, black brown; female, ash yellow.

Ornamented Uromastyx. Uromastyx Ocellatus, Licht. *Uromastyx Ornatus,* Heyden, Ruppel, Trav.

Dull green, body brown, ringed subcaudal segment consists of a single series of armless scales. Africa. Dongola.

61

Black Uromastyx. *Uromastyx Niger,* Merrem. *Stellio Niger,* Daud.

Tail twice as long as the body, with twenty-seven rings of spines ; toes long, sub-depressed ; scales small, rhombic.

Hardwick's Uromastyx. *Uromastyx Hardwickii,* Gray, Zool. Jour. *U. Reticulatus,* Cuv. R.A.

Greenish, marbled with black dots; a large black spot in front of each thigh. Hindostan.

Cuvier indicates *Uromastyx Grisseus,* from New Holland,

** *Tail with rings of small armless spines. Lecolepis,* Cuv. *Cynosaura,* Schegel.

Bell's Uromastyx. *Uromastyx Bellii,* Gray, Zool. Jour. *Lecolepis Guttatus,* Cuv. Guerin. Icon, t. 7. *L. Guttato Lineatus,* Mus. Paris. *Cyn. Punctatus,* Schegel. Mus. Leyd.

Olive; back, with three lines alternating with black edged white spots ; limbs white-eyed ; sides black spotted.

Reeves Uromastyx. *Uromastyx Revesii,* Gray, Mus. Brit. Olive, with a series of bright red spots down each side.

Spotted Uromastyx. *Uromastyx Maculatus,* Gray. *Lecolepis Maculatus,* Cuv. Mus. Par.

Pale brown ; head, with a black streak from the eye to the neck ; back, legs and tail with six rows of oblong black spots, sides of the legs black netted.

Order II. OPHIOSAURI.

Mouth not dilatable, skin covered with regular equal scales.

Section I.

Body with equal similar scales, above and below, and a compressed line of small scales on each side ; tongue short contractile, two cut. Ptygopleura.

62

Gen. I. Zonurus.

Teeth placed on the side of the jaws; legs four, moderate; ears, exposed; vent medial.

1. *Toes 5, 5; femoral pores distinct; tail with spinose whorled scales.* Africa. Zonurus.

Common Zonurus. Lac. Cordylus, Linn. *Cordylus Grisseus,* Seba, i. t. 48. f. 4.

Brown with a yellow line down the back; the scales nearly equal. *Cor. Niger,* Cuv., is from a badly preserved specimen, and *Cor. Dorsalis,* from a good coloured specimen. Cape.

Armed Zonurus. Zonurus Cataphractus. Bonn. Trans.

Head with large scales; scales of the body large and hard, of the sides three toothed; of the thighs, neck, and tail, ending in a trigonal spine.

Small scaled Zonurus. Zonurus Microlepidotus. Gray.

Dorsal scales about half the size of ventral, bluntly keeled.

2. *Toes 5, 5, femoral pores distinct; tail unarmed.* Cicigna, Gray, 1815. Gerrhosaurus, Wiegmann.

Common Cicigna. Cicigna Sepiformis. Scincus, Merrem. *Lac. Seps,* Lin. *Gerrhosaurus Flavigularis,* Weigmann.

Olive brown, beneath white, chin, throat, and upper lateral scales, yellow; fore, middle, and hinder toes longest.

Smooth Cicigna. Cordylus Lævigatus, Cuv.

Brown, with four black and yellow lines on each side the back, and two series of black and yellow spots; scales obliquely four sided. keeled. Perhaps the same as former.

63

Madagascar Cicigna. Cicigna Madagascariensis. Gray.
Green, with yellow lines on each side the back; back and
sides brown spotted; scales smooth; when young, scales
keeled; variety. β *Ornata.* Back with five yellow and six
black lines, and sides black, yellow spotted. Madagascar.
Mus. Brit.

C. *Toes 5, 5; femoral pores, none; tail with whorled,
unarmed scales.* America. Gerrhonotus, Wiedman.

Burnet's Gerrhonote. Gerrhonotus Burnettii, Gray.

Scales of back, and sides of body and tail, obliquely four
sided, slightly keeled; dark brown; sides, with dark band
and cross rows of white spots beneath.

Depp's Gerrhonote. Gerr. Deppii, Wiegm. Isis. xxi.

Dorsal scales four angular, smooth; olive black, irregularly
white spotted, beneath white; tail round, white ringed.
Mexico.

Retired Gerrhonote. Gerrhonotus Tæniatus. Wiegm.

Dorsal scales four angular, smooth; bluish, with black
angular cross bands; tail black, ringed.

Blue Gerrhonote. Gerrhonotus Cæruleus, Wiegm.

Dorsal scales, four angular; olive bluish, with three series
of black spots; tail round, hemeolate. Brazils.

Rough-necked Gerrhonote. Gerrhonotus Rudicollis, Wiegm.

Head rough, with elevated shields; scales four, angular;
placed in cross series. Grey green, base of tail, four angular.

Imbricate Gerrhonote. Gerrhonotus Imbricatus, Wiegm.

Head shields elevated; dorsal scales keeled, four angular,
olive grey, tail, hemiolate. Mexico. Brit. Mus.

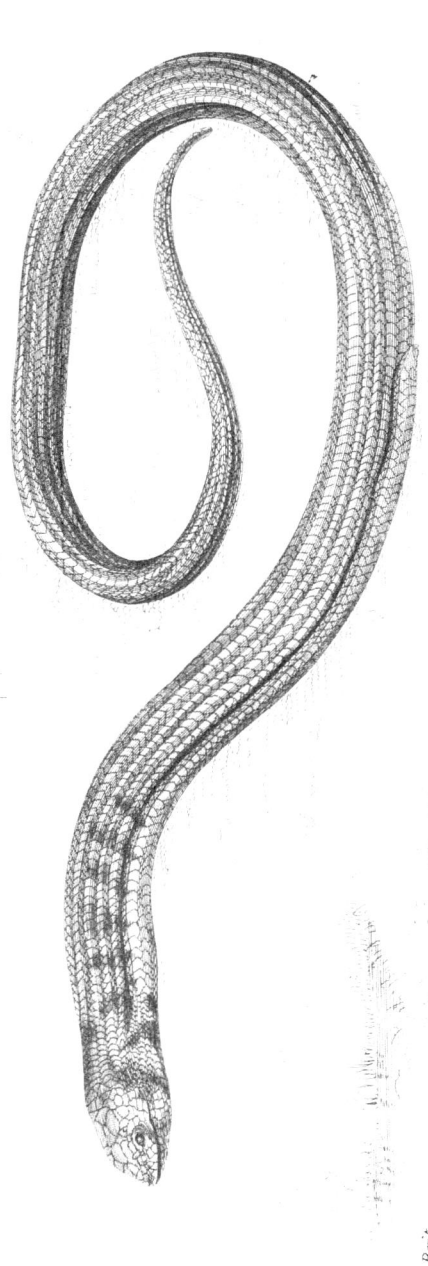

Mus. Brit.

DURVILLE'S PSEUDOPUS. PSEUDOPUS DURVILLII.

London. Published by Whittaker & Cᵒ Ave Maria Lane. Novʳ 1830.

Smooth-headed Gerrhenote. *Gerr. Leiocephalus,* Weigm.

Head smooth, scales four angular, in cross series, of middle of back keeled, of na pe and sides smooth ; grey green, bluish beneath, with nine longitudinal lines. Mexico.

Toes 4, 4, femoral pores — Saurophis *Fitz.* Tetradactylus, *Merrem.*

New Holland Saurophis. Saurophis Lacepedii, Lac. Tetradactylus, Lacep. Ann. Mus. ij. t. 59. f. 2. *Tetradactylus Chalcidices,* Merrem.

Gen. II. OPHISAURUS.

Teeth on the side of the jaw ; legs rudimentary ; ears distinct ; lateral line distinct ; vent medial.

a. Hinder extremities only rudimentary, on the side of the vent, undivided. Pseudopus, *Merrem.*

Pallas's Scheltopusik. Pseu. Serpentinus, Merrem. *Ps. Pallasii,* Cuv. *Lac. apoda,* Pall. Nov. Com. xix. t. 9.

Scales smooth ; caudal scales square ; front feet rudimentary, length two feet. Russia.

Durville's Pseudopus. Pseutopus Durvillii, Cuv.

Brown ; dorsal and caudal scales keeled, rough. Brit. Mus.

b. Legs none. Ophisaurus, *Daud.* Hyalinus, *Merrem.*

Common Glass Snake. O. Ventralis, Daud. *Anguis,* Linn.

Yellow green, black spotted. Sometimes the spots form continued lines. *Oph. Striatalus* and *Oph. Punctatus,* Cuv. MSS. differ very little, if at all, from this species.

Gen. III. CHALIDES.

Teeth on the side of the jaws ? Legs four, distinct ; ears hid under the skin ; vent medial, linear.

a. Toes 5, 5. Chalcides, *Fitz.*
Common Chalcides.

Ferruginous with six dorsal lines.

b. Toes 4, 5 ; *femoral pores distinct.* Heterodactylus.
Spix's Heterodactyle. Het. Imbricatus, Spix, Braz. t. 27.

Olive brown ; sides of body and tail yellow-lined ; tail long ; central dorsal scales acutely keeled. Brazils.

c. Toes 4, 4. Brachypus, *Fitzinger.*

Common Brachypus. Bra. Abdominalis, Thunb.

Brown, with stripes on the belly.

d. Toes very small, rudimentary, 5 *before,* 3 *behind. Chalchis* and *Colobus*, Merrem. *Cophias*, Fitz.

Annulated Chalchides, Shaw. *Cophias Flavescens*, Gray. *Le Challide*, Lacep. t. 32. *Chamœsaura Cophias*, Schn.— *Chalchide Menodactylus*, Daud. *Chalchydes Flavescens*, Bonnat. *Chalchide Tridactylus*, Daud. *Colobus Daudini*, Merrem. Guiana.

Gen. IV. AMPHISBŒNA.

Teeth in sockets ; ears hid ; vent roundish, subterminal.

a. Legs two, very small ; subanal pores distinct Chirotes.

Worm-like Bimane. Lacerta Lumbricoides, Shaw. *Chamasaurus Propus*, Schn. Lacep. t. 41. Mexico.

b. Legs none, subanal pores distinct ; thoracic rings regular. Amphisbœna, *Lin.*

White Amphisbœna. Am. Alba, Lin. Lacep. ij. t. 21. f. 1. Eyes small ; occiput smooth. Perhaps same as next.

Sooty Amphisbœna. Amph. Fuliginosa, Linn. Seba, t. 18. f. 2. t. 100. f. 3. t. 83. f. 4.

Eyes small ; occiput grooved ; black, white-varied.

See *Amp. flavescens.* Pr. Max. t. *Amp. pathyura*, Wolf.— *Amp. punctata*, Bell.—*Amp. vermicularis*, Wagler.— *Ampt. Cinerea*, Vandelle. *Amp. cxyura*, Wagler, Braz. t. 25. f. 1. *Blanus*, Wagler.

66

Blind Amphisbœna. Amph. Cœca, Cuv.

Eyes none.   Martinique.

c. Legs and subanal pores none ; thoracic rings irregular ;
the plates united together in front. Leposternon, *Spix.*

Dotted Leposternon-   Leposternon Microcephalus, Spix
Amph. Punctata, Pr. Max.

Blue gray, with numerous black dots.

Sharp-nosed Leposternon. Leposternon Oxyrhychus. Dekay
Siliman Jour. xiv. 907.

Yellowish ;   back with three longitudinal grooves ;   nose
pointed, not mucronate.

11. *Body covered with imbricate scales.*

Gen. V. SCINCUS.

Body covered with uniform imbricate scales;   head long ;
tongue short, contractile, two-cut.

A. Legs four ; ears distinct.

a. Toes 5, 5, margined ; muzzle acute, produced. Scincus.

Shop. Scinc. Scincus Officinalis, Linn.   Egypt.

b. Toes 5, 5 ; muzzle rounded ; body moderate ; scales
very hard, bony ; tail broad, depressed.   Trachydosaurus.

Rugose. Trachydosaurus. Tra. Rugosa, Gray. King's Voy.
Scincus Pachyurus.   Peron MSS.

c. Toes 5, 5 ; muzzle rounded ;   body moderate ; scales
thin, silvery ; tail conical   Tiliqua, *Gray.*

New Holland Tiliqua.   Tiliqua Whitii.   Lacerta Scin-
coides, ˌShaw.   White Jour. at p. 242. *Cyclodus,* Wagler. *Sc. Tuberculatus,* Merrem.

Large ;   pale ;   scales large, whitish ;   six series on the
back of neck.   New Holland.   Brit. Mus. *Cyclodeus*
Flacigularis, Wagler, t. 6, appears to differ little from this.

Black and yellow Tiliqua. Tiliqua Nigroluteus. Scincus, Quoy and Gaim. Frey. Voy. t. 41. New Holland.

New Holland Tiliqua. Tiliqua Crotaphomelas. New Holland.

Indian Tiliqua. Tiliqua carinatus, Schn. Gray, Zool, Jour. *Scincus rufescens,* Cuv. *S. multicarinatus, Sc. lineatus, and Sc. nigro fasciatus,* Kuhl. Mus. Par.

Greenish, with a yellow line along each side, scales, three keeled. India, Java, &c.

Cape Tiliqua. Tiliqua Capensis, Gray. *Sc. trivittatus,* Cuv.

Brown, with three longitudinal pale lines, on the back and tail, with black spot between the lines. Cape.

Three Streaked Tiliqua. T. trivittatus, Gray, Zool, Jour.

Pale brown, with three broad black edged yellow streaks on back and tail; sides, pale black spotted. India.

Ribbon Galley-Wasp. T. tæniolata, Lac. tæniolata, White, Jour. t. ap. p. 245. *Sc. octolineatus,* Daud. *Sc. decemlineatus,* Lacep. *Sc. undecemstriatus,* Kuhl.

Brown, back with ten or twelve white bands, with black sides, the two dorsal ones on each side uniting into one over the eyes, scales thin polished. New Holland.

Many scaled Galley-Wasp. Tiliqua multiscutatus Sc. Cuv. *Anolis parcé,* Geoff. Rept. Egypt, t. 4. f. 4.

Large Galley-Wasp. Tiliqua Cyprinus Sc. Cuv. *Anolis gigantesque,* Geoff. Rept. Egypt, t. 3. f. 3.

Scales smooth, tail longer than the body, brown, with a pale line on each side. Egypt.

Eyed Galley-Wasp. Tiliqua ocellatus Sc. Schn. *Sc. variegatus,* Cuv. Daud. t. 56. *Anolis marbré,* Geoff.
68

Rept. Egypt, t. 5. f. 1. Savigny, Rept. Egypt, t. 2. f. 2. *Sc. Tiligugus*, Merrem. Oliv. Voy. 1. t. 16.

Pale brown, with semilunar, black spot divided in the centre, by a short whitish band. South of Europe, and Egypt. See also *Sc. vittatus*, Oliver, Voy. t. 29. f. 1. Varied with a pale line on each side the back. Brit. Mus.

Common Galley-Wasp. Tiliqua occidua, Lacerta occidua, Shaw, Sloane, Jam. y. t. 273. f. 9. *Sc. fossar.* Merrem.

Large ; pale brown ; scales moderate, about twelve series of scales on the back of the neck. Brit. Mus. Jamaica.

Lacepedes Galley-Wasp. Sc. Mabouya, Sh. Lacep. t. 24.

Smooth, greenish brown, back black dotted, and a brown band from the temple over the shoulder.

Doubled Galley-Wasp. Tiliqua bistriatus Sc. Spix, t. 26. f. 1.

Coppery brown with a broad black streak from the nostrils to the nose, edged with white on each side. Brazils.

Five lined Galley-Wasp. Tiliqua quinquelineatus, Lacerta quinquelineatus, Lin.

Greenish blue, with five dorsal lines, the central one forked over the head. Tail reproduced, blue. *Lacerta fasciata,* Lin. Catesby, Car. t. 67.

Seven Striped Galley-Wasp. Tiliqua homolocephalus, Wiegmann.

Olive brown, with seven black longitudinal streaks.

Black spotted Galley-Wasp. Tiliqua nigro-punctatus Sc. Spix, t. 26. f. 2.

Black brown, golden, with an unconspicuous black band along each side to the thigh, back, black punctate, tail short.

69

Coppery Galley-Wasp, Tiliqna ænea, Gray.

Golden green ; back with five lines of black spots, sides from the eyes, with a dark brown band, beneath pale, sides of belly black spotted.　Brazils.　Brit. Mus.

Sloan's Galley-Wasp. Tiliqua Sloanii, Sc. Daud. t. 55. f. 2.
Pr. Max. x. 11. t. 1. f. 1.

Golden green, with four longitudinal black streaks extended over the base of the tail, paler beneath, scales thin, smooth, Toes short thick.　Brit. Mus.

Red-headed Galley Wasp. Ti. Erythrocephala. Sc. Ery-throcephalus, Gilliams, Jour. Acad. N.S.　Phil. i.

Reddish brown cupreous, beneath whitish, head above red.

Two Coloured Galley Wasp. Ti. Bicolor. Sc. Harlan. Acad. N. S. Phil. iv. t. 18. f. 1.

Dusky brown, darkest on the head, beneath silvery, two longitudinal whitish lines on each side the body ; tail tapering with two obsolete lines on the hinder part of the thighs.

Lateral Galley Wasp. Tiliqua Lateralis. Scincus, Say
Sc. unicolor, Harlan.

Light brown, with blackish lateral lines, beneath greenish white ; a cross row of scales behind the plates larger than the remaining cervical scales.

Bell's Galley Wasp. Tiliqua Bellii, Gray.

Pale brown, with irregular cross bands of small white scales, with central black bands, sides varied with dark brown, scales small, smooth, about four series on back of neck.

Double-streaked Galley Wasp. Tiliqua bistriatus, Gray.

Brown, sides with two narrow streaks, back with five series of black spots, with a central yellow triangle in each.

70

Thin Scaled Galley Wasp. Tiliqua Tenuis, Gray.

Pale brown, back with irregular black spots, forming a band on each side the back, scales thin polished.

Fine Scaled Galley Wasp. Tiliqua Microlepis, Gray.

Pale, with numerous irregular transverse brown bands; head very depressed, forehead flat, eyes large, tail slender, toes very unequal. Scales very thin, closely adpressed keeled.

Banded Galley Wasp. Tiliqua Fasciatus.

Silvery blue, head with three, neck with one, back and tail with six broad black bands, toes, hinder rather unequal, five last in Brit. Mus.

Spix has placed his genus *Lepidosoma* near the Scinks. In one of my notes made on the animal at Paris I have re-marked that it is nearly allied to *Polychrus*, but has no pores or keeled scales.

d. *Toes* 4, 5, *rest like Tiliqua, but no eyelids.* Gymno-tholamus, *Merrem.*

Four-lined Galley Wasp. Gym. quadrilineatus, Merren. *Lacerta lineata*, Lin. Seba, y. t. 41. f. 6. *Lac. quadrilineatus*, Lin. Pr. Max. Beytr. xiii. f. 2. *Sc. Cyanurus*, Sching.

Blackish, with four yellowish dorsal bands.

e. *Toes* 5, 5, *muzzle rounded, body very long slender, feet small, far apart, ears distinct.* Lygosoma, Gray.

Short-footed Lygosoma. Scincus Brachypus, Schn. Geoff. Rept. Egypt, t. f. 9, 10. *Spœnops Capistrata*, Wagler.

Hinder toes unequal, long; pale brown, with narrow brown longitudinal band, and a black streak through each eye.

Dotted Lygosoma. Lygosoma Punctata, Gray.

Hind toes unequal, long; pale brown, with many series of minute dark spots; head brown; tail long, dotted.

71

Scincord Lygosoma. Lygosoma Scincoides, Seps. Cuv.

Ears large; hind toes unequal, long. Pale brown, with black slender bands formed of spots on the centre of each scale. Mus. Paris.

Lygosoma Serpens, Lacerta Serpens, Gmel. Blo. Nutur-
fosh. y. t. 2.

Hinder toes subequal short.

Golden Ligosoma. Ly. Aurata, Gray. *L. Serpens,* Gray.

Golden brown, with many longitudinal black lines, one between each series of scales; hind toes unequal, rather short.

f. Toes 4, 4, *or* 3, 3; *body very long and slender; feet far apart: ears distinct.* Seps.

Four-toed Seps. Seps Peronii, Fitz.

Toes 4, 4, hinder unequal; golden brown, with streaks; sides brown and white spotted; beneath white, with cross black spots.

Peron's Seps. Tridactylus Decresiensis, Peron.

Toes, 3. 3; hinder unequal, one short and two long, subequal; pale brown, with long dark lines; beneath netted.

Common Seps. Seps Chalcides Zygnis, Fitz.

Grey, with four longitudinal lines on the back. Brit. Mus.

Striated Seps. Seps Striata. Zygnis Striata, Fitz.

Grey, with eight or nine equal spaced dorsal lines.

g. Toes 3, 3; *body long; feet far apart, slender; ears* 0; *femoral pores* 0. Saiphos, *Gray.* Cophias, *Fitz.*

Lacertine Saiphos. Saiphos Equalis, Gray. *Seps Equalis,*
Gray Ann. Phil. *Anguis Lacertina,* E. W. Gray, MSS.

Grey; scales of the head equal; toes short, subequal. Brit. Mus.

72

h. Feet four, undivided, oblong, scaly ; scales of body and tail keeled, pointed subverticulate. Monodactylus, *Merrem.* Chamæ Saura, *Fitzinger.*

Anguine Monodactyle. Monodactylus Anguinus. Lacerta Anguinus, Linn. *Lac. Monodactyle,* Lacep. Ann. Mus. ij. t. 49. f. 1. The scales of the back in straight, and those of the belly in oblique whorles.

Gen. VI. BIPES.

Feet two, posterior imperfect ; scales uniform, imbricate, the ventral rather the largest ; head long; tongue short, contractile, two-cut.

a. Feet undivided, oblong ; scales smooth ; femoral pores none ; ears o ; Ophoides, *Wagler.* Pygopus, *Spix.*

Brazilian Bipes. Ophiodes Striatus, Wagler. *Pygopus Cariococca,* Spix, t. 28. f. 1., adult. *Pygopus Striatus,* Spix, t. 28. f. 2. Young. *Seps fragilis,* Raddi. *Pseudopus Olfersii,* Licht. Brazils. Mus. Brit. *Bipes Lineata,* Cuv. Mus. Paris. The lines become double near the head.

Wagler cites another species, *O. Gronovii,* Wagler. *Pygopus Striatus,* Fitzinger. *Pygodactylus Gronovii,* Merrem.; but I have not seen it.

b. Feet undivided, oblong ; scales smooth ; femoral pores none ; ears distinct. Delma, *Gray.*

Frasers Bipes. Bipes Fraseri, Gray. Brit. Mus.

Tail one-third longer than the body ; black ; beneath paler ; head with four narrow pale cross bands. New Holland.

c. Feet undivided, oblong ; scales of back keeled; femoral pores distinct. Pygopus, *Merrem.*

New Holland Bipes. Pygopus Lepidopus, Merrem. *Bipes*

73

Lepidopus, Lacep. Ann. Mus. iv. t. 65. Tail when perfect, longer than the body.

d. *Feet with two unequal toes ; scales keeled ; femoral pores none.* Bipes, Merrem. *Scelotes*, Fitz. *Pygodactylus*, Merrem. *Zignis*, Wagler.

Cape Bipes. Bipes Anguinus, Merrem. *Anguis Bipes*, Linn. *Lacerta Bipes*, Gmel. Seba, j. t. 85. f. 3.

Golden brown ; back with ten longitudinal black lines, of which the four lateral are the most distinct. The feet are considered as the generative organs by Seba. *Seps. Sex lineata.* Harlan. Jour. Acad. N. S. Philad. iv. t. 10. f. 1., according to Cuvier is a variety.

Gen. VII. ANGUIS.

Legs o. ; body and tail covered with smooth imbricate scales ; femoral pores none.

a. Ears distinct. Siguana.

Otto's Blind-worm. Siguana Ottonis, Gray.

Brown, with darker lines ; beneath paler ; tail rather longer than the body. Breslau. Discovered by Dr. Otto.

b. Ears hidden by the skin. Anguis.

Common Blind-worm. Anguis Fragilis, Lin. *Anguis Eryx*, Daud., the young. *Anguis clivica*, Wolf. *A. Lineata*, Laur.

Silvery grey, when young with a black dorsal line and black sides. Europe.

American Blind-worm. Anguis Eryx, Linn., said to have four nostrils, I have not been able to see them.

Gen. VIII. TORTRIX.

Body long, cylindrical ; back covered with imbricate scales ;
74

beneath with a series of larger scales ; tongue short, contractile, two-cut ; head long.

a. Tail blunt, with a single series of larger scales beneath ; Anilius *Oken.* Tortrix *Oppel.*, not *Linn.* Ilysia, *Hemprick.* Torquatrix, *Haworth.* Cylindrophis, *Wagler.*

Scytale Coral Snake. Anilius Scytale. Anguis, Linn. Mus. Adolph. t. 6. Daud. vii. t. 87. Seba, ji. t. 20.

With irregular black and white rings. S. America.

Common Coral Snake. Anilius Corallinus. Anguis, *Laur.* Seba, ij. t. 73. f. 1. 2, 3. S. America.

Black Coral Snake. Anilius Rufus. Anguis Ater, Cuv. Seba, ij. t. 25. f. 1. t. 7. f. 3. *Anguis Rufa,* Laur. *A. Striatus,* Gmel. *Eryx Rufus,* Daud. *Tortrix Rufa,* Merr. *Illysia Rufa,* Mempr. *Cylindrophis Resplendens,* Wagler. Russel Ind. Serp. ij. t. 27. t. 28.

Copper colour, with white spot on tail.

Spotted Coral Snake. Anilius Maculatus. Anguis, Linn. Mus. Adolph. t. 21. f. 3. and *A. Tessellata,* Laur. Seba, j. t. 53. f. 7. ? ij. t. 100. f. 2 Russell, ij. t. 89.

Yellow, with black dorsal band.

Dotted Coral Snake. An. Melanostictus. Anguis, Schn. *Eryx,* Daud. *Tortrix Punctatus,* Cuv., from Seba, ij. t. 2. f. 1, 4. Ind. Serp. j. t. 42.

Pale, yellowish punctated, with different coloured spots.

Black-eared Coral Snake. Anilius Melanotis. Cylindrophis, *Wagler.* Tortrix, Remw. MSS.

Like L. Resplendens ; but tip of tail black ; occipital band red ; forehead black, and bands of body nearer together.

Indian Coral Snake, Anilius Latta, Tortrix, Cuv. Seba, ij. t. 30. f. 3. Russel, t. 46.

75

b. Tail obliquely truncated, with a double series of larger scales beneath ; muzzle pointed. Uropeltis, *Cuv.*

Cuvier has indicated, but not described, two species of this genus.

Gen. IX. ACONTIAS.

Body sub-cylindrical, covered with uniform imbricate scales. Tongue ———— ? ; head short, blunt, masked; ears hid ; tail short.

Cape Pintado Snake. Acontias Meleagris, Merrem. *Anguis,* Linn. *Eryx.* Daud.

Eyes distinct ; back with eight lines of black spots. Cape. *Netted Pintado Snake. Acontias Reticulata,* Merrem. Eyes distinct ; scales rough.

      Blind Pintado Snake. Acontias Cæcus, Cuv.

Eyes covered. Africa.

Gen. X. TYPHLOPS.

Body sub-cylindrical, covered with uniform imbricate scales ; tongue long, forked ; muzzle produced ; vent subterminal.

a. Head same size as body, blunt.

Bramin Typhlops. Typhlops Braminus, Cuv. from Russel. t. 43. *Eryx,* Daud., and *Tortrix Russelli,* Merrem.

b. Muzzle depressed and blunt, with many plates in front, Stenostoma, *Spix.*

Netted Typhlops. Typhlops Reticulatus, Merrem. *Anguis.* Cuv. Sch. Phys. Scer. t. 757. f. 4.

    Ash Typhlops. Ty. Cinereus, Merrem.
Ash coloured.

Banded Typhlops. Ty. Fasciatus, Merrem.

Body equal, tail ending in a spine.

Yellow Typhlops. Ty. Crocotatus. Anguis, Schn. *T. Rostralis,* Merrem.

Body depressed, and then compressed, tail acute.

White-bellied Typhlops. Typhlops Leucogaster, Prin. Max. Beytr. j. 495.

Olive brown, beneath white.

Cuvier here also indicates, by name, two undescribed species. *T. Undecimstriatus* and *T. Leucorrhous.*

c. Muzzle covered with a single sharp edged plate in front. Typhlops.

Worm-like Typhlops. Ty. Lumbricalis. T. Vermicularis, Merrem. *Le Lombric. Anguis,* Lacep. ij. t. 20. f. 1. Seba, t. 86. f. 2.

Tail conical, blunt.

White-fronted Typhlops. Typhlops Albifrons. Stenosoma Albifrons, Spix, Braz. t. 25. f. 3.

Forehead white.

d. Muzzle ending in a conical point ; eyes quite hid ; tip of tail enveloped in an oval horny shield. Rhinophis, *Wagler.*

Phillipine Typhlops. Typhlops Phillipinus, Cuv.

Blackish ; length eight inches. The Phillipines.

Sharp-nosed Typhlops. Ty. Oxyrhynchus, Merrem.

e. Muzzle covered with a single large convex shield ; eyes none. Typhlina, *Wagler.*

Seven-streaked Typhlops. Typhlops Septemstriatus, Merr.

Tail ending in a spine.

Lined Typhlops. Typhlops Lineatus Acontias, Reimw.

77

Order OPHIDII.   *Serpents.**

The jaws very dilatile; the mastoïd bones free from the skull, and the branches of the lower jaws free in front; legs o, body with a row of larger scales benenth.

I. *Upper jaw toothless, fangs large, distinct vertebral plates broad ; tail short, conical.*   Venemous.

Gen. 1. CROTALUS.   Rattle-snakes.

Head broadly triangular, with a large pit before the eyes.

A. Tail with a rattle, head covered with scales like the back.
Crotalus.   New World.

Common Rattle Snake.   Crotalus Horridus, Lin. Catesby, ij. t. 41.   Crot. Catesbæi.   Hempr.——Lozenge spotted Rattle-snake.   Crotalus Durissus.   Cr. Horridus.   Pr. Max. Ch. Rhombifer, Boie.   Cr. Cascavella, Spix, t. 24.——add Cr. Confluentus, as North America, and Boie has indicated, in the Leyden Mus.   C. Atricaudatus, and C. Drijinus ; and Humboldt in his Zoological observation describes Crotalus Loeflingii.   Crot. Strepitans, Daud.   Seba, ij. t. 96. f. 2, is said to be Boa Canina ; and Crot. triseriatus, Weigmann.   Is referred to Mexico.

B. Tail with a rattle, head with shields.   Crotalophorus, *Gray.*   Caudisona, *Fitz.*

Miliar Rattle-snake.   Crotalophorus Miliaris.   Crotalus, Lin. Catesby, ij. t. 42.   Mus Brit.   See also Crot. Tergeminus, Say.   North America.

---

* It has been found absolutely necessary to omit the specific characters of the ensuing divisions of this class.   ED.

The Catalogue on the Species, is greatly dependant on the one published by Boie, in the Iris, Vol. xx.   The species are separated by a short dash, thus ——.   J. E. G.

CLASS REPTILIA.

C. Tail ending in a spine; subcaudal plates simple, one rowed, head with shields behind the eyes. Tisiphone *Fitz.* Cenchris. *Daud.*—Ancistodan, *Pallas.*

Catesby's Tisiphone. Tisiphone Shausii, Gray. Coluber Tisiphone, Shaw. Catesby Car. ij. t. 45. Peleas Niger fusens, Merrem. Cenchris Mockeson, Daud, t. 60, and t. 70. f. 3, 4.—See also Tisiphone Cupoea, Fitz. South America.

D. Tail ending in a spine; subcaudal plates double and single; head with scales. Lachesis.

L. Rhombeata. Pr. Max. Crotalus Mutus, Lin. Boa Mutus, Lacep. Scytale Catenata, Licht. Lachesis Mutus, Daud. Cophias Crotalinus, Merrem. Curuacea, Margr. Bothrops Suruacea, Wagler. Braz. t. 23.

E. Tail simple; sub-caudal plates double; head scaly. Cophias, *Merrem.* Crassedocephalus, *Fitzenger.* Bothrops, *Spix.* Trigenocephalus, *Oppel.* Alecto, Megæra, Atropos, and Tropidolæmus, *Wagler.*

American.

Cophias Lanceolatus. Coluber, Lacep. ij. t. 5. f. 1. Vipera, Daud. Col. Megæra, Shaw. Brazils.——Cophias Mægera. Bothrops, Spix, t. 19.——Cophias Furia, Boie. Bothrops, Spix, t. 20.——Cophias Leucostigma. Bothrops, Spix, t. 21. f. 1.——Cophias Tessalatus. Bothrops, Spix, t. 21. f. 2.——Cophias Tæniatus. Bothrops, Spix, t. 21. f. 3.—— Cophias Triangulum, Boie. Mus. Leyd. Vip. Brasiliana, Lacep.——Cophias Jacaraca, Pr. Max. adult and jun. Col. Atrox, Pr. Max. not Merrem.——Cophias Atrox, Merrem. Col. Atrox, Linn. Mus. Adolph. j. t. 23. f. 9. Vip. Atrox, Laur. Col. Ambiguum, Weigl. Vip. Weigelii, Daud. Vip. Tigrina, Daud.——Vipera Tigrina, Licht. Cophias Neuweidii. C. Bothrops, Spix, t. 21. f. 1.——

79

Cophias Leucurus, Boie, Bothrops, Spix, t. 22. f. 2. appears to be the young. Cophias Bilineatus Pr. Max. Vipera Chloris, Gravenhorst.

Asiatic,

Cophias Trigonocephalus, Merrem. Col. Capite Triangularis, Lacep. t. 5. f. 2. Col. Trigonocephalus, Dau. Vip. Trigonocephalus, Daud. Sch. P. S. t. 749. f. 11. Inhab. Isle of St. Eustach. Asia. Boie ——Cophias Viridis, Merrem. Trimesurus Viridis, Lacep. Ann. Mus. iv. t. 56. f. 2. Coluber gramineus, Shaw. Brodro. Pam. Russel, t. 9. Green with a narrow yellowish line on each side. Coph. Punicea, Boie. Atropos, Wagler. Java.——Cophias Sumatrana, n. Col. Sumatranous, Raffles. Coph. Wagleri, Boie. Seba, ij. t. 68. f. 4. Tropidolæmus, Wagleri.—— Cophias Russelii, n. Russel, ij. t. 22. Inhab. Phillipine Islands. Mus. Paris.

F. Tail, end simple ; subcaudal plates double ; head shielded Trigonocephalus, *Fitz.*

Old World.

Trigonocephalus Rhodostoma, Reinw. Vipera Pretextata, Gray. India.——Trigon. Hypnale, Boie. Cophias Hypnale, Merrem. Col. Lebetinus, Linn. Vipera Lebetina, Daud. Inhab. North Africa and Southern Europe and Asia.—— Trigonocephalus Haly., Boie. Col. Halys, Pall. Vipera Halys, Daud. Vipera Aspis, B. Pallasii, Merrem. Lichten, Reese. Siberia.——Trigonocephalus Blomhoffi, Boie. Iris. 1826, 214. Bul. Sci. Nat. x. 151. Japan.——Trigonocephalus Orophyas, Oppel. Boie.——Trigonocephalus Boeii, Boie. Iris. xx. 561. Inhab. Levant. Olivier.

American.

Trigonocephalus Cacodæmon, Boie. Col. Cocadæmon.
80

Shaw. Pelais Niger, Merrem. 2 Catesby, Car. t. 44. Scytale Niger, Daud.——Trigonocephalus Boiei. Iris, xx. 561. Inhab. Carolina.

Gen. II. VIPERA.

Head broad, without any pit ; covered with scales or small shields.

A. Head covered with scales like the back ; subcaudal plates double ; nostrils large. Vipera. Echidna, *Merrem.* (not Geoffrey.) Cerastes, *Wagler !*

Vipera Nasicornis, Daud. Coluber, Shaw. Nat. Misc. t. 94. Zool. t. 104. Brit. Mus.——Vipera Russelii. Coluber, Shaw. Zool. t. 108. Russel Ind. Serp t. 7. Col. Trinoculus, Schn. Vipera Elegans, Daud. Col. Triseriatus, Herm. Bengal. Brit. Mus.——Vipera Cerastes; male with horn before the eyes. Col. Cerastes, Hasselq. Ellis Phil Trans 4. vi. t. 14. Shaw Zool. t. 103. Lacep. ij. t. 1. f. 9. Daud, vi. t. 47. f. 1. Col. Cornutus, Hasselq ; female hornless. Col. Vipera, Hasselq. Aspis Cleopatra, Laur. Col. Egypticus, Lacep. Vipera Ægytia, Latr. Vipera Ægyptica, Daud. North Africa. Brit. Mus.——Vipera Lophophris, Cuv. Vipera Flava, Merrem. ? Paterson, Africa. t. 15.——Vipera Arietans, Echidna, Merrem. Vipera Inflata, Burchel. Seba, ij. t. 30. f. 1. Col. Intumescens, Donnd. Vipera Severa, Daud. Col. Dubius, Gmel. Col. Hebriacus, Lacep. Col. Bitis, Bonnat. Vip. Brachyura, Cuv.—— Vipera Daboia, Daud. Daboia Lacep. t. 13. f. 2. Col. Braziliensis, Lacep. t. 4. f. 2.——Vipera Ocellata, Latr. Aspic, Lacep. ij. t. 2. f. 1.——Vipera Atropos. Coluber, Linn. Mus. Adolph. t. 13. f. 1.

B. Head covered with granular scales ; sub-caudal plates double ; nostrils moderate, Berus.

Berus Vulgaris. Col. Berus, Lin. Var. Col. Aspis, Linn. Sturm Fauna, t.——Berus Ammodytes. Col. Ammodytes, Jacq. Coll. iv. t. 24, 25. Vip. Illyrica, Ald. Sturm Fauna, t.

C. *Head covered with scales, with three larger scales on the forehead ; sub-caudal plates double ; nostrils moderate.* Pelias, *Merrem.*

Pelias Chersea. Col. Berus, Laur. and Daud. Col. Chersia, Linn. Sturm. Fauna, Pelias Berus, Merrem. Vipera Berus, Fitz. Var. black. Col. Prester, Linn. Sturm Fauna.

D. *Head covered with small scales ; sub-caudal plates simple.* Echis, *Merrem.* Scytale, *Daud.* Pseudoboa, *Schn.*

Echis Carinata, Merrem. Pseudoboa Carinata, Schn. Scytale Bizonata, Daud. t. 70. Boa Horatta, Shaw. Russel, t. 2. India. —— Echis Krait, Merrem. Boa Krait, Williams. Pseudoboa, Schn. Scytale, Daud. India.—— Echis Arenicola, Boie. Geoff. Rept. Egypt, t. 7. f. 1. Savigny, t. 4. f. 1—4. North Africa.

E. *Head covered with scales in front ; tail ending in a hook ; sub-caudal plates simple and double.* Acanthopis, *Daud.* Ophrias, *Merrem.*

Acanthopis Cerastinus, Daud. v. t. 77. Merrem. Beytr. ij. t. 9. Boa Palpebrosa, Shaw. Ophria Acanthropis, Merrem. Boa Aculeata, E. W. Gray MSS. 1796. Mus. Brit. New Holland.——Acanthopis Brownii, Leach Zool. Misc. j. t. 3. Boa Antartica, Shaw. New Holland.

Gen. 3. NAIA.

Head moderate, without any pit, covered with large regular shields.

A. Head broad ; neck not dilatile ; sub-caudal plates double.
Sepedon, *Merrem.* Ophis, *Spix.* Naia, *Boie.* Hæmaca-
thus, *Flem.* Causus, *Wagler.*

Sepedon Rhombeata, Licht. Col. Vip. Nigrum, Licht.
Scheush. Ph. Sac. iv. t. 717. Naia Rhombeata, Boie.
Mus. Leyd.——Sepedon Hœmachates, Merrem. Hœma-
chate, Lacep. L. O. ij. t. 3. f. 2. Vipera Hœmachates,
Seba, ij. t. 58. f. 1, 3. Naie Hœmachate, Boie.——Sepe-
don Porphyraceus. Coluber Shaw, Zool. iij. 423.——
Sepedon Merremii. Ophis Merremi Spix, Braz. t. 17.

*B. Head narrow ; neck dilatile ; dorsal scales linear ; tail
conical; sub-caudal plates two-rowed.* Naja, *Daud.*
Uræus and Aspis, *Wagler.*

Naia Tricuspidans, Merrem. Col. Naja. Latr. Rus-
sel, I. S. j. t. 5 and 6. ij. t. 1. Daud iv. t. 60. Lacep. ij.
t. 3. f. 1. Has been divided into two several species by
Laurente.——Naja Haje, Merrem. Col. Haje, Hasselq.
Vipera Haje. Daud. Geoff. Rept. Egypt, t. 7. Savigny,
t. 3. Africa.——Naja Nivea, Cuv. Boie. Isis, xx. 557.
Vipera Flava, Merrem. Cape of Good Hope.——Naja
Spectatrix. Reinw. Boie. Iris, xx. 557. Elaps Fuscus,
Merrem. Col. Castaneus, Oppel. Russel. ij. t. 36.——Naie
Latratus. Boie. Col. Latratus, Kuhl. Beytr. 69.

*C. Head narrow ; neck not dilatile ; dorsal scales equal;
tail conical, moderate ; sub-caudal plate two-rowed*
Elaps.

Elaps Anguiformis, Schn.——Elaps Lacteus. Col. Do-
micella, Lin. Col. Domicellarum, Lac. Col. Lacteus, Lin.
Mus. Adolph. t. 17. f. 1. Seba, ij. t. 35. f. 2. 54. f. 1.
——Elaps Lemniscatus, Cuv. Col. Lemniscatus, Lin.
Seba, j. t. 10 ij. t. 76. f. 2.——Elaps Surinamensis, Cuv.

Seba, ij. t. 6. f. 2, and t. 86. f. 1.——Col. Surinamensis, Merrem, n. 184. Inhab. Surinam. Mus. Leyd. (Perhaps var. of former.)——Elaps Latonius. Col. Latonius, Merr. j. t. 2. Seba, 3. ij. t. 34. f. 4. t. 43. f. 3. Col. Lubricus, Merrem. Cape of Good Hope.——Elaps Flavius. Col. Flavius.——Elaps Micurus. Micrurus Spixii, Wagler, Spix, t. 18. ——Elaps Ibibaboca, Merr. Inhab. Brazils.——Elaps Corallinus, Merrem, 10. Pr. Max. ——Elaps Corallinus Merrem, 61. Vipera Psyches, Daud. t. 100. f. 1. Inhab. Brazils. Mus. Leyd.——Elaps Coccineus, Merrem.——E. Collaris, Boie. n. 5.——Elaps Furcatus, Schneider. Russel, ij. t. 19. Java——Elaps Bivirgatus, Kuhl. Boie. Isis, xx. 556. Java.——Elaps Boie nob. Elaphoides Fusca, Boie. Isis, xx. Java.—Elaps Chalybeum, Latostoma, Wagler.

D. *Head indistinct; body cylindrical; dorsal scales equal; tail compressed, two-edged; caudal plates two-rowed.* Platurus, *Latr.*

Platurus Laticaudatus. Col. Laticaudatus, Lin. Hydrus Colubrinus, Shaw. Platurus Faciatus, Boie. Daud. viii. t. 85. Lin. Mus. Alolph. j. t. 16. f. 1.——Platurus Semifaciatus, Reinw. Boie. (Mus. Leyden,) adult. Laticauda Scutale, Laur. Bechst. ij. t. 20. f. 1. Shaw, Zool. ij. t. 233.

E. *Head distinct; body fusiform; dorsal scales* ———— ; *tail conical; caudal plates partly double, partly simple.* Trimesurus, *Lacep.*

Trimesurus Leptocephalus, Lacep. Ann. Mus. iv. t. 56. f. 1.

F. *Head distinct; body fusiform; dorsal scales* ————— ; *tail conical; caudal plates all simple.* Oplocephalus, *Cuvier.*

Oplocephalus Cuvieri, nob. Mus. Paris.

84

II. Upper jaw toothed, fangs none, or small.

Gen. 3.   COLUBER.

Ventral shields broad, anal spurs none, tail, usually conical, and elongate.

A. Body subcylindrical, scales equal, subcaudal plates, double or single, head moderate, above shielded.

a. Head regularly shielded, nose acute, recurved. Heterodon.

Coluber constrictor, Lin. (Ed. 10.) Boa contortrix, Lin. (Ed. 12.) Col. Heterodon, Daud. Heterodon platyrhincus, Latr. Boa porcaria, Lacep. Catesby, 7. t. 56.——Heterodon simus. Col. simus, Lin. Merrem. 96. Col. borealis, Schoeff. ——Heterodon nasua, Vipera nasua, Oppel. Rhinostoma nasua, Wagler. Rhinostoma proboscideum, Fitz.

b. Head regularly shielded, nose rounded, Coluber.

A. Head distinct, oblong, ovate before, slightly truncated, gape, very wide, body long, cylindrical, tail half as long as the body, continued, acute, scales imbricate lanceolate, mostly keeled, placed in longitudinal series, abdominal shields, simple, arched. Tropidonotus, *Kuhl.*

European.

Trepidomurus natrix, Col. natrix, Lin. Col. murorum. Col. mur. Fitz.——Col. Oppelii, Dumeril. C. Murorum, Mus.——Col. Viperinus, Daud. Latr. Col. pseudo echidna, Herm.——Col. hybridus, Merrem.——Tr. scaber, Boie. Col. scaber, Lin. Mus. Adolph. t. 10, f. 1. Dacypeltis scaber, Wagler.

Asiatic.

Tr. melanozostus, Gravenh. Iris, 1826, 206.——Tr. funebris, Oppel.——Tr. chrysargus, Kuhl.——Trop. sub-

miniatus. Col. subminiatus, Reinw. n. 5.——Col. stola-
tus, Lin. Mus. Adolph. t. 22. f. 1. Coronella cervina,
Laur. Col. malpolon, Lacep. C. cervinus. Gm. Vipera
stolata, Lat. Col. sibitans, Lat.——Tr. cyanocephalus,
Boie, n. 5.——Col. palustris, Mer. 121. Hydrus palustris,
Schn. Col. Braminus, Daud. Russel, t. 20.——Tr. spilo-
gaster, Boie, n. 5.—— Tr. rhodomelas, Kuhl. and V. H.
n. 5.——Tr. trianguliferus.   Col. trianguligerus, Reinw.
——Tr. Mortuarius, Kuhl. Russel, t. 28.——Tropidonotus
hypostictus, Boie, n. 5.——Tropidonotus piscator, Mer. Hydrus piscator, Seb. Enchydus piscator, Latr. Col. anastomo-
satus, Daud. Russel, t. 33.——Tropidonotus dora. Col. dora,
Merrem, 104. Russel. ——Tropidonotus, t. 5, lugubris, Oppel.
——Tropidonotus lacrimans, Oppel.——Tropidonotus vibi-
kari, Boie, Iris, 1826. 207.——Tropidonotus tigrinus, Boie,
Do. 206.

American.

Col. variabilis, Pr. Max.——Tropodontus tesselata. Coronella tesselata. Laur. Col. viperinus, Daud. Col. Hydrus, Pallas. Natrix cherscoides. Var. bivittatus and Natrix occellata,
Wag.——Trop. æstirus, Boie. Col. æstirus, Lin. Catesb. Col.
subrirides, Lac.——Tr. poecilostoma. Col. poecilostoma,
Pr. Max. Natrix sculptura, Wag.——Col. saurita, Lin.
Ribon Snake, Cates. Daud. vi. t. 81. f. 2.——Tr. vittatus.
Col. vittatus, Lin. Mus. Adolph. t. 18. f. 2. Scheuchz,
t. 66. f. 8. Seba, t. 15. f. 3.——Col. fasciatus, Lin. Wam-
pum Snake, Cates. t. 58.   Dirty yellow with two white lines
down the back, and cross black bars.——Tr. Nattereri, Pr.
Max.——Tr. parietalis. Col. parietalis, Say, Isis, 1824, 255.
——Tr. proximus. Col. proximus, Say, Isis, 1824.——Tr.
ordinatus. Col. ordinatus, Lin.——Tr. dimidiatus, Boie.
——Tr. porcatus, Boie. Col. porcatus, Latr. Cates. Col.
erythrogaster, Sh.

B. *Head distinct ; gape very small ; tail half the length of the body ; scales placed in longitudinal series, mostly smooth, abdominal shields angularly recurved.* Coluber.

American.

Col. punctatus, Lin.——Col. cyaneus, Lin. Col. viridi-cæruleus, Lac. Seba, 17. t. 48. f. 2.——Col. azurus, Lac. ——Col. caninana, Merr. Seba, ij. t. 20. f. 1. &c. C. pulla-tus, Lin. Mus. Adolph. Col. peruvianus. Sh. Col. plutoni-cus, Daud. C. humanus and C. coronatus, Gm. Spiletes pullatus, Wag. Col. Nova Hispaniæ, Gm. Col. leucomelas, Gm. Col. variabilis, Pr. Max.——Col. Lichtenstenii, Pr. Max. Wag. Col. capistratus, Lich.——Col. bifossatus, Radd.——Col. pantherinus, Merr.——Col. scurrula, Wag. Natrix scurrula, Wag. Braz. t. 8.——Col. pileatus, Pr. Max. i. 344. Col. olfersii, Hemp. Col. olivaceus, Olfers. Philodryas olfersii, Wag.——Col. plumbeus, Pr. Max.—— Col. triangulum, Lac.——Col. corais, Daud. Boie. Isis, 1826, 538.——Col. getulus, Daud.——Col. bahaiensis, Spix. Braz. t. 10. f. 1.——Col. variabilis, Kuhl.

Africa.

Col. Hippocrepis, Lin. Col. canus, Lin. Periops hippo-crepis, Wag. Lin. Mus. Adolph. —— Col. barbarus, Boie. ——Col. Geoffroyii. Coulévre aux raies parelleles, Geoff. Rept. Egypt. Periops, Wag.——Col. trabalis, Pall.—— Var..*a.* fuliginosus, Oppel. *b.* minochronis, Cuv.

European.

Col. Elaphis.——Col. scalaris, Sh. Col. Meifrenii, Oppel Col. bilineatus, Dumeril.——Col. flavescens, Scop.——Col. viridiflavus, Scop. Col. personatus, Daud. young, Col. atro-virens, Metaxa. Zamerinus viridiflavus, Wag.——Col. Æscu-lapii, Lacep. not Lin. Metaxa. Zamenis Æsculapii, Wag.

Asiatic.

Col. Oppelii, Wag. Russel, ii. t. 30.——Col. melanopis, Oppel. Col. flavolineatus, Rienw.——Col. trivergatus, Boie. Iris, 1826, 209. —— Col. clamacophorus, Do. 210. ——Col. vulneratus, Do. 212.——Col, conspicillatus, Do. 211. Col. canus, Lin. Merrem. i. t. 7. C. cinarascens, Lacep. C. ammobates, Shaw, Seba, ii. t. 3. f. 2. t. 78. f. 2. young C. margaritaceus, Daud. B. ij. t. 9. C. crucifer, Sh. C. Parias, Hem.——Col. radiatus, Russel, t. 42.——Col. mucosus, Lin. ——Col. melanurus, Oppel.—— Col. cancellatus, Oppel. Col. korros, Rein. Col. Saturninus, Merr. 84. ——Col. tricolor, Boie.——Col. geminiatus, Oppel. Isis, 1826.——Col. oxycephalus, Rein. Gonyosoma viride, Wag. t. 9. ——Col. trabalis, Pallas.——Col. Helena, Daud. Rus. t. 32.——Col. obscurus, Daud. Rus. t. 18.

C. Head large, distinct, depressed ; gape very wide ; nostrils large, open ; body fuciform, thick ; tail shorter than half the length of the body ; scales equal, placed in oblique transverse series. Xenodon, *Boie.*

American, Ophis, *Wagler.*

Col. severus, Lin. Col. versicolor, Lin. Sch. Phy. Sac. t. 661. f. 7. Seba, j. t. 85. f. 1.——X. Merremii, Ophis Merremii, Wag. Spix, Braz. t. 17.——Col. rhapdocephalus, Pr. Max. Braz. i. 351.——Dipsas scholti, Fitz.—— X. aneus, Boie, n. 5.——X. ocellatus, Boie.——X. saurocephalus, Boie. Col. saurocephalus, Boie, Pr. Max.

D. Asiatic, Xenodon, *Wagler.*

Xenodon inornatus, Boie. Col. inornatus, Kuhl.

Head short, blunt, and narrow : tail short, continued, Oligodon, *Boie.*

O. bitorquatus, Boie.——O. atriventus, Boie. Col. atri-

88

ventus, Daud. Merr. Seba, ij. t. 86. f. 5.   Col. lutrix, Lin. ?
Homalosoma arctiventris, Wag. Elaps Lutonia and E. Duber-
ria, Sch.   Duberria   arctiventris, Fitz.——O. punctatus.
Col. punctatus, Latr.   Homolosoma punctatum, Wagler.

*E. Head, very long, very distinct ; gape very wide ; tail,
very long ; scales small and oblique series, partly
keeled.* Herpotodryas, *Wag.* Erpetodryas, *Boie.* Inhab.
South America.

C. carinatus, Boie.   Col. carinatus, Lin. Merr. Seba, ii·
t. 54, t. 56, f. 3, t. 71, f. 12.   C. fuscus, Lin.   Col. sub-
fuscus, Lacep.   Natrix bicarinatus, Wag.——E. lævicollis,
de Col. lævicollis, Pr. Max.——E. exoletus, Lin. Mus.
Adolph. Col. bicarinatus, Pr. Max.——Col. carinatus, var.
Merr. Natrix exoleta, Lau.——Col. quadricarinatus, Fitz.
Erpet, Boie. Col. pyrrhopogon. Pr. Max.——Col. sex-
carinatus, Pr. Max.   Natrix cinnamomea, Wag.   Without
epidermis, back with six series of keeled scales.

*F. Tail conical, sub-caudal plates first simple, then two
rowed, dorsal scales uniform.* Hurria, *Daud.*

Hurria bilineatus, Daud, Rup.——Hurria Nympha, Mer.
Col. Nympha, Daud. from Russel, t. 36, 37.

*G. Head scarcely distinct from the body, gape small or
moderate, teeth colubrine, the hinder rather larger, scales
imbricate, very smooth, abdomen arcuate,* Coronella *Laur.*
Erythrolampus Liophis *and* Cloelia, *Wag.*

American.

Cor. venustissima. Col. venustissima, Pr. Max. Col. bina-
tus, Lich.——Col. agilis, Lin.   Col. Æsculapii, Lin. Mus.
Adolph. t. 21.——Col. poecelogyrus, Pr. Max. Liophis poe.
Wag. Body marbled the whole length.——Col. formosus,
Pr. Max. Ery. formosus, Wag.——Col. clœlia, Daud.

t. 78.——Col. melanocephalus, Lin. Mus. Adolph. t. 15, f. 2. Col. capite niger, Lacep. Clœlia melanocephalus, Wag.——Col. cobella, Lin. Elaps Cob. Sch. Cerastes Cob. Laur. Merr. Col. serpentinus, Daud. Cates. Leophis Cobella, Wag.——Col. miliaris, Lin. Col. Merremii, Pr. Max. Col. Dictyodes, Id. Natrix chiametla, Wag. t. 2, f. 2. Col. ammobates, Shaw. Liophis miliaris, Wag.—— Natrix Forsteri Wag. t. 4, f. 1. Liophis Forsteri, Id.—— Col. typhlus, Lin. Lac.——Col. viridissimus, Lin. Mer. 1, t. 12. Col. Ianthinius, Daud. Chlorosoma viridissima, Wag. Col. crassicaudatus, Mer. C. Africanus, Bonat. Seba, ij. 35, f. 4. Col. Reginæ, Lin. Liophis Reginæ, Wag. Natrix semilinea, Wag. t. 11, f. 2. Natrix alamensis, jun. Wag. t. 10, f. 3.——Col. Boddartii, Scitz.——Col. doliatus, Daud. Pr. Max. 368.——Col. meleagris, Sh. from Seba, ij. 32, f. 1, 56 ——Col. orythrogastra, Pr. Max.—— Natrix occipitalis, Wag. Spix, Braz. t. 6. Cloelia occipitalis, Wag. Col. melanocephalus, Boie. Elaps melanocephalus, Wag. t. 2, f 1. Cloelia dorsata, Wagler.——Col. lineatus, Lin.——Col. Nicandri, Mer.——Col. bicinctus, Herm.—— Col. raninus, Boie, from Seba, ij. t. 9.——Elaps triscalis, Mer.

Africa.

Col. annulatus, Lin. Mus. Adol. t. 8. Col. ignobilis. Laur. Col. atrofuscus, Lacep. C. caudiolus, Lac. C. orientalis, Gm. C. epidaurius, Herm. Seba, Thes. i. t. 72. ij. t. 9, t. 13, t. 52, 57, 82, f. 2. — Col. rufulus, Lich. Col. 1823.——Col. rufescens, Gm. Coron. Kotamboija, Laur. C. Hitambocia, Gm. Seba, t. 33, f. 6.——Col. Aurora, Lin. Cloelia aurora, Wag. Lin. Merr. Adol. 19. Seb. ij. 22.—— Col. annulatus, Boie. Lin. Mus. Adolph. i. t. 8. Seba, i. t. 54. Merrem, t. 11. Oxyrhopus annulatus, Wagler.——Cor. Petolarius. Coluber pet. Lin. Ib. t. 6. Col. pethola, Bech.

90

Lacep. t. 3. Oxyrhopus pet. Wagler.——Col. rhombeatus, Boie. Col. rhom. Lin. Ib. t. 24. f. 1.

Asiatic.

Col. Baliedeira, Boie. Col. bal., Kuhl.——Col. octolineatus, Boie. Elaps oct., Schneid. Russel, 4, 38.——Col. Bilineatus, Boie. Col. bilineatus, Lac. ——Col. tæniolatus, Boie. Col. tæn. Russel, t. 19.

European.

Col. meridionalis, Daud.——Col. lævis, Lacep.——Col. tessalata, Lat.

H. Head not distinct from body ; no hind frontal scale ; body cylindrical, tail very short, blunt abdominal shields, entire caudal shields, in two rows, scales smooth. Calamaria, *Boie.*

Col. Linnæ, Boie. Col. calamarius, Lin. Mus. Adol. 1. t. 6. f. 1. Anguis calamaria, Laur.——Col. lumbricoides, Boie. ——Col. tessalata, Boie.——Col. maculosa, Boie.——Col. multipunctata, Boie.——Col. virgulata, Boie. ——Col. reticulata, Boie.

I. Head not distinct ; eyes small ; tail short, acute. Brachyorrhos, *Kuhl.* Atractos, *Wagler.*

Br. Albus, Kuhl. Col. albus, Lin. Mus. Adol. t. 14, f. 2. Anguis alba, Laur. Mer. t. 7. Col. brachyurus, Shaw.—— Br. dimidiatus, Kuhl. Col. dimidiatus, Oppel.——Br. Kuhlii, Boie. Br. brachyurus, Kuhl. ——Br. Decussata, Kuhl.——Br. torquatus.——Br. badius, Boie.——Br. flammigerus, Boie.——Br. trilineatus, Atractus trilineatus, *Wagler. Isis,* 1828.——Br. Schach, Boie.

K. Body, long ; sub-compressed ; tail continued short, abdominal shields convex ; scales rhomboidal, nearly square, imbricate. Lycodon, *Boie.*

91

Col. Maximiliana, Merr. Col. audax, Daud. Col. cancellatus, Merr. Col. cattenularis, Daud. Tar tutta, Russel, 1, t. 15.——Lycedon Hebe, Boie. Col. Hebe, Daud. Col. fasciolatus, Sh. Russel, 1, t. 21.——Lycodon aulicus, Boie. Col. aulicus, Lin. Mus. Adol. t. 12, f. 2.; Seba, t. 91. f. 5. Nat. aulica, Laur. Lappiata, Lacep.——Lycodon subcinctus, Boie. Col. sub. Rein; Seba, t. 109, f. 7; Russel, ij. t. 41. Ophites, sub. Wagler.——Lyc. capucinus, Boie; Russel, 11, t. 37. Col. aulicus, Kuhl.——Lyc. fuliginosus, Boie.—— Lyc. pethola. Col. pethola, Lin.; Seba, t. 28, f. 2, t. 54; f. 4, t. 110; f. 3.: ii. t. 38, f. 2. Cor. pethola, Laur.—— Lyc. leucocephalus. Col. leucocephalus, Merr.——Lyc. Russelii. Col. Russelius, Daud. Katla tutta, Russel, 1, t, 38.——Lyc. unicolor, Boie; Russel, ij. 39.——Lyc. malignus, Boie. Col. Malignus, Merr.——Lyc. macrochinus. Col. macro, Boie.——Lyc. galathea, Boie. Col. galathea, Daud. Russel, 1, t. 26.

L. Nose rounded; head shielded; shield triangular, larger than the dorsal scales; tail conical. Xenopeltis.

Xen. Concolor Rein. Boie. Isis. 564.——Xen. Unicolor. Rein. Boie. Ib. 564. Col. alvearius Oppel.——Xen. leucocephalus, Rien. Boie. Isis, 564.

M. Head ovate indistinct, shielded; body cylindrical, scales equal, tail conical, with a single series of subcaudal scales. Scytale. *Merr. not Daud.* Pseudoboa. *Schn.*

Scytale coronata. Mer. Seba, ij. t. 41. f. 1. Pseudoboa coronata. Schn.——Scy. coronata. Pr. Max.——Hurria carinata. Kuhl. Ceraspis carinata. Wag.——Scy. Brachyorrhos, Boie. Isis, 20. Aspidura Brachyorrhos. Wag.

N. Head regularly shielded; body long; slender scales, equal; head long; muzzle acute; tail very long; subcau-

92

dal plates in two rows. Dryophis. *Fitz.* Dryinus, *Mer.* Oxybelis, *Wag.*

Col. fulgidus, Daud., t. 80. Seba, ij. t. 53 Col. nasutus, Shaw, and Col. purpurascens, Shaw. Dryophis fulgidus, Wag.——Col. acuminatus. Pr. Max. Dryinus Æneus. Spix. Braz. Oxybelis Œneus, Wag. Dryinus auratus. Bell.——Dryophis Panthoraria, Kuhl.

O. Head very long, regularly shielded; body very long; dorsal scales triangular, lateral scales linear.

Muzzle acute appendaged. Passerita.

Dryinus oxyrhynchus, Bell. Dryophis nasutus, Boie. Col. nasutus, Russel, 1. t. 12. Tragops, Wagler. Dryinus nasutus, Merr. Dryophis Russelii, Bottla. Passericki, Russel, t. 13. ——Dry. pavoninus. Cuv. Boie. Tragops, Wag. ——Dry. xanthozonica, Kuhl. Tragops, Wag.——Dry. prasina, Rein. Russel, ij. t. 25. Seba, ij. t. 53. f. 4. Dry. nasutus, Bell. Tragops, Wag.——D. rostratus, Reinw.

Muzzle blunt, rounded : ventral plates angularly keeled on the sides. Ahœtula. Leptophis, *Wagler.*

Ahæ Linnei. Col. ahætuta, Lin. Natrix ahætula, Laur. Boiga. Lacep. ij. t. 2. Dendrophis ahætula, Boie.——Ahæ. Richardi. Col. Richardi, Bory.——Ahæ. liocercus. Col. liocercus, Pr. Max. Dendrophis lio, Boie. These two are considered by Wagler as varieties of the first.——Col. caracarus, Gm. Seb. ij. Bungarus filiformis, Oppel.——Daud. periophthalmica, Boie.

Psamnophis, *Boie.*

Psam. Girondicus, Boie. Col. Girond, Daud. Col. Gallicus, Hem. Zacholus Girondicus, Wagler.——Coronella Austriaca, Laur. Zacholus Austriacus, Wagler.——Col. Crucifer, Daud. Mer. Seba, j. t. 109. f. 8.——Col. moniliger, Lac. Col. sibelans, Seba, ij. t. 56.——Macrosoma elegans,

93

Leach. Col. elegans, Sch.—Psam. pulverulenta, Kuhl. Boie.
Isis. xx. 547.——Psam. schokari, Boie. Col. schokari,
Forsh.——Psam. lacertina, Boie. Col. lac. Spix, t. 5. Co-
clopeltes, Lac. Wag. Geoff. Egypt, t. 5. f. 2.——Psam. rhombeatus. Col. Rhom. Lin. Mus. Ad. t. 24. f. 2. Col. Tyria,
Kuhl. Coclopeltis, Wagler.——Col. condanarus, Mer. Russel,
t. 27. j. t. 27.——Psam. bicolor, Wag.

Dendrophis, *Boie.*

Den. chaireacos, Boie. Russel, ij. t. 26. Elaps bilineatus,
Sch.——Den. maniar, Russel, ij. 25. Leptophis mancas,
Bell. Zool. Jour.——Den. tristis. Col. tristis, Daud. Col.
scandens, Mer. Russel, t. 31.——Den. formosa, Rien. Boie.
——Den. picta, Boie. Col. pictus, Gm. Col. decorus, Sh.
Col. cœruleus. Bonnat. Dips. Schokari, Kuhl. Bungarus
filum, Oppel. Seba, j. t. 99.——Den polychroa, Boie.——
Den scandens, Col. scandens, Mer. Russel, j. t. 31.

Ventral plates two keeled. Chysopelea, *Boie.*

Col. ornatus, Mer. Col. Ibibcea, Daud. Russel, ij. f. 2.
Seba, j. t. 94. ij. t. 7. f. 1. t. 56. f. 1. t. 61. f. 2.——Chrys.
rhodopleura, Rern.——Chup. pardalis, Boie.——Chry. para-
disi, Boie. Isis, 20. 547. Seba, i. 94. f. 7. i. j. 61. f. 2.

*P. Head very high, above flat, shielded before, truncate ; body
compressed, with one or three series of longer vertebral
scales, lateral scales imbricate.* Amblicephalus, *Kuhl.*
Pareas, *Wag.*

A. carinata, Boie. Dipsas carinata, Kuhl. Pareas· car.
Wag.——A. mikani. Col. mikani, Fitz.——A. coccineus.
Col. coccineus, Blum.——A. lœvis, Khul.——A. Nattereri.
Col. Natereri, Mikan. Dryophyllax Nattereri, Wag.——
Natrix punctatissima, Wag. t. 14. Col. lineolatus, Oppel.
Thammodynastes punctatissima, Wag.—·—Rhinostomus pro-
biscidens, Fitz.

94

*Q. Head short and broad, above shielded ; body long, com-
pressed ; vertebral scales square; lateral scales linear.*
Dipsas.

a. Subcaudal scales double. Dipsas, *Laur.* Bungarus,
Oppel.

Dipsas Indica, Laur. Col. bucephalus, Shaw, Seba, i.
t. 43. Col. atrox, Gmel. Dipsas bucephalus, Boie.——
Dipsas dendrophila, Rein. Wagler, t. 8. Col. peruvianus, Sh.
——Dip. Drapiezii, Boie, Isis, xx. 549.——Dipsas multo-
maculata, Rein. Russ. t. 23.——Dipsas cynodon, Cuv. Boie,
Isis.——Dipsas trigonatus, Boie. ——Dipsas irregularis.
Col. irreg, Mer. t. 4. Hurria pseudotriga, Daud.—— Dip. cen-
choa, Boie. Col. cenchoa, Lin. Seba, ij. t. 16. f. 2.——Dip.
Catesbœi. Col. Cates. Weig. Bungarus, Oppel. ——Dip.
nebulatus. Col. nebulatus, Lin.——Dip. compressus, Boie.
Col. compressus, Daud.——Dip. Savignii, Boie.

b. Sub-caudal plates simple, fangs, with the teeth. Bungarus, *Daud.*

Bungarus cæruleus, Daud, Boie, Russel, t. 1. Boa latotecta
Herm. —— Bungarus annularis, Daud. Russel, t. 3. ——
Bung. semifasciatus, Oppel. Seba. Col. candidus, Lin.

*R. Head, scales, with small plates over the face and between
the eyes, with plates, simple, double, a simple and double.
Cerebus, Cuv.* Homolopsis, *Boie.*

Col. cerberus, Daud. Hydrus rhynchops, Schn. Euby-
dris rhynchops, Latr. Hydrus cinereus, Sh. Python rhynchops, Mer. Homalophis rhinchops, Boie.——Homolopsis
obtusatus, Boie.——Hern. erythrogammus, Boie. Col. erythrogammus, Daud.——Hern. carinicaudatus, Boie. Col.
carinicaudatus, Pr. Max.——Col. monilis, Lin. Col. buc-
catus, Lin. Mus. t. 19. Vipera buccata, Daud. Vipera

95

semifasciata, Merr. Col. varius, Merr. Seba.——Col. angu-
latus, Lin. Merr. t. 9. Natrix aspera, Spix. Braz.——
Homolophis aer, Boie. Col. aer, Oppel.——Col. plicatus,
Lin. Mus. t. 6. f. 1. Bali. Lacep. Elaps plicatilis, Schn.
Cerastes plicatilis, Laur.——Col. Æneus, Spix.——Hom.
moluroides, Boie. Col. molurus, Lac. t. 10.—— Hom. plumbea,
Kuhl, Boie.——Hom. Jara, Boie. Col. Jara, Shaw, Russel,
t. 14. Col. Linnæi, Merrem.

Genus. BOA.

Ventral shields narrow ; body thick ; anal spurs distinct ;
nostrils opical ; subcaudal plates simple, a double tail
conical.

*Body compressed fusiform ; head depressed ; tail prehensile ;
back of the head scaly ; subcaudal plates simple.* America. Boa.

Muzzle covered with scales ; labial scales flat. Boa, Wagler.
Boa constrictor, Lin. Devin or Boa, Empereur, Daud.

Muzzle covered with scales ; labial scales pitted. Epicrates,
Wagler.

Epicrates Cenechria. Boa cenchris, Lin. Daud. Seba. Boa
cenchrya, Pr. Max.

Muzzle covered with plates ; labial scales flat. Eunectes,
Wag.

Eunectes murina. Boa murina. Boa scytale, Linn. Seba.
Boa anacondo, Daud. Boa aquatica, Pr. Max.——Eunectus
lateristrigota. Boa lateristriga, Boie.

Muzzle covered with plates ; labial scales pitted. Xiphosoma, Wag.

Xiphosoma canina. Boa canina. Lin. Boa hypnale, Lin.
Xip. araramboya, Wag.——Xiph. hortulana. Boa hortulana.

Lin. Sel. Vipera et Echidna cærulescens. Mer. Vip. moderans and Boa elegans, Daud. Coluber glaucus, Gm. Xiphes dorsale, and X. ornatum, Wag.——Xip. Merrem. Corallus obtusirostris, Daud. accidental deformity.

Body much compressed, keeled ; head obliquely cut behind and truncated in front, covered with small shields in front ; sub-caudal plates simple. Cenchris, *Gray*, Engyrus, *Wagler.* India.

Cenchris regia, Boa regia, Shaw. Boa carinata, Merrem. Wetter, Amer. t. 9. —— Cenchris ocellata, Boa ocellata, Oppel. Boa carinata, Var. C. Mer.

Body fusiform ; head covered with scales ; muzzle with small plates ; sub-caudal plates, two rowed. Python. Asia.

a. Labial scales with deep pits, dorsal scales rhombic. Constrictor, Wagler.

Python Schneideri, Mer. Boa reticulata and Boa rhombeata, Schn. Python Javanicus, Kuhl. Seba, t. 62. Col. Javanicus, Shaw.——Python Poda, Boie, Pedda Poda. Russel. t. 23. Col. boaformis, Sh. Python Tigris, Daud. ——Python bivittatus, Kuhl. Col. Sebœ, Lin.——Python amethystina, Schn.——Boa orbiculata, Schn. Bora, Russel, t. 39.

a. Labial scales equal, flat. Python, *Wagler.*

Python Peronii, Cuv. Wag. Python punctatus, Mer. K. H.

Body fusiform ; head covered with keeled scales like the back ; tail short ; sub-caudal plates simple ; spurs distinct. Gongylophis, *Wagler,* Boa, part *Cuv.* Asia.

Boa conica, Schn. Boa ornata, Daud. Boa viperina, Shaw. Padani Cootoo, Russel, t. 4. Mus. Brit.

Body cylindrical ; head covered like the back with small

I   97

scales ; muzzle shielded, tail short. Eryx and Clothonia, *Daud.*

Eryx jaculus and E. turcica, Daud. Boa Tartarica, Lich. but Cuvier states that this is without spurs.——Eryx angui-formis. Boa anguiformis, Schn. Eryx Indica, Cuv.

HYDRUS, WATER OR SEA SNAKES.

Ventral shields narrow, formed of two united scales; body compressed ; anal spurs none ; tail compressed, except in Achrocordus ; nostrils vertical operculated.

A. Head small, shielded ; body covered with scales. Hydro-phis, Hydrus, *Wagler.*

    a. Abdomen with two rows of larger scales.

Hydrus major, Sh.——Hyd. nigrocinctus, Merr. from Russel, t. 6. Anguis Xiphura, Herm.——Hyd. chloris, Mer. Russel, t. 7.——Hyd. obscurus, Merr. Russel, t. 8. Leio-seloma obscura. Var. a Fitz.——Hyd. schistosus, Mer. from Russel, t. 10.——Hyd. cyanocinatus, Mer. from Russel.

b. Abdomen with two rows of larger scales united into a band.

Hydrus Shawii, Hyd. major, var. Sh. MSS.——Hyd. cœrulescens, Sh. Mus. Brit.--— Hyd. leiolepis, Gray.—— Hyd. lanceolatus, Gray. Mus. Brit.——Hyd. spiralis, Shaw. ——Hyd. doliatus, Mer. Distera, Lacep. Pelamis, Wag. ——Hyd. striatus, Mer.—— Hyd. lævis, Mer. —— Hyd. Brugmansii, Boie.——Hyd. atricapillus, Rein.

B. Head large, depressed, dilated behind, shielded ; body with square plates. Pelamis, Hydrophis, *Wag.*

a. The central abdominal line with a series of scales on each side.

Pelamis bicolor, Daud. Hyd. platura, Lin.——Pelamis ornatus, Gray, Mus. Brit.--—Pelamis gracilis, Hydrus gra-

98

cilis, Sh. Pel. Valakadin, Boie, Russ. t. 11. Disteria, Fitz. ——Pelamis Rupelii, Hyd. Schiddil, Boie. Russ. t. 12.—— Pel. melanurus, Hyd. melanurus, Wag. t. 3.

b. The central abdominal line formed of a series of long plates.

Pelamis fasciatus, Daud. Hydrus, Sh. Russ. t. 44. Anguis laticauda, Lin.——Pel. Kadll, Russ. t. 13.——Pelamis curtus, Hyd. curtus, Shaw.——Pelamis carinatus, Hyd. carinatus, Cuv. Pel. Lindsayi. Gray. Mus. Brit.

c. Head and body covered with small keeled scales ; tail compressed. Chersydrus, *Cuv.*

Ch. granulatus, Mer. Achrocordus fasciatus, Sh. Hydrus granulatus, Schn.

d. Head and body covered with small keeled scales ; tail conical. Achrocordus, *Horst.*

Ach. dubius, Sh. Ach. Javanicus, Horst.

e. Head shielded, with two soft scaly tentacula, ventral plates narrow ; tail long. Herpeton, *Wagler.*

Herpeton tentaculatus, Daud. Erpeton, Lacep. Merrem.

AMPHIBIA.

Body with a naked skin.

Order 1. MUTABILIA.

Undergoing a transformation; gills deciduous, covered with a deciduous operculum.

Gen. 1. RANA.

Body thick, tailless ; feet four, long ; sternum and clavicles distinct. Larva elongate, fishlike, tailed and without legs ; gills four on each side.

A. Tongue and tympanum distinct ; mouth toothed ; skin smooth ; toes clawless. Rana.

a. Toes simple, eyebrows rounded.

Green Frog. Rana esculenta, Lin.——Common Frog. Rana temporaria, Lin.——Rana cultripes, Cuv.——Rana punctata, Daud. R. Daudini, Mer. Hypscaphelia, Boie.——R. plicata, Daud. R. Daudini, β Mer.——R. paradoxa, Lin.——R. tigrina, Daud. R. pipens, Daud. t. 18 ——R. cyanophlyctis, Schn. Bufo cyanophlyctis, Latr. ——R. rudibunda, Pallas. Bufo, Sch.——R. vespertina, Pal. Bufo, Schn.——R pipiens, Lin. not Daud. R. maxima, Cates. R. Catesbiana, Shaw. R. mugiens, Mer.——R. palmipes, Spix.——R. Virginica, Gm. R. Halecina, Daud. R. pipiens, Mer. Shad Frog, Bartram.——R. clamitans, Latr. Daud. t. 16. Grunting Frog, Bartram. Argus Frog, Shaw. R. ocellata, Lin. R. pentadactyla, Laur. R. sonans, Bonnat. Cystygnathus, Wag. R. gruniens, Daud. is thought by Harlan to be the same.——R. melanota, Raffinesque.——R. cutricularis, Harlan.——R. scapularis, Harlan.——R. flaviventris, Har. R. fontinalis, Leconte. Spring Frog, Bartram. ——R. sylvatica, Leconte. —— R. palustris, Leconte. R. pardalis, Harl.——R. pumilla, Leconte.——R. Gryllus, Id.——R. dorsalis, Harlan. Savannah Cricket of the Americans.——R. nigrita, Leconte.—— R. typhonia, Daud. R. Virginica, Mer. Cystignathus, Wagler.——R. Gigas, Spix. R. coriacea, Id. R. pachypus, Spix Cystignathus pachypus, Wag.——R. sibilatrix, Pr. Max. R. pygmœa, Spix. Cystignathus, Wagler.—— R. rubella, Latr. R. Daudini, Mer.——R. mystacea, Spix. Cystignathus, Wagler.——R. miliaris, Spix.——R. labyrinthica, Spix. Cystignathus, Wagler.——R. maculata, Latr. Daud.

100

b. Toes simple ; eyebrows horned. Ceratophys.

America. Ceratophys cornuta, Pr. Max. C. varius, Boie. C. cornuta, Lin. Seba, j. t. 72. f. 1. 2. Ceratophys dorsata, Pr. Max. Back with separate shield-like plates, the front one the largest.——Ceratophys Spixii, Cuv. R. megastoma, Spix, perhaps the same species.——Cer. clypeata, Cuv. Mus. Par. ——Cer. Boiei, Pr. Max. Cer. granosus, Cuv.——Cer. scutata. R. scutata, Spix. Hemiphractus Spixii, Wagler.—— Asia. Cer. montanus, Megalophrys Montana, Kuhl. Wagler.

c. End of toes dilated. Hyla, *Laurent.* Calamita, *Schn.*
Tree Frogs.

Hyla arborea. R. arborea, Lin. Hyas, Wagler.—— Hyla ranœformis, Lau. Calamita, Mer. La Bossue, Lacep. Seba. ——Hyla Indica, Gray.

Hy. bilineata Daud.——Hy. Blochii, Daud. Calamita cinereus, Schn. Cal. bilineatus, β Mer.——Hyl. Scychellerii, Peron.——Hy. vermiculata, Peron.——Hy. Leschenaultii, Gray.——Hy. Peronii, Gray.——H. pulchra, Gray. ——Hy. bicolor, Daud. Phyllomedusa, Wagler.——Hy. boans, R. Boans, Lin. R. maxima, Lau. Calamita maximus, Schn. Hyla palmata, Latr. Rana Zebra, Shaw. Auletris, Wagler.——Hy. venulosa, Latr. Rana ven., Laur. Hy. viridi-fusca, Laur. Calamita Boans, Schn. Hyla Boans, Mer. Rana Meriana, Shaw. Rana arborea, Var. Gm. R. squamigera, Walb.——Hy. marmorata, Latr. Rana gibbosa, β Gm.——Hy. tinctoria, Latr. Daud. Rana, Shaw. Rametta a tapiner, Lacep.

* Some of the Indian species deposit their eggs on the under side of leaves, hanging over water. General Hardwicke has observed them place the eggs on a leaf which stood over a pail of water, so that the young dropt into the water.

Hy. Surinamensis, Daud. Calamita, Mer.——Hy. hypochondrialis, Latr. Daud. Calamita, Mer.——Hy. intermixta.——Hy. ruber, Laur. La Rouge, Lacep. Hy. sceleton, Laur. —— Hy. aurantiaca, Daud. Auletris, Wagler.——Hy. lactea, Laur. Rana lactea, Lin. Calam. melanorabdotus, Sch. Hy. Boans, Daud.——Hy. tibiatrix, Laur. H. aurantiaca, Laur.——Hy. lactea, Latr. Daud. Hy. verrucosa, Daud. Hy. variegata, Daud. Hy. lecophyllatus, Beirus. Hy. frontalis, Latr.——Hy. bufonia, Spix.——Hy. geographica, Id.——Hy. albomarginata, Id——Hy. papillaris, Id.——Hy. pardalii, Id. ——Hy. cinerascens. Id——Hy. affinis, Id.——Hy. trivitata, Id.——Hy. abbreviata, Id.——Hy. lateralis, Daud. Rana arborea, β Gm. Hyla viridis, Laur. Calamita Carolinensis, Penn. C. cinerea, Schn. Rana bilineata, Shaw.—— Hy. femoralis, Daud.— Hy. syuirella, Daud. H. occularis, var Leconte.——Hy. delitescens, Leconte.——Hy. versicolor, Id. —Hy. occularis, Latr.——Hy. Quoyii, Bory.——Hy. Gaimardii, Id.——Hy. Seuerii, Desm. Hy. bifasciata, Gray, Mus. Paris.——Blue tree-frog, White's Journal, Hy. Cyanea, Daud. R. austraciæ, Schn. R. cærulea, Daud.

White's specimen of the last mentioned seems to have wanted a toe on its hind feet, and from this Fitzinger has made of it a genus, under the name Calamita. Hyla violacepoda, and Hyla erythropoda, of the French museum, appear to be nearly applied to, and probably the young of this species. Hyla bufonoides of the Paris Museum, of which H. centripoda of the same Museum may be a variety. Hy. ocellata, and Hy. rubeola, all also of the same Museum, are from New Holland.

d Tongue and tympanum distinct; mouth toothless. Nose rounded. Bufo.

Common Toad, Rana Bufo, Lin. Bufo cinereus, Schn.

102

Bufo vulgaris, Laur. Bufo Roeselii, Latr. Daud. t. 27?——
Bufo roseus, Wasserkrote, Meyer.——Bufo fuscus, Laur.
Rana bombina, Gm. Roesel, Fro. t. 15. Pelobates fuscus,
Wagler.——Bufo variabilis, Mer. R. variabilis, Pallas. t. 6,
f. 12. Bufo viridis, Laur. Rana viridis, Shaw. Bufo Schre-
berianus, Laur. Rana sitibunda, Pallas. Bufo sitibundus,
Lepeche. See also Geoffroy Reptiles of Egypt, t. 4. Bufo
obstetricans, Laur. R. obstetricans, Sturm. Bombina ob-
stetricans, Mer. Alytes, Wagler.——Bufo calamita, Laur.
R. protentosa, Blumen. R. Bufo, var. Gm. B. cruciatus,
Schn. R. fœtidissima, Hern. R. mephitica, Shaw. Bufo
salsus Schrank, Daud. Common on heaths near London.——
Bufo marina, Daud. t. 37. B. maculiventris, Spix.——Rana
marina, Lin. Bufo marinus, Sch., is from Seba; t. 76, f. 7, it
has toes like an Hyla. Copied R. maxima, Shaw.——Bufo
ictericus, Spix, Brazils. ——B. lazarus, Id.——B. stellateus,
Id. ——B. scaber, Id.——Bufo cinctus, Pr. Max.——B.
Agua, Id.——Bufo scaber, Daud.——Bufo pustulosus, Laur.
Bufo melanostrictus, Schn. Seba.——Bufo cognatus, Say,
Brickcoloured toad, Bartram.——Bufo musicus, Latr. Daud.
——Rana musica, Lin? Bufo clamosus, Schn. Land-frog,
Bartram.——Bufo Bengalensis, Daud. Bufo marinus, Mer.
R. dubia, Shaw.——Bufo flaviventris, Daud. B. chlorogaster,
Daud.——Bufo prætextatus, Boie, MSS.

e. *Nose acute, head rounded behind.* Rhinella, *Fitz.* Oxyrhynchus, *Spix.*

Rhinella proboscideus, Bufo, Spix. ——Rhi. semilineatus,
Bufo, Spix.——Rhi. granulosus.——Rhi. acutirostris——
Rhi. nasicus——all Bufo, Spix, t. 21, and 24.

f. *Nose acute; head with a crest extending on each side, to
the paratoids.* Otilopha, Les Otilophes, *Cuv.*

Otilopha typhonia, R. typhonia, Lin. Seba. R. margira-

103

tifera, Laur. Bufo margaritifera, Daud.—— Oli. ocellata, Tympanum, obscurely seen through the skin. Dark brown, with white edged brown spots.

B. *Tongue distinct; tympanum hidden; toes* 3, *inner clawed; mouth toothed.* Dactylethra, Cuv. Engystoma, *Fitz.* not *Merrem.* Xenopus, *Wagler.*

Xenopus Boiei, Wag. Pipra Buffonia, Mer. Bufo lævis, Daud. Pipa lævis, Mer. Engystoma ovalis, Fitz. ?—Rana fasciata of Burchel's Travels, may belong to this genus.—— Xenopus ovalis, Rana ovalis, Schn. Microps unicolor, Wag.

C. *Tongue distinct, tympanum hidden; toes clawless; mouth large, eyebrows rounded.* Bombinator.

B. igneus. Rana bombina, Gm. variegata, Lin.——Bombinator Daudini, Bufo ventricosus, Daud. Engystoma ventricosa, Fitz.

D. *Mouth large; eyebrows acute.* Strombus, *Gray.*

Ceratophys granulata, Cuv. C. Boiei, Pr. Max. Physalamis Cuvieri, Fitz.

E. *Mouth and head small, paratoides none; toes,* 5, 4. Breviceps. Engystoma, part Fitz.

Breviceps gibbosus, Bufo gibbosus, Daud. Engystoma dorsatum, Cuv. Seba. R. systoma, Schn.——Breviceps marmoratum, Engyst. marmoratum, Cuv. not described. —— Bre. granosus, Engyst. granosus, Cuv.—— Bre. Surinamensis, R. Suninamensis, Daud. R. Bufonia, Mer.——Bre. globulosus, Bufo, glo. Spix, t. 19, f. 1. Chaunus marmoratus, Wagler. ——Bre. albifrons, Bufo albifrons, Spix, t. 19, f. 2. Paludicola albifrons, Wagler.——Bufo ephippium, Spix. This species has only three toes on all the feet, and Fitzinger has made a genus of it, therefore, under the name of Brachycephalus.

104

F. *Tongue indistinct, tympanum, hidden, end of toes, four, cut.* Pipa, *Daud.* Asterodactylus, *Wagler.*

Common Pipa, R. Pipa, Lin. Seba, j. t. 77. The females of this species hatch their eggs on the back.——Pipa Curucura, Spix, t. 22. The female of this species do not carry their eggs on the back. Inhabits the bottom of the lakes in Brazil.

Gen. 2. SALAMANDRA.

Body subcylindrical, tailed, feet four, short; sternum clavicules none. Larva, with four feet, gills, three on each side.

* *Toes* 4, 4. *Salamandrina.*

Salamandrina perspicilleta. Savi; les trois doights, Lacep. ij. t. 36. Molge tridactylus, Merrem. Salam. tridactylus, Daud. See also, Salam. Savi, Gosse, with the toes half webbed. Mus. Paris.

** *Toes* 4, 5. *Salamandra.*

a. Skin smooth, paratoids glandular, with group of tubercles on the sides, in two series down the back and tail, each ending in a large pore, lateral line, simple.

Common Salamandra. Salamandra maculosa, Laur. Lac. Salamandra. Lin. Black, yellow spotted.——Black Salamandra. Salamandra atra, Laur. Sal. fusca Gesner, Sturm fauna. Black may be only a variety.

b. Skin granular, granicles ending in a group of minute pores, paratoids porous, with a series of pores on each side, forming a line between the legs.

Common Salamander. Salamandra palustris. Salam. cristata, Latr.——The Marbled Salamander. Salem marmorata, Latr. Triton Gesneri, Laur., and the Alpine Salamander.

Salam. alpestris, Bechst. Lacep. t. 20, perhaps belong to this section.

c. *Skin and paratoids granular, each granule ending in a large pore, lateral line porus.* Pleurodeles ?

Spanish Salamander. Salamandra major, nob, Mus. Brit. Salam. fenestrata, Mus. Par. Inhab. Gibraltar, Mus. Col. Surg. and Brit. length ten, or twelve inches. This species is very like Pleurodeles Waltl. Michaellis, Isis, 1830, t. 2, said to be found in the South of Spain, and said to differ from the other Salamanders, by having fourteen perfect ribs.

d. *Skin and paratoids smooth, with scattered granules, each ending in a pore, place of lateral line occupied by a large blood vessel.*

Common Eft. Salamandra vulgaris. Sal. punctata, Latr. ——The Webfooted Eft. Salam. palmata, Latr., scarcely appears to be distinct.

e. *Skin and paratoids smooth, minutely punctulated, the pustules along the upper surface of the tail rather larger, and more distinct.*

Violet Salamander. Salamandra subviolocea, Barton. Scargus Valeno. Mus. Par. Sal. Venenosa, Barton, Amer. Phil. Trans. vi. t. 1. Brit. Mus.

f. *Skin and paratoids smooth, minutely punctated, lateral line none, toes long and free,* Molge. nob.

Glutinous Salamander. M. glutinosa. Salamandra gluti-nosa, Green. Mus. Brit.——Red backed Salamander. Sala-mandra Erythronota, Green. Brit. Mus.—— Cuvier says that Salam. Japonica, Bechst. Lacep. ij. t. 10, f. 1, is allied to this species.——Ashy Salamander. Salamandra cinerea, Green.——Banded Salamander. Salamandra fasciata, Green, J. A. N. S. Phil. Mus. Brit.

Probably the following American species belong to this section.

Punctate Salamander. Salam. symetrica. Sal. punctata, Bechst. Lacep. ij. t. 10, f. 2.——Tigerine Salamander. Salam. tigrina, Green. N. America.——Two lined Salamander. Salam. bislineata, Green, New Jersey.——Red Salamander. Salem. rubra, Daud. Salam. rubiventris, Green. North America——Variolated Salamander. Salam. variolata, Gilliams. Jour. Sic. Nat. Philad. 1, t. 18, f. 1, Pensylvania.——Cylindrical Salamander. Salam. cylindracea, Harlan. South Carolina.——White faced Salamander. Sala mandra frontalis. Sal sinciput-alba, Green.——Brown Salamander. Salamandra fusca, Green, Jour. Acad. Philad. New Jersey.——Dorsal Salamander. Salamandra dorsalis, Harlan. South Carolina.——Painted Salamander. Salamandra picta, Harlan. Pensylvania.——Yellow Salamander. Salamandra flavissima, Harlan. Pensylvania.—— Capt. Beechy's Salamander. Salamandra Beechii, Gray. ——Spotted Salamander. Salamandra maculata, Green. New Jersey. Mus. Brit.——Brownish Salamander, Salamandra subfusca, Green. New Jersey.——Long Tail Salamander. Salamandra longicaudata, Green.——Black Salamander. Salamandra nigra, Green. Pennsylvania.—— Green's Salamander. Salamandra Greeni, Gray.

g. Skin and paratoids smooth, minutely punctuated; lateral line none; toes very short webbed.

Salamandra variegata. Gray. Sal. platydactylus. Cuv. Mus. Paris. Brown, with irregular dorsal bands. Mexico.

Sect. II. Amphipneusta. *Not undergoing transformation.*

Gen. I. Proteus.

Gills free, persistent; skull formed of several distinct bones;

107

body rather depressed; tail compressed; legs four; muzzle depressed.

    a. Toes 3 before, 2 behind. Hydochthon.

Common Proteus. Proteus anguinus, Laur. Siren anguinus, Schn. See Configliachi, Monografra and Schrieber's, Monograph, Carniola.

    b. Toes 4 *before,* 4 *behind.* Menobranchus, *Harlan.* Necturus, *Raffin.* Phanerobranchus, *Fitz.*

Say's Menobranchus. Menobranchus lateralis, Harlan. Triton lateralis. Acad. Nat. Sci. Philad. iv. t. 21. Ann. Lyc. N. ij. 1. 16. Proteus of the Lakes, Mitchell, Silliman Jour, iv. and vii.——Salamandra alleghaniensis, Jun. Say. ——Lacepedes Menobranchus. Menobranchus Lacepedii. Proteus tetradactylus. Lacep. Ann. Mus. x. t.

    c. Toes 4 *before,* 5 *behind.* Phyllhydrus, *Brooks,* Gyrinus.   *Shaw* not *Lin.* Siredon, *Wagler.*

Common Axolotl. Phyllhydrus pisceformormis. Siren pisciformis. Shaw Zool. iij. t. 140. Misc. t. Axolotl Humboldt. Obs. Zool. t. 12. Home Phil. Trans. and Lacep. Anat. Siredon Axolotl, Wagler. Mexico. Consult also Proteus Nove Cæsarcenris, Green. Jour Acad. Phil. and Siren operculata, Pal de Beauv. Amer. Phil. Trans. iv.

    Gen. II. Siren.

Gills free, persistent; skull of distinct bones : body subcylindrical; tail compressed; legs two, anterior.

    a. Toes 4, 4. Siren, *Linn.* Muræna, *Gmel.*

Common Siren. Siren lacertina, Linn. Mud Iguana. Ellis Phil. Trans.——Intermediate Siren. Siren intermedia, Leconte Ann. Lyc. New York ij. t. 1.
  108

b. Toes 3 . 3. Pseudobranchus, *Gray.*

Striated Siren. Pseudobranchus striatus. Siren striata, Leconte Ann. Lyc. New York, j. t. 4. Captain Leconte has unjustly and unscientifically attacked this genus. See Ann. Lyc. New York, ij. 133, though it was established originally on the authority of his erroneous account of the Opercula.

Gen. III. AMPHIUMA.

Gills none ; skull formed of a solid bony substance ; body subcylindrical ; tail compressed ; legs four.

a. Legs rather strong ; toes 4, 5, *outer edge of the feet fringed.* Abranchus.

Large Hellbender. Abranchus horridus. Salamandra gigantea. Salamandra horrida, and Protonopsis horrida, Barton. Menopoma Alleghaniensis, Harlan. Triton Alleghaniensis, Daud. Molge gigantea, Merrem. Salamandra Alleghaniensis, Michaux. Abranchus alleghaniensis, Leconte, Ann. Lyc. j. t. 17.

b. Legs weak, boneless. Amphiuma.

Common Congo Snake. Amphiuma means. Garden. Florida and Georgia. Mus. Brit. Chrysodonta larvaformis, Mitchel.——Three Toes Congo Snake. Amphiuma tridactylum. Cuv. Mem. Mus. xiv. t. 1. Mus. Col. Surg. Four footed. Siren. Barton.

Gen. IV. CÆCILIA. · Linn.

Gills none, head depressed, formed of a solid bony substance ; legs none ; body cylindrical ; tail short, blunt ; vent roundish, nearly terminal—perhaps this should be placed near Amphisbœna.

a. Muzzle blunt ; nostrils bearded ; body distinctly ringed.
Cæcilia. Siphonops, *Wagler.*

Ringed Cœcilia. Cæcilia annulata, Spix, t. 27. f. 1.——
Interrupted Ringed Cæcilia. Cæcilia interrupta, Cuv.——
Bearded Cæcilia. Cæcilia tentaculata, Linn. Amon. Acad. j.
t. 17. f. 1. Cæcilia albiventris, Daud. vii. t. 122. f. 2.——
Glutinous Cæcilia. Cæcilia glutinosa, Lin. Mus. Adolph, t. 4.
f. 1. Seba, t. 25. f. 2. Ceylon.——Two Banded Cæcilia.
Cæcilia bivittata, Cuv. America.——Sharp-nosed Cæcilia.
Cæcilia rostrata, Cuv.

b. Muzzle prominent, body slender, very long, smooth, ten-
tacula distinct. Ichthyophis, *Fitz.* Cæcilia, *Wagler.*

Wormlike Cæcilia. Cæcilia lumbricoides, Cuv. Daud.
t. 92, f. 2, Lin. Mus. Adolph. t. 5, f. 2.

c. Muzzle depressed, bearded ; body, rather fusiform, very
closely ringed, eyes small. Epicrium, *Wagler.*

Javanesse Cæcilia. Cæcilia hypocyana, Van. Hasselt, Isis.
1827. Epicrium Hasseltii, Wagler, Java. Cæcilia nasuta,
is said to be a fish, the Sphagebranchus Coromandelicus of
Russel, t. 37.